Practical Linear Algebra

Practical Linear Algebra
A Geometry Toolbox
Third Edition

Gerald Farin
Dianne Hansford

CRC Press
Taylor & Francis Group
Boca Raton London New York

CRC Press is an imprint of the
Taylor & Francis Group, an **informa** business
AN A K PETERS BOOK

CRC Press
Taylor & Francis Group
6000 Broken Sound Parkway NW, Suite 300
Boca Raton, FL 33487-2742

© 2014 by Taylor & Francis Group, LLC
CRC Press is an imprint of Taylor & Francis Group, an Informa business

No claim to original U.S. Government works

Printed on acid-free paper
Version Date: 20130524

International Standard Book Number-13: 978-1-4665-7956-9 (Hardback)

Library of Congress Cataloging-in-Publication Data

Farin, Gerald E.
 Practical linear algebra : a geometry toolbox / Gerald Farin, Dianne Hansford. -- Third edition.
  pages cm
 Summary: "Practical Linear Algebra covers all the concepts in a traditional undergraduate-level linear algebra course, but with a focus on practical applications. The book develops these fundamental concepts in 2D and 3D with a strong emphasis on geometric understanding before presenting the general (n-dimensional) concept. The book does not employ a theorem/proof structure, and it spends very little time on tedious, by-hand calculations (e.g., reduction to row-echelon form), which in most job applications are performed by products such as Mathematica. Instead the book presents concepts through examples and applications. "-- Provided by publisher.
 Includes bibliographical references and index.
 ISBN 978-1-4665-7956-9 (hardback)
  1. Algebras, Linear--Study and teaching. 2. Geometry, Analytic--Study and teaching. 3. Linear operators. I. Hansford, Dianne. II. Title.

 QA184.2.F37 2013
 512'.5--dc23   2013016125

Visit the Taylor & Francis Web site at
http://www.taylorandfrancis.com

and the CRC Press Web site at
http://www.crcpress.com

With gratitude to
Dr. James Slack and Norman Banemann.

Contents

Preface

Just about everyone has watched animated movies, such as *Toy Story* or *Shrek*, or is familiar with the latest three-dimensional computer games. Enjoying 3D entertainment sounds like more fun than studying a linear algebra book. But it is because of linear algebra that those movies and games can be brought to a TV or computer screen. When you see a character move on the screen, it's animated using some equation straight out of this book. In this sense, linear algebra is a driving force of our new digital world: it is powering the software behind modern visual entertainment and communication.

But this is not a book on entertainment. We start with the fundamentals of linear algebra and proceed to various applications. So it doesn't become too dry, we replaced mathematical proofs with motivations, examples, or graphics. For a beginning student, this will result in a deeper level of understanding than standard theorem-proof approaches. The book covers all of undergraduate-level linear algebra in the classical sense—except it is not delivered in a classical way. Since it relies heavily on examples and pointers to applications, we chose the title *Practical Linear Algebra*, or *PLA* for short.

The subtitle of this book is *A Geometry Toolbox*; this is meant to emphasize that we approach linear algebra in a geometric and algorithmic way. Our goal is to bring the material of this book to a broader audience, motivated in a large part by our observations of how little engineers and scientists (non-math majors) retain from classical linear algebra classes. Thus, we set out to fill a void in the linear algebra textbook market. We feel that we have achieved this, presenting the material in an intuitive, geometric manner that will lend itself to retention of the ideas and methods.

Review of Contents

As stated previously, one clear motivation we had for writing PLA was to present the material so that the reader would retain the information. In our experience, approaching the material first in two and then in three dimensions lends itself to visualizing and then to understanding. Incorporating many illustrations, Chapters 1–7 introduce the fundamentals of linear algebra in a 2D setting. These same concepts are revisited in Chapters 8–11 in a 3D setting. The 3D world lends itself to concepts that do not exist in 2D, and these are explored there too.

Higher dimensions, necessary for many real-life applications and the development of abstract thought, are visited in Chapters 12–16. The focus of these chapters includes linear system solvers (Gauss elimination, LU decomposition, the Householder method, and iterative methods), determinants, inverse matrices, revisiting "eigen things," linear spaces, inner products, and the Gram-Schmidt process. Singular value decomposition, the pseudoinverse, and principal components analysis are new additions.

Conics, discussed in Chapter 19, are a fundamental geometric entity, and since their development provides a wonderful application for affine maps, "eigen things," and symmetric matrices, they really shouldn't be missed. Triangles in Chapter 17 and polygons in Chapter 18 are discussed because they are fundamental geometric entities and are important in generating computer images.

Several of the chapters have an "Application" section, giving a real-world use of the tools developed thus far. We have made an effort to choose applications that many readers will enjoy by staying away from in-depth domain-specific language. Chapter 20 may be viewed as an application chapter as a whole. Various linear algebra ingredients are applied to the techniques of curve design and analysis.

The illustrations in the book come in two forms: figures and sketches. The figures are computer generated and tend to be complex. The sketches are hand-drawn and illustrate the core of a concept. Both are great teaching and learning tools! We made all of them available on the book's website http://www.farinhansford.com/books/pla/. Many of the figures were generated using PostScript, an easy-to-use geometric language, or Mathematica.

At the end of each chapter, we have included a list of topics, *What You Should Know* (*WYSK*), marked by the icon on the left. This list is intended to encapsulate the main points of each chapter. It is not uncommon for a topic to appear in more than one chapter. We have made an effort to revisit some key ideas more than once. Repetition is useful for retention!

Exercises are listed at the end of each chapter. Solutions to selected exercises are given in Appendix B. All solutions are available to instructors and instructions for accessing these may be found on the book's website.

Appendix A provides an extensive glossary that can serve as a review tool. We give brief definitions without equations so as to present a different presentation than that in the text. Also notable is the robust index, which we hope will be very helpful, particularly since we revisit topics throughout the text.

Classroom Use

PLA is meant to be used at the undergraduate level. It serves as an introduction to linear algebra for engineers or computer scientists, as well as a general introduction to geometry. It is also an ideal preparation for computer graphics and geometric modeling. We would argue that it is also a perfect linear algebra entry point for mathematics majors.

As a one-semester course, we recommend choosing a subset of the material that meets the needs of the students. In the table below, LA refers to an introductory linear algebra course and CG refers to a course tailored to those planning to work in computer graphics or geometric modeling.

	Chapter	LA	CG
1	Descartes' Discovery	•	•
2	Here and There: Points and Vectors in 2D	•	•
3	Lining Up: 2D Lines		•
4	Changing Shapes: Linear Maps in 2D	•	•
5	2×2 Linear Systems	•	•
6	Moving Things Around: Affine Maps in 2D	•	•
7	Eigen Things	•	
8	3D Geometry	•	•
9	Linear Maps in 3D	•	•
10	Affine Maps in 3D	•	•
11	Interactions in 3D		•

	Chapter	LA	CG
12	Gauss for Linear Systems	•	•
13	Alternative System Solvers	•	
14	General Linear Spaces	•	·
15	Eigen Things Revisited	•	
16	The Singular Value Decomposition	•	
17	Breaking It Up: Triangles		•
18	Putting Lines Together: Polylines and Polygons		•
19	Conics		•
20	Curves		•

Website

Practical Linear Algebra, A Geometry Toolbox has a website:
http://www.farinhansford.com/books/pla/

This website provides:

- teaching materials,

- additional material,

- the PostScript files illustrated in the book,

- Mathematica code,

- errata,

- and more!

Gerald Farin   March, 2013
Dianne Hansford   Arizona State University

Descartes' Discovery

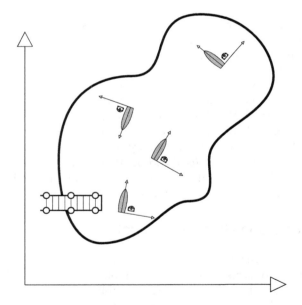

Figure 1.1.
Local and global coordinate systems: the treasure's local coordinates with respect to
the boat do not change as the boat moves. However, the treasure's global coordinates,
defined relative to the lake, do change as the boat moves.

There is a collection of old German tales that take place sometime in
the 17th century, and they are about an alleged town called Schilda,

whose inhabitants were not known for their intelligence. Here is one story [12]:

> An army was approaching Schilda and would likely conquer it. The town council, in charge of the town treasure, had to hide it from the invaders. What better way than to sink it in the nearby town lake? So the town council members board the town boat, head for the middle of the lake, and sink the treasure. The town treasurer gets out his pocket knife and cuts a deep notch in the boat's rim, right where the treasure went down. Why would he do that, the other council members wonder? "So that we will remember where we sunk the treasure, otherwise we'll never find it later!" replies the treasurer. Everyone is duly impressed at such planning genius!
>
> Eventually, the war is over and the town council embarks on the town boat again, this time to reclaim the treasure from the lake. Once out on the lake, the treasurer's plan suddenly does not seem so smart anymore. No matter where they went, the notch in the boat's rim told them they had found the treasure!

The French philosopher René Descartes (1596–1650) would have known better: he invented the theory of *coordinate systems*. The treasurer recorded the sinking of the treasure accurately by marking it on the boat. That is, he recorded the treasure's position relative to a *local* coordinate system. But by neglecting the boat's position relative to the lake, the *global* coordinate system, he lost it all! (See Figure 1.1.) The remainder of this chapter is about the interplay of local and global coordinate systems.

1.1   Local and Global Coordinates: 2D

This book is written using the LaTeX typesetting system (see [9] or [13]), which converts every page to be output to a page description language called PostScript (see [1]). It tells a laser printer where to position all the characters and symbols that go on a particular page. For the first page of this chapter, there is a PostScript command that positions the letter **D** in the chapter heading.

In order to do this, one needs a two-dimensional, or 2D, coordinate system. Its origin is simply the lower left corner of the page, and the x- and y-axes are formed by the horizontal and vertical paper edges meeting there. Once we are given this coordinate system, we can position objects in it, such as our letter **D**.

The **D**, on the other hand, was designed by font designers who obviously did not know about its position on this page or of its actual size. They used their own coordinate system, and in it, the letter **D** is described by a set of points, each having coordinates relative to **D**'s coordinate system, as shown in Sketch 1.1.

We call this system a *local coordinate system*, as opposed to the *global coordinate system*, which is used for the whole page. Positioning letters on a page thus requires mastering the interplay of the global and local systems.

Following Sketch 1.2, let's make things more formal: Let (x_1, x_2) be coordinates in a global coordinate system, called the $[\mathbf{e}_1, \mathbf{e}_2]$-system. The boldface notation will be explained in the next chapter. You may be used to calling coordinates (x, y); however, the (x_1, x_2) notation will streamline the material in this book, and it also makes writing programs easier. Let (u_1, u_2) be coordinates in a local system called the $[\mathbf{d}_1, \mathbf{d}_2]$-system. Let an object in the local system be enclosed by a box with lower left corner $(0, 0)$ and upper right corner $(1, 1)$. This means that the object "lives" in the *unit square* of the local system, i.e., a square of edge length one, and with its lower left corner at the origin. Restricting ourselves to the unit square for the local system makes this first chapter easy—we will later relax this restriction.

We wish to position our object into the global system so that it fits into a box with lower left corner $(\min_1, \min_2)$ and upper right corner $(\max_1, \max_2)$ called the *target box* (drawn with heavy lines in Sketch 1.2). This is accomplished by assigning to coordinates (u_1, u_2) in the local system the corresponding target coordinates (x_1, x_2) in the global system. This correspondence is characterized by preserving each coordinate value with respect to their extents. The local coordinates are also known as *parameters*. In terms of formulas, these parameters are written as quotients,

$$\frac{u_1 - 0}{1 - 0} = \frac{x_1 - \min_1}{\max_1 - \min_1}$$
$$\frac{u_2 - 0}{1 - 0} = \frac{x_2 - \min_2}{\max_2 - \min_2}.$$

Thus, the corresponding formulas for x_1 and x_2 are quite simple:

$$x_1 = (1 - u_1)\min_1 + u_1\max_1, \tag{1.1}$$
$$x_2 = (1 - u_2)\min_2 + u_2\max_2. \tag{1.2}$$

We say that the coordinates (u_1, u_2) are *mapped* to the coordinates (x_1, x_2). Sketch 1.3 illustrates how the letter **D** is mapped. This concept of a parameter is reintroduced in Section 2.5.

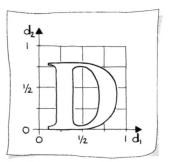

Sketch 1.1.
A local coordinate system.

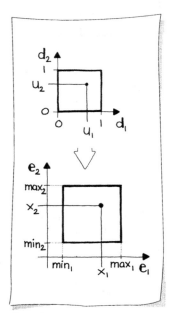

Sketch 1.2.
Global and local systems.

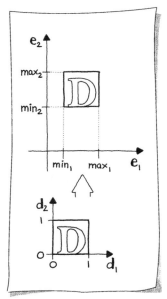

Sketch 1.3.

Local and global **D**.

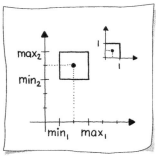

Sketch 1.4.

Map local unit square to a target box.

Let's check that this actually works: the coordinates $(u_1, u_2) = (0, 0)$ in the local system must go to the coordinates $(x_1, x_2) = (\min_1, \min_2)$ in the global system. We obtain

$$x_1 = (1 - 0) \cdot \min_1 + 0 \cdot \max_1 = \min_1,$$
$$x_2 = (1 - 0) \cdot \min_2 + 0 \cdot \max_2 = \min_2.$$

Similarly, the coordinates $(u_1, u_2) = (1, 0)$ in the local system must go to the coordinates $(x_1, x_2) = (\max_1, \min_2)$ in the global system. We obtain

$$x_1 = (1 - 1) \cdot \min_1 + 1 \cdot \max_1 = \max_1,$$
$$x_2 = (1 - 0) \cdot \min_2 + 0 \cdot \max_2 = \min_2.$$

Example 1.1

Let the target box be given by

$$(\min_1, \min_2) = (1, 3) \quad \text{and} \quad (\max_1, \max_2) = (3, 5),$$

see Sketch 1.4. The coordinates $(1/2, 1/2)$ can be thought of as the "midpoint" of the local unit square. Let's look at the result of the mapping:

$$x_1 = (1 - \frac{1}{2}) \cdot 1 + \frac{1}{2} \cdot 3 = 2,$$
$$x_2 = (1 - \frac{1}{2}) \cdot 3 + \frac{1}{2} \cdot 5 = 4.$$

This is the "midpoint" of the target box. You see here how the geometry in the unit square is replicated in the target box.

A different way of writing (1.1) and (1.2) is as follows: Define $\Delta_1 = \max_1 - \min_1$ and $\Delta_2 = \max_2 - \min_2$. Now we have

$$x_1 = \min_1 + u_1 \Delta_1, \tag{1.3}$$
$$x_2 = \min_2 + u_2 \Delta_2. \tag{1.4}$$

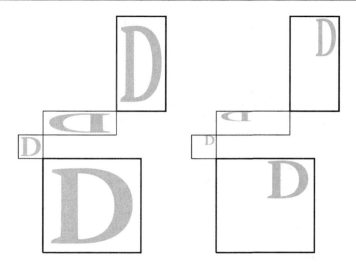

Figure 1.2.
Target boxes: the letter D is mapped several times. Left: centered in the unit square. Right: not centered.

A note of caution: if the target box is not a square, then the object from the local system will be distorted. We see this in the following example, illustrated by Sketch 1.5. The target box is given by

$$(\min_1, \min_2) = (-1, 1) \quad \text{and} \quad (\max_1, \max_2) = (2, 2).$$

You can see how the local object is stretched in the $\mathbf{e}_1$-direction by being put into the global system. Check for yourself that the corners of the unit square (local) still get mapped to the corners of the target box (global).

In general, if $\Delta_1 > 1$, then the object will be stretched in the $\mathbf{e}_1$-direction, and it will be shrunk if $0 < \Delta_1 < 1$. The case of $\max_1$ smaller than $\min_1$ is not often encountered: it would result in a reversal of the object in the $\mathbf{e}_1$-direction. The same applies, of course, to the $\mathbf{e}_2$-direction if $\max_2$ is smaller than $\min_2$. An example of several boxes containing the letter $\mathbf{D}$ is shown in Figure 1.2. Just for fun, we have included one target box with $\max_1$ smaller than $\min_1$!

Another characterization of the change of shape of the object may be made by looking at the change in *aspect ratio*, which is the ratio of the width to the height, Δ_1/Δ_2, for the target box. This is also written as $\Delta_1 : \Delta_2$. The aspect ratio in the local system is one. Revisiting

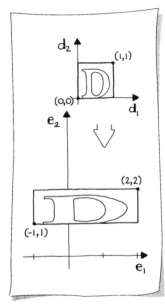

Sketch 1.5.
A distortion.

Example 1.1, the aspect ratio of the target box is one, therefore there is no distortion of the letter **D**, although it is stretched uniformly in both coordinates. In Sketch 1.5, a target box is given that has aspect ratio 3, therefore the letter **D** is distorted.

Aspect ratios are encountered many times in everyday life. Televisions and computer screens have recently changed from nearly square 4 : 3 to 16 : 9. Sketch 1.5 illustrates the kind of distortion that occurs when an old format program is stretched to fill a new format screen. (Normally a better solution is to not stretch the image and allow for vertical black bars on either side of the image.) All international (ISO A series) paper, regardless of size, has an aspect ratio of $1 : \sqrt{2}$. Golden rectangles, formed based on the golden ratio $\phi = (1 + \sqrt{5})/2$ with an aspect ratio of $1 : \phi$, provide a pleasing and functional shape, and found their way into art and architecture. Credit cards have an aspect ratio of 8 : 5, but to fit into your wallet and card readers the size is important as well.

This principle, by the way, acts strictly on a "don't need to know" basis: we do not need to know the relationship between the local and global systems. In many cases (as in the typesetting example), there actually isn't a known correspondence at the time the object in the local system is created. Of course, one must know where the actual object is located in the local unit square. If it is not nicely centered, we might have the situation shown in Figure 1.2 (right).

You experience this "unit square to target box" mapping whenever you use a computer. When you open a window, you might want to view a particular image in it. The image is stored in a local coordinate system; if it is stored with extents $(0, 0)$ and $(1, 1)$, then it utilizes *normalized coordinates*. The target box is now given by the extents of your window, which are given in terms of *screen coordinates* and the image is mapped to it using (1.1) and (1.2). Screen coordinates are typically given in terms of *pixels*;[1] a typical computer screen would have about 1440×900 pixels, which has an aspect ratio of 8 : 5 or 1.6.

1.2   Going from Global to Local

When discussing global and local systems in 2D, we used a target box to position (and possibly distort) the unit square in a local $[\mathbf{d}_1, \mathbf{d}_2]$-system. For given coordinates (u_1, u_2), we could find coordinates (x_1, x_2) in the global system using (1.1) and (1.2), or (1.3) and (1.4).

---

[1]The term is short for "picture element."

How about the inverse problem: given coordinates (x_1, x_2) in the global system, what are its local (u_1, u_2) coordinates? The answer is relatively easy: compute u_1 from (1.3), and u_2 from (1.4), resulting in

$$u_1 = \frac{x_1 - \min_1}{\Delta_1}, \tag{1.5}$$

$$u_2 = \frac{x_2 - \min_2}{\Delta_2}. \tag{1.6}$$

Applications for this process arise any time you use a mouse to communicate with a computer. Suppose several icons are displayed in a window. When you click on one of them, how does your computer actually know which one? The answer: it uses Equations (1.5) and (1.6) to determine its position.

Example 1.2

Let a window on a computer screen have screen coordinates

$$(\min_1, \min_2) = (120, 300) \quad \text{and} \quad (\max_1, \max_2) = (600, 820).$$

The window is filled with 21 icons, arranged in a 7×3 pattern (see Figure 1.3). A mouse click returns screen coordinates (200, 709). Which icon was clicked? The computations that take place are as follows:

$$u_1 = \frac{200 - 120}{480} \approx 0.17,$$

$$u_2 = \frac{709 - 300}{520} \approx 0.79,$$

according to (1.5) and (1.6).

The u_1-partition of normalized coordinates is

$$0, \quad 0.33, \quad 0.67, \quad 1.$$

The value 0.17 for u_1 is between 0.0 and 0.33, so an icon in the first column was picked. The u_2-partition of normalized coordinates is

$$0, \quad 0.14, \quad 0.29, \quad 0.43, \quad 0.57, \quad 0.71, \quad 0.86, \quad 1.$$

The value 0.79 for u_2 is between 0.71 and 0.86, so the "Display" icon in the second row of the first column was picked.

Figure 1.3.
Selecting an icon: global to local coordinates.

1.3 Local and Global Coordinates: 3D

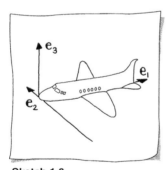

Sketch 1.6.
Airplane coordinates.

These days, almost all engineering objects are designed using a *Computer-Aided Design* (*CAD*) system. Every object is defined in a coordinate system, and usually many individual objects need to be integrated into one coordinate system. Take designing a large commercial airplane, for example. It is defined in a three-dimensional (or 3D) coordinate system with its origin at the frontmost part of the plane, the e_1-axis pointing toward the rear, the e_2-axis pointing to the right (that is, if you're sitting in the plane), and the e_3-axis is pointing upward. See Sketch 1.6.

Before the plane is built, it undergoes intense computer simulation in order to find its optimal shape. As an example, consider the engines: these may vary in size, and their exact locations under the wings need to be specified. An engine is defined in a local coordinate system, and it is then moved to its proper location. This process will have to be repeated for all engines. Another example would be the seats in the plane: the manufacturer would design just one—then multiple copies of it are put at the right locations in the plane's design.

Following Sketch 1.7, and making things more formal again, we are given a local 3D coordinate system, called the $[\mathbf{d}_1, \mathbf{d}_2, \mathbf{d}_3]$-system, with coordinates (u_1, u_2, u_3). We assume that the object under consideration is located inside the *unit cube*, i.e., all of its defining points satisfy

$$0 \leq u_1, u_2, u_3 \leq 1.$$

This cube is to be mapped onto a *3D target box* in the global $[\mathbf{e}_1, \mathbf{e}_2, \mathbf{e}_3]$-system. Let the target box be given by its lower corner $(\min_1, \min_2, \min_3)$ and its upper corner $(\max_1, \max_2, \max_3)$. How do we map coordinates (u_1, u_2, u_3) from the local unit cube into the corresponding target coordinates (x_1, x_2, x_3) in the target box? Exactly as in the 2D case, with just one more equation:

$$x_1 = (1 - u_1)\min_1 + u_1\max_1, \tag{1.7}$$
$$x_2 = (1 - u_2)\min_2 + u_2\max_2, \tag{1.8}$$
$$x_3 = (1 - u_3)\min_3 + u_3\max_3. \tag{1.9}$$

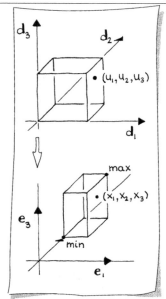

As an easy exercise, check that the corners of the unit cube are mapped to the corners of the target box.

The analog to (1.3) and (1.4) is given by the rather obvious

$$x_1 = \min_1 + u_1\Delta_1, \tag{1.10}$$
$$x_2 = \min_2 + u_2\Delta_2, \tag{1.11}$$
$$x_3 = \min_3 + u_3\Delta_3. \tag{1.12}$$

As in the 2D case, if the target box is not a cube, object distortions will result—this may be desired or not.

1.4 Stepping Outside the Box

We have restricted all objects to be within the unit square or cube; as a consequence, their images were inside the respective target boxes. This notion helps with an initial understanding, but it is not at all essential. Let's look at a 2D example, with the target box given by

$$(\min_1, \min_2) = (1, 1) \quad \text{and} \quad (\max_1, \max_2) = (2, 3),$$

and illustrated in Sketch 1.8.

The coordinates $(u_1, u_2) = (2, 3/2)$ are not inside the $[\mathbf{d}_1, \mathbf{d}_2]$-system unit square. Yet we can map it using (1.1) and (1.2):

$$x_1 = -\min_1 + 2\max_1 = 3,$$
$$x_2 = -\frac{1}{2}\min_2 + \frac{3}{2}\max_2 = 4.$$

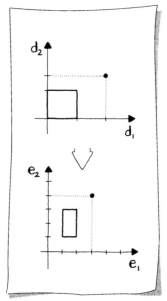

Sketch 1.8.
A 2D coordinate outside the target box.

Since the initial coordinates (u_1, u_2) were not inside the unit square, the mapped coordinates (x_1, x_2) are not inside the target box. The notion of mapping a square to a target box is a useful concept for mentally visualizing what is happening—but it is not actually a restriction to the coordinates that we can map!

Example 1.3

Without much belaboring, it is clear the same holds for 3D. An example should suffice: the target box is given by

$$(\min_1, \min_2, \min_3) = (0, 1, 0) \quad \text{and}$$
$$(\max_1, \max_2, \max_3) = (0.5, 2, 1),$$

and we want to map the coordinates

$$(u_1, u_2, u_3) = (1.5, 1.5, 0.5).$$

The result, illustrated by Sketch 1.9, is computed using (1.7)–(1.9): it is

$$(x_1, x_2, x_3) = (0.75, 2.5, 0.5). \qquad \text{}$$

1.5 Application: Creating Coordinates

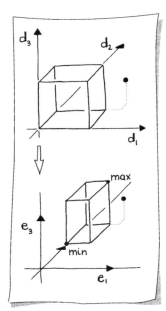

Sketch 1.9.

A 3D coordinate outside the 3D target box.

Suppose you have an interesting real object, like a model of a cat. A friend of yours in Hollywood would like to use this cat in her latest hi-tech animated movie. Such movies use only mathematical descriptions of objects—everything must have coordinates! You might recall the movie *Toy Story.* It is a computer-animated movie, meaning that the characters and objects in every scene have a mathematical representation.

So how do you give your cat model coordinates? This is done with a *CMM,* or *coordinate measuring machine;* see Figure 1.4. The CMM is essentially an arm that is able to record the position of its tip by keeping track of the angles of its joints.

Your cat model is placed on a table and somehow fixed so it does not move during digitizing. You let the CMM's arm touch three points on the table; they will be converted to the origin and the $\mathbf{e}_1$- and $\mathbf{e}_2$-coordinate axes of a 3D coordinate system. The $\mathbf{e}_3$-axis (vertical to the table) is computed automatically.[2] Now when you touch your

---

[2] Just how we convert the three points on the table to the three axes is covered in Section 11.8.

Figure 1.4.
Creating coordinates: a cat is turned into math. (Microscribe-3D from Immersion Corporation, http://www.immersion.com.)

cat model with the tip of the CMM's arm, it will associate three coordinates with that position and record them. You repeat this for several hundred points, and you have your cat in the box! This process is called *digitizing*. In the end, the cat has been "discretized," or turned into a finite number of coordinate triples. This set of points is called a *point cloud*.

Someone else will now have to build a mathematical model of your cat.[3] Next, the mathematical model will have to be put into scenes of the movie—but all that's needed for that are 3D coordinate transformations! (See Chapters 9 and 10.)

- unit square
- 2D and 3D local coordinates
- 2D and 3D global coordinates
- coordinate transformation
- parameter
- aspect ratio
- normalized coordinates
- digitizing
- point cloud

---

[3] This type of work is called *geometric modeling* or *computer-aided geometric design*, see [7] and Chapter 20.

1.6 Exercises

1. Let the coordinates of triangle vertices in the local $[\mathbf{d}_1, \mathbf{d}_2]$-system unit square be given by

   $$(u_1, u_2) = (0.1, 0.1), \quad (v_1, v_2) = (0.9, 0.2), \quad (w_1, w_2) = (0.4, 0.7).$$

   (a) If the $[\mathbf{d}_1, \mathbf{d}_2]$-system unit square is mapped to the target box with

   $$(\min_1, \min_2) = (1, 2) \quad \text{and} \quad (\max_1, \max_2) = (3, 3),$$

   where are the coordinates of the triangle vertices mapped?

   (b) What local coordinates correspond to $(x_1, x_2) = (2, 2)$ in the $[\mathbf{e}_1, \mathbf{e}_2]$-system?

2. Given local coordinates $(2, 2)$ and $(-1, -1)$, find the global coordinates with respect to the target box with

   $$(\min_1, \min_2) = (1, 1) \quad \text{and} \quad (\max_1, \max_2) = (7, 3).$$

   Make a sketch of the local and global systems. Connect the coordinates in each system with a line and compare.

3. Let the $[\mathbf{d}_1, \mathbf{d}_2, \mathbf{d}_3]$-system unit cube be mapped to the 3D target box with

   $$(\min_1, \min_2, \min_3) = (1, 1, 1) \quad \text{and} \quad (\Delta_1, \Delta_2, \Delta_3) = (1, 2, 4).$$

   Where will the coordinates $(u_1, u_2, u_3) = (0.5, 0, 0.7)$ be mapped?

4. Let the coordinates of triangle vertices in the local $[\mathbf{d}_1, \mathbf{d}_2, \mathbf{d}_3]$-system unit square be given by

   $$(u_1, u_2, u_3) = (0.1, 0.1, 0), \quad (v_1, v_2, v_3) = (0.9, 0.2, 1),$$
   $$(w_1, w_2, w_3) = (0.4, 0.7, 0).$$

   If the $[\mathbf{d}_1, \mathbf{d}_2, \mathbf{d}_3]$-system unit square is mapped to the target box with

   $$(\min_1, \min_2) = (1, 2, 4) \quad \text{and} \quad (\max_1, \max_2) = (3, 3, 8),$$

   where are the coordinates of the triangle vertices mapped? *Hint: See Exercise 1a.*

5. Suppose we are given a global frame defined by $(\min_1, \min_2, \min_3) = (0, 0, 3)$ and $(\max_1, \max_2, \max_3) = (4, 4, 4)$. For the coordinates $(1, 1, 3)$ and $(0, 0, 0)$ in this frame, what are the corresponding coordinates in the $[\mathbf{d}_1, \mathbf{d}_2, \mathbf{d}_3]$-system?

6. Assume you have an image in a local frame of 20 mm^2. If you enlarge the frame and the image such that the new frame covers 40 mm^2, by how much does the image size change?

7. Suppose that local frame coordinates $(v_1, v_2) = (1/2, 1)$ are mapped to global frame coordinates $(5, 2)$ and similarly, $(w_1, w_2) = (1, 1)$ are mapped to $(8, 8)$. In the local frame, $(u_1, u_2) = (3/4, 1/2)$ lies at the midpoint of two local coordinate sets. What are its coordinates in the global frame?

8. The size of a TV set is specified by its monitor's diagonal. In order to determine the width and height of this TV, we must know the relevant aspect ratio. What are the width and height dimensions of a $32''$ standard TV with aspect ratio $4 : 3$? What are the dimensions of a $32''$ HD TV with aspect ratio $16 : 9$? (This is an easy one to find on the web, but the point is to calculate it yourself!)

9. Suppose we are given coordinates $(1/2, 1/2)$ in the $[\mathbf{d}_1, \mathbf{d}_2]$-system. What are the corresponding coordinates in the global frame with aspect ratio $4 : 3$, $(\min_1, \min_2) = (0, 0)$, and $\Delta_2 = 2$?

10. In some implementations of the computer graphics viewing pipeline, *normalized device coordinates*, or NDC, are defined as the cube with extents $(-1, -1, -1)$ and $(1, 1, 1)$. The next step in the pipeline maps the (u_1, u_2) coordinates from NDC (u_3 is ignored) to the *viewport*, the area of the screen where the image will appear. Give equations for (x_1, x_2) in the viewport defined by extents $(\min_1, \min_2)$ and $(\max_1, \max_2)$ that correspond to (u_1, u_2) in NDC.

2

Here and There: Points and Vectors in 2D

Figure 2.1.
Hurricane Katrina: the hurricane is shown here approaching south Louisiana. (Image courtesy of NOAA, katrina.noaa.gov.)

In 2005 Hurricane Katrina caused flooding and deaths as it made its way from the Bahamas to south Florida as a category 1 hurricane. Over the warm waters of the Gulf, it grew into a category 5 hurricane, and even though at landfall in southeast Louisiana it had weakened to a category 3 hurricane, the storm surges and destruction it created rates it as the most expensive hurricane to date, causing more than $45 billion of damage. Sadly it was also one of the deadliest, particularly for residents of New Orleans. In the hurricane image (Figure 2.1), air is moving rapidly, spiraling in a counterclockwise fashion. What isn't so clear from this image is that the air moves faster as it approaches the eye of the hurricane. This air movement is best described by points and vectors: at any location (point), air moves in a certain direction and with a certain speed (velocity vector).

This hurricane image is a good example of how helpful 2D geometry can be in a 3D world. Of course a hurricane is a 3D phenomenon; however, by analyzing 2D slices, or cross sections, we can develop a very informative analysis. Many other applications call for 2D geometry only. The purpose of this chapter is to define the two most fundamental tools we need to work in a 2D world: points and vectors.

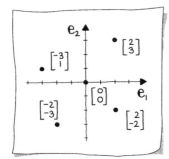

Sketch 2.1.
Points and their coordinates.

2.1   Points and Vectors

The most basic geometric entity is the *point*. A point is a reference to a *location*. Sketch 2.1 illustrates examples of points. In the text, boldface lowercase letters represent points, e.g.,

$$\mathbf{p} = \begin{bmatrix} p_1 \\ p_2 \end{bmatrix}. \tag{2.1}$$

The location of $\mathbf{p}$ is p_1-units along the $\mathbf{e}_1$-axis and p_2-units along the $\mathbf{e}_2$-axis. Thus a point's *coordinates*, p_1 and p_2, are dependent upon the location of the coordinate origin. We use the boldface notation so there is a noticeable difference between a one-dimensional (1D) number, or *scalar p*. To clearly identify $\mathbf{p}$ as a point, the notation $\mathbf{p} \in \mathbb{E}^2$ is used. This means that a 2D point "lives" in 2D Euclidean space $\mathbb{E}^2$.

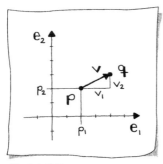

Sketch 2.2.
Two points and a vector.

Now let's move away from our reference point. Following Sketch 2.2, suppose the reference point is $\mathbf{p}$, and when moving along a straight path, our target point is $\mathbf{q}$. The directions from $\mathbf{p}$ would be to follow the *vector* $\mathbf{v}$. Our notation for a vector is the same as for a point: boldface lowercase letters. To get to $\mathbf{q}$ we say,

$$\mathbf{q} = \mathbf{p} + \mathbf{v}. \tag{2.2}$$

To calculate this, add each component separately; that is,

$$\begin{bmatrix} q_1 \\ q_2 \end{bmatrix} = \begin{bmatrix} p_1 \\ p_2 \end{bmatrix} + \begin{bmatrix} v_1 \\ v_2 \end{bmatrix} = \begin{bmatrix} p_1 + v_1 \\ p_2 + v_2 \end{bmatrix}.$$

For example, in Sketch 2.2, we have

$$\begin{bmatrix} 4 \\ 3 \end{bmatrix} = \begin{bmatrix} 2 \\ 2 \end{bmatrix} + \begin{bmatrix} 2 \\ 1 \end{bmatrix}.$$

The *components* of $\mathbf{v}$, v_1 and v_2, indicate how many units to move along the $\mathbf{e}_1$- and $\mathbf{e}_2$-axis, respectively. This means that $\mathbf{v}$ can be defined as

$$\mathbf{v} = \mathbf{q} - \mathbf{p}. \qquad (2.3)$$

This defines a vector as a difference of two points, which describes a *direction and a distance*, or a *displacement*. Examples of vectors are illustrated in Sketch 2.3.

How to determine a vector's length is covered in Section 2.4. Above we described this length as a distance. Alternatively, this length can be described as speed: then we have a *velocity vector*.[1] Yet another interpretation is that the length represents acceleration: then we have a *force vector*.

A vector has a *tail* and a *head*. As in Sketch 2.2, the tail is typically displayed positioned at a point, or *bound to a point* in order to indicate the geometric significance of the vector. However, unlike a point, a vector does *not* define a position. Two vectors are equal if they have the same component values, just as points are equal if they have the same coordinate values. Thus, considering a vector as a difference of two points, there are any number of vectors with the same direction and length. See Sketch 2.4 for an illustration.

A special vector worth mentioning is the *zero vector*,

$$\mathbf{0} = \begin{bmatrix} 0 \\ 0 \end{bmatrix}.$$

This vector has no direction or length. Other somewhat special vectors include

$$\mathbf{e}_1 = \begin{bmatrix} 1 \\ 0 \end{bmatrix} \quad \text{and} \quad \mathbf{e}_2 = \begin{bmatrix} 0 \\ 1 \end{bmatrix}.$$

In the sketches, these vectors are not always drawn true to length to prevent them from obscuring the main idea.

To clearly identify $\mathbf{v}$ as a vector, we write $\mathbf{v} \in \mathbb{R}^2$. This means that a 2D vector "lives" in a 2D *linear space* $\mathbb{R}^2$. (Other names for $\mathbb{R}^2$ are *real* or *vector spaces*.)

---

[1] This is what we'll use to continue the Hurricane Katrina example.

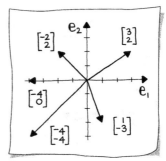

Sketch 2.3.
Vectors and their components.

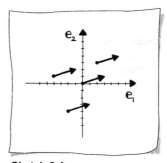

Sketch 2.4.
Instances of one vector.

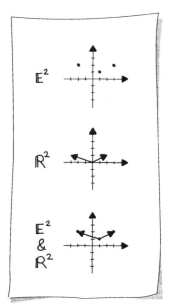

Sketch 2.5.

Euclidean and linear spaces illustrated separately and together.

2.2   What's the Difference?

When writing a point or a vector we use boldface lowercase letters; when programming we use the same data structure, e.g., arrays. This makes it appear that points and vectors can be treated in the same manner. Not so!

Points and vectors are different geometric entities. This is reiterated by saying they live in different spaces, $\mathbb{E}^2$ and $\mathbb{R}^2$. As shown in Sketch 2.5, for convenience and clarity elements of Euclidean and linear spaces are typically displayed together.

The primary reason for differentiating between points and vectors is to achieve geometric constructions that are *coordinate independent*. Such constructions are manipulations applied to geometric objects that produce the same result regardless of the location of the coordinate origin (for example, the midpoint of two points). This idea becomes clearer by analyzing some fundamental manipulations of points and vectors. In what follows, let's use $\mathbf{p}, \mathbf{q} \in \mathbb{E}^2$ and $\mathbf{v}, \mathbf{w} \in \mathbb{R}^2$.

Coordinate Independent Operations:

- Subtracting a point from another ($\mathbf{p} - \mathbf{q}$) yields a vector, as depicted in Sketch 2.2 and Equation (2.3).

- Adding or subtracting two vectors yields another vector. See Sketch 2.6, which illustrates the *parallelogram rule:* the vectors $\mathbf{v} - \mathbf{w}$ and $\mathbf{v} + \mathbf{w}$ are the diagonals of the parallelogram defined by $\mathbf{v}$ and $\mathbf{w}$. This is a coordinate independent operation since vectors are defined as a difference of points.

- Multiplying by a scalar s is called *scaling*. Scaling a vector is a well-defined operation. The result $s\mathbf{v}$ adjusts the length by the scaling factor. The direction is unchanged if $s > 0$ and reversed for $s < 0$. If $s = 0$ then the result is the zero vector. Sketch 2.7 illustrates some examples of scaling a vector.

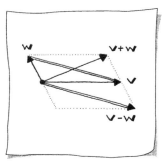

Sketch 2.6.

Parallelogram rule.

- Adding a vector to a point $(\mathbf{p} + \mathbf{v})$ yields another point, as in Sketch 2.2 and Equation (2.2).

Any coordinate independent combination of two or more points and/or vectors can be grouped to fall into one or more of the items above. See the Exercises for examples.

Coordinate Dependent Operations:

- Scaling a point (s**p**) is not a well-defined operation because it is not coordinate independent. Sketch 2.8 illustrates that the result of scaling the solid black point by one-half with respect to two different coordinate systems results in two different points.

- Adding two points (**p**+**q**) is not a well-defined operation because it is not coordinate independent. As depicted in Sketch 2.9, the result of adding the two solid black points is dependent on the coordinate origin. (The parallelogram rule is used here to construct the results of the additions.)

Some special combinations of points are allowed; they are defined in Section 2.5.

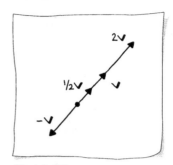

Sketch 2.7.
Scaling a vector.

2.3   Vector Fields

Figure 2.2.
Vector field: simulating hurricane air velocity. Lighter gray indicates greater velocity.

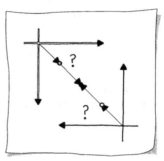

Sketch 2.8.
Scaling of points is ambiguous.

A good way to visualize the interplay between points and vectors is through the example of *vector fields*. In general, we speak of a vector field if every point in a given region is assigned a vector. We have already encountered an example of this in Figure 2.1: Hurricane Katrina! Recall that at each location (point) we could describe the air velocity (vector). Our previous image did not actually tell us anything about the air speed, although we could presume something about the direction. This is where a vector field is helpful. Shown in Figure 2.2 is a vector field simulating Hurricane Katrina. By plotting all the

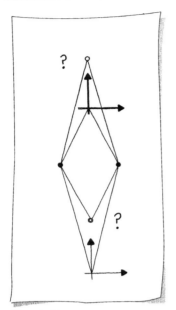

Sketch 2.9.
Addition of points is ambiguous.

vectors the same length and using *gray scale* or varying shades of gray to indicate speed, the vector field can be more informative than the photograph. (Visualization of a vector field requires *discretizing* it: a finite number of point and vector pairs are selected from a continuous field or from sampled measurements.)

Other important applications of vector fields arise in the areas of automotive and aerospace design: before a car or an airplane is built, it undergoes extensive aerodynamic simulations. In these simulations, the vectors that characterize the flow around an object are computed from complex differential equations. In Figure 2.3 we have another example of a vector field.

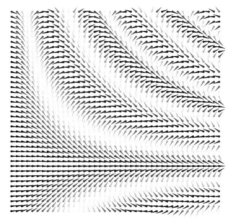

Figure 2.3.
Vector field: every sampled point has an associated vector. Lighter gray indicates greater vector length.

2.4   Length of a Vector

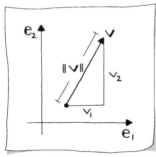

Sketch 2.10.
Length of a vector.

As mentioned in Section 2.1, the length of a vector can represent distance, velocity, or acceleration. We need a method for finding the length of a vector, or the *magnitude*. As illustrated in Sketch 2.10, a vector defines the displacement necessary (with respect to the $\mathbf{e}_1$- and $\mathbf{e}_2$-axis) to get from a point at the tail of the vector to a point at the head of the vector.

In Sketch 2.10 we have formed a right triangle. The square of the length of the hypotenuse of a right triangle is well known from the *Pythagorean theorem*. Denote the *length* of a vector $\mathbf{v}$ as $\|\mathbf{v}\|$. Then

$$\|\mathbf{v}\|^2 = v_1^2 + v_2^2.$$

Therefore, the magnitude of $\mathbf{v}$ is

$$\|\mathbf{v}\| = \sqrt{v_1^2 + v_2^2}. \tag{2.4}$$

This is also called the *Euclidean norm*. Notice that if we scale the vector by an amount k then

$$\|k\mathbf{v}\| = |k|\|\mathbf{v}\|. \tag{2.5}$$

A *normalized vector* $\mathbf{w}$ has *unit length*, that is

$$\|\mathbf{w}\| = 1.$$

Normalized vectors are also known as *unit vectors*. To *normalize* a vector simply means to scale a vector so that it has unit length. If $\mathbf{w}$ is to be our unit length version of $\mathbf{v}$ then

$$\mathbf{w} = \frac{\mathbf{v}}{\|\mathbf{v}\|}.$$

Each component of $\mathbf{v}$ is divided by the scalar value $\|\mathbf{v}\|$. This scalar value is always *nonnegative*, which means that its value is zero or greater. It can be zero! You must check the value before dividing to be sure it is greater than your *zero divide tolerance*. The zero divide tolerance is the absolute value of the smallest number by which you can divide confidently. (When we refer to checking that a value is greater than this number, it means to check the absolute value.)

In Figures 2.2 and 2.3, we display vectors of varying magnitudes. But instead of plotting them using different lengths, their magnitude is indicated by gray scales.

Example 2.1

Start with

$$\mathbf{v} = \begin{bmatrix} 5 \\ 0 \end{bmatrix}.$$

Applying (2.4), $\|\mathbf{v}\| = \sqrt{5^2 + 0^2} = 5$. Then the normalized version of $\mathbf{v}$ is defined as

$$\mathbf{w} = \begin{bmatrix} 5/5 \\ 0/5 \end{bmatrix} = \begin{bmatrix} 1 \\ 0 \end{bmatrix}.$$

Clearly $\|\mathbf{w}\| = 1$, so this is a normalized vector. Since we have only scaled $\mathbf{v}$ by a positive amount, the direction of $\mathbf{w}$ is the same as $\mathbf{v}$.

Figure 2.4.
Unit vectors: they define a circle.

There are infinitely many unit vectors. Imagine drawing them all, emanating from the origin. The figure that you will get is a circle of radius one! See Figure 2.4.

To find the *distance between two points* we simply form a vector defined by the two points, e.g., $\mathbf{v} = \mathbf{q} - \mathbf{p}$, and apply (2.4).

Example 2.2

Let
$$\mathbf{q} = \begin{bmatrix} -1 \\ 2 \end{bmatrix} \quad \text{and} \quad \mathbf{p} = \begin{bmatrix} 1 \\ 0 \end{bmatrix}.$$

Then
$$\mathbf{q} - \mathbf{p} = \begin{bmatrix} -2 \\ 2 \end{bmatrix}$$

and
$$\|\mathbf{q} - \mathbf{p}\| = \sqrt{(-2)^2 + 2^2} = \sqrt{8} \approx 2.83.$$

Sketch 2.11 illustrates this example.

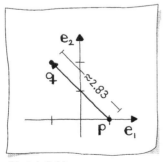

Sketch 2.11.
Distance between two points.

2.5 Combining Points

Seemingly contrary to Section 2.2, there actually is a way to combine two points such that we get a (meaningful) third one. Take the example of the midpoint $\mathbf{r}$ of two points $\mathbf{p}$ and $\mathbf{q}$; more specifically, take

$$\mathbf{p} = \begin{bmatrix} 1 \\ 6 \end{bmatrix}, \quad \mathbf{r} = \begin{bmatrix} 2 \\ 3 \end{bmatrix}, \quad \mathbf{q} = \begin{bmatrix} 3 \\ 0 \end{bmatrix},$$

as shown in Sketch 2.12.

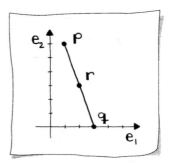

Sketch 2.12.
The midpoint of two points.

Let's start with the known coordinate independent operation of adding a vector to a point. Define $\mathbf{r}$ by adding an appropriately scaled version of the vector $\mathbf{v} = \mathbf{q} - \mathbf{p}$ to the point $\mathbf{p}$:

$$\mathbf{r} = \mathbf{p} + \frac{1}{2}\mathbf{v}$$

$$\begin{bmatrix} 2 \\ 3 \end{bmatrix} = \begin{bmatrix} 1 \\ 6 \end{bmatrix} + \frac{1}{2}\begin{bmatrix} 2 \\ -6 \end{bmatrix}.$$

Expanding, this shows that $\mathbf{r}$ can also be defined as

$$\mathbf{r} = \frac{1}{2}\mathbf{p} + \frac{1}{2}\mathbf{q}$$

$$\begin{bmatrix} 2 \\ 3 \end{bmatrix} = \frac{1}{2}\begin{bmatrix} 1 \\ 6 \end{bmatrix} + \frac{1}{2}\begin{bmatrix} 3 \\ 0 \end{bmatrix}.$$

This is a legal expression for a combination of points.

There is nothing magical about the factor $1/2$, however. Adding a (scaled) vector to a point is a well-defined, coordinate independent operation that yields another point. Any point of the form

$$\mathbf{r} = \mathbf{p} + t\mathbf{v} \qquad (2.6)$$

is on the line through $\mathbf{p}$ and $\mathbf{q}$. Again, we may rewrite this as

$$\mathbf{r} = \mathbf{p} + t(\mathbf{q} - \mathbf{p})$$

and then

$$\mathbf{r} = (1 - t)\mathbf{p} + t\mathbf{q}. \qquad (2.7)$$

Sketch 2.13 gives an example with $t = 1/3$.

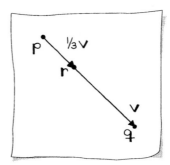

Sketch 2.13.
Barycentric combinations:
$t = 1/3$.

The scalar values $(1 - t)$ and t are *coefficients*. A weighted sum of points where the coefficients sum to one is called a *barycentric combination*. In this special case, where one point $\mathbf{r}$ is being expressed in terms of two others, $\mathbf{p}$ and $\mathbf{q}$, the coefficients $1 - t$ and t are called the *barycentric coordinates* of $\mathbf{r}$.

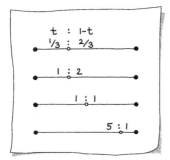

Sketch 2.14.

Examples of ratios.

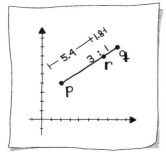

Sketch 2.15.

Barycentric coordinates in relation to lengths.

A barycentric combination allows us to construct $\mathbf{r}$ anywhere on the line defined by $\mathbf{p}$ and $\mathbf{q}$. This is why (2.7) is also called linear interpolation. If we would like to restrict $\mathbf{r}$'s position to the *line segment* between $\mathbf{p}$ and $\mathbf{q}$, then we allow only *convex combinations*: t must satisfy $0 \le t \le 1$. To define points outside of the line segment between $\mathbf{p}$ and $\mathbf{q}$, we need values of $t < 0$ or $t > 1$.

The position of $\mathbf{r}$ is said to be in the *ratio* of $t : (1 - t)$ or $t/(1 - t)$. In physics, $\mathbf{r}$ is known as the *center of gravity* of two points $\mathbf{p}$ and $\mathbf{q}$ with weights $1 - t$ and t, respectively. From a constructive approach, the ratio is formed from the quotient

$$\text{ratio} = \frac{||\mathbf{r} - \mathbf{p}||}{||\mathbf{q} - \mathbf{r}||}.$$

Some examples are illustrated in Sketch 2.14.

Example 2.3

Suppose we have three collinear points, $\mathbf{p}$, $\mathbf{q}$, and $\mathbf{r}$ as illustrated in Sketch 2.15. The points have the following locations.

$$\mathbf{p} = \begin{bmatrix} 2 \\ 4 \end{bmatrix}, \quad \mathbf{r} = \begin{bmatrix} 6.5 \\ 7 \end{bmatrix}, \quad \mathbf{q} = \begin{bmatrix} 8 \\ 8 \end{bmatrix}.$$

What are the barycentric coordinates of $\mathbf{r}$ with respect to $\mathbf{p}$ and $\mathbf{q}$?

To answer this, recall the relationship between the ratio and the barycentric coordinates. The barycentric coordinates t and $(1 - t)$ define $\mathbf{r}$ as

$$\begin{bmatrix} 6.5 \\ 7 \end{bmatrix} = (1 - t) \begin{bmatrix} 2 \\ 4 \end{bmatrix} + t \begin{bmatrix} 8 \\ 8 \end{bmatrix}.$$

The ratio indicates the location of $\mathbf{r}$ relative to $\mathbf{p}$ and $\mathbf{q}$ in terms of relative distances. Suppose the ratio is $s_1 : s_2$. If we scale s_1 and s_2 such that they sum to one, then s_1 and s_2 are the barycentric coordinates t and $(1 - t)$, respectively. By calculating the distances between points:

$$l_1 = ||\mathbf{r} - \mathbf{p}|| \approx 5.4,$$
$$l_2 = ||\mathbf{q} - \mathbf{r}|| \approx 1.8,$$
$$l_3 = l_1 + l_2 \approx 7.2,$$

we find that

$$t = l_1/l_3 = 0.75 \text{ and}$$
$$(1 - t) = l_2/l_3 = 0.25.$$

These are the barycentric coordinates. Let's verify this:

$$\begin{bmatrix} 6.5 \\ 7 \end{bmatrix} = 0.25 \times \begin{bmatrix} 2 \\ 4 \end{bmatrix} + 0.75 \times \begin{bmatrix} 8 \\ 8 \end{bmatrix}.$$

The barycentric coordinate t is also called a *parameter*. (See Section 3.2 for more details.) This parameter is defined by the quotient

$$t = \frac{\|\mathbf{r} - \mathbf{p}\|}{\|\mathbf{q} - \mathbf{p}\|}.$$

We have seen how useful this quotient can be in Section 1.1 for the construction of a point in the global system that corresponded to a point with parameter t in the local system.

We can create barycentric combinations with *more than two points*. Let's look at three points $\mathbf{p}$, $\mathbf{q}$, and $\mathbf{r}$, which are not collinear. Any point $\mathbf{s}$ can be formed from

$$\mathbf{s} = \mathbf{r} + t_1(\mathbf{p} - \mathbf{r}) + t_2(\mathbf{q} - \mathbf{r}).$$

This is a coordinate independent operation of point + vector + vector. Expanding and regrouping, we can also define $\mathbf{s}$ as

$$\begin{aligned} \mathbf{s} &= t_1\mathbf{p} + t_2\mathbf{q} + (1 - t_1 - t_2)\mathbf{r} \\ &= t_1\mathbf{p} + t_2\mathbf{q} + t_3\mathbf{r}. \end{aligned} \tag{2.8}$$

Thus, the point $\mathbf{s}$ is defined by a barycentric combination with coefficients t_1, t_2, and $t_3 = 1 - t_1 - t_2$ with respect to $\mathbf{p}$, $\mathbf{q}$, and $\mathbf{r}$, respectively. This is another special case where the barycentric combination coefficients correspond to *barycentric coordinates*. Sketch 2.16 illustrates this. We will encounter barycentric coordinates in more detail in Chapter 17.

We can also *combine points* so that the result is a vector. For this, we need the coefficients to sum to zero. We encountered a simple case of this in (2.3). Suppose we have the equation

$$\mathbf{e} = \mathbf{r} - 2\mathbf{p} + \mathbf{q}, \qquad \mathbf{r}, \mathbf{p}, \mathbf{q} \in \mathbb{E}^2.$$

Does $\mathbf{e}$ have a geometric meaning? Looking at the sum of the coefficients, $1 - 2 + 1 = 0$, we would conclude by the rule above that $\mathbf{e}$ is a vector. How to see this? By rewriting the equation as

$$\mathbf{e} = (\mathbf{r} - \mathbf{p}) + (\mathbf{q} - \mathbf{p}),$$

it is clear that $\mathbf{e}$ is a vector formed from (vector + vector).

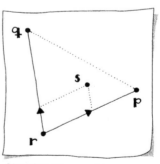

Sketch 2.16.
A barycentric combination of three points.

2.6   Independence

Two vectors **v** and **w** describe a parallelogram, as shown in Sketch 2.6. It may happen that this parallelogram has zero area; then the two vectors are parallel. In this case, we have a relationship of the form **v** = *c***w**. If two vectors are parallel, then we call them *linearly dependent*. Otherwise, we say that they are *linearly independent*.

Two linearly independent vectors may be used to write any other vector **u** as a *linear combination*:

$$\mathbf{u} = r\mathbf{v} + s\mathbf{w}.$$

How to find r and s is described in Chapter 5. Two linearly independent vectors in 2D are also called a *basis* for $\mathbb{R}^2$. If **v** and **w** are linearly dependent, then you cannot write all vectors as a linear combination of them, as the following example shows.

Example 2.4

Let

$$\mathbf{v} = \begin{bmatrix} 1 \\ 2 \end{bmatrix} \quad \text{and} \quad \mathbf{w} = \begin{bmatrix} 2 \\ 4 \end{bmatrix}.$$

If we tried to write the vector

$$\mathbf{u} = \begin{bmatrix} 1 \\ 0 \end{bmatrix}$$

as $\mathbf{u} = r\mathbf{v} + s\mathbf{w}$, then this would lead to

$$1 = r + 2s, \tag{2.9}$$
$$0 = 2r + 4s. \tag{2.10}$$

If we multiply the first equation by a factor of 2, the two right-hand sides will be equal. Equating the new left-hand sides now results in the expression $2 = 0$. This shows that **u** cannot be written as a linear combination of **v** and **w**. (See Sketch 2.17.)

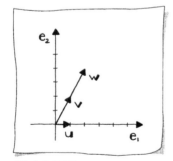

Sketch 2.17.
Dependent vectors.

2.7   Dot Product

Given two vectors **v** and **w**, we might ask:

- Are they the *same* vector?

- Are they *perpendicular* to each other?

- What *angle* do they form?

The *dot product* is the tool to resolve these questions. Assume that $\mathbf{v}$ and $\mathbf{w}$ are not the zero vector.

To motivate the dot product, let's start with the Pythagorean theorem and Sketch 2.18. There, we see two perpendicular vectors $\mathbf{v}$ and $\mathbf{w}$; we conclude

$$\|\mathbf{v} - \mathbf{w}\|^2 = \|\mathbf{v}\|^2 + \|\mathbf{w}\|^2. \tag{2.11}$$

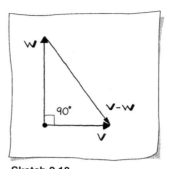

Sketch 2.18.
Perpendicular vectors.

Writing the components in (2.11) explicitly

$$(v_1 - w_1)^2 + (v_2 - w_2)^2 = (v_1^2 + v_2^2) + (w_1^2 + w_2^2),$$

and then expanding, bringing all terms to the left-hand side of the equation yields

$$(v_1^2 - 2v_1w_1 + w_1^2) + (v_2^2 - 2v_2w_2 + w_2^2) - (v_1^2 + v_2^2) - (w_1^2 + w_2^2) = 0,$$

which reduces to

$$v_1w_1 + v_2w_2 = 0. \tag{2.12}$$

We find that perpendicular vectors have the property that the sum of the products of their components is zero. The short-hand vector notation for (2.12) is

$$\mathbf{v} \cdot \mathbf{w} = 0. \tag{2.13}$$

This result has an immediate application: a vector $\mathbf{w}$ perpendicular to a given vector $\mathbf{v}$ can be formed as

$$\mathbf{w} = \begin{bmatrix} -v_2 \\ v_1 \end{bmatrix}$$

(switching components and negating the sign of one). Then $\mathbf{v} \cdot \mathbf{w}$ becomes $v_1(-v_2) + v_2v_1 = 0$.

If we take two arbitrary vectors $\mathbf{v}$ and $\mathbf{w}$, then $\mathbf{v} \cdot \mathbf{w}$ will in general not be zero. But we can compute it anyway, and define

$$s = \mathbf{v} \cdot \mathbf{w} = v_1w_1 + v_2w_2 \tag{2.14}$$

to be the *dot product* of $\mathbf{v}$ and $\mathbf{w}$. Notice that the dot product returns a scalar s, which is why it is also called a *scalar product*. (Mathematicians have yet another name for the dot product—an *inner product*. See Section 14.3 for more on these.) From (2.14) it is clear that

$$\mathbf{v} \cdot \mathbf{w} = \mathbf{w} \cdot \mathbf{v}.$$

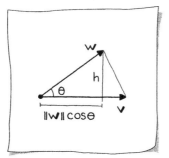

Sketch 2.19.
Geometry of the dot product.

This is called the *symmetry property*. Other properties of the dot product are given in the Exercises.

In order to understand the geometric meaning of the dot product of two vectors, let's construct a triangle from two vectors **v** and **w** as illustrated in Sketch 2.19.

From trigonometry, we know that the height h of the triangle can be expressed as

$$h = \|\mathbf{w}\| \sin(\theta).$$

Squaring both sides results in

$$h^2 = \|\mathbf{w}\|^2 \sin^2(\theta).$$

Using the identity

$$\sin^2(\theta) + \cos^2(\theta) = 1,$$

we have

$$h^2 = \|\mathbf{w}\|^2 (1 - \cos^2(\theta)). \tag{2.15}$$

We can also express the height h with respect to the other right triangle in Sketch 2.19 and by using the Pythagorean theorem:

$$h^2 = \|\mathbf{v} - \mathbf{w}\|^2 - (\|\mathbf{v}\| - \|\mathbf{w}\| \cos\theta)^2. \tag{2.16}$$

Equating (2.15) and (2.16) and simplifying, we have the expression,

$$\|\mathbf{v} - \mathbf{w}\|^2 = \|\mathbf{v}\|^2 + \|\mathbf{w}\|^2 - 2\|\mathbf{v}\|\|\mathbf{w}\| \cos\theta. \tag{2.17}$$

We have just proved the *Law of Cosines*, which generalizes the Pythagorean theorem by correcting it for triangles with an opposing angle different from 90°.

We can formulate another expression for $\|\mathbf{v} - \mathbf{w}\|^2$ by explicitly writing out

$$\begin{aligned}
\|\mathbf{v} - \mathbf{w}\|^2 &= (\mathbf{v} - \mathbf{w}) \cdot (\mathbf{v} - \mathbf{w}) \\
&= \|\mathbf{v}\|^2 - 2\mathbf{v} \cdot \mathbf{w} + \|\mathbf{w}\|^2.
\end{aligned} \tag{2.18}$$

By equating (2.17) and (2.18) we find that

$$\mathbf{v} \cdot \mathbf{w} = \|\mathbf{v}\|\|\mathbf{w}\| \cos\theta. \tag{2.19}$$

Here is another expression for the *dot product*—it is a very useful one! Rearranging (2.19), the cosine of the angle between the two vectors can be determined as

$$\cos\theta = \frac{\mathbf{v} \cdot \mathbf{w}}{\|\mathbf{v}\|\|\mathbf{w}\|}. \tag{2.20}$$

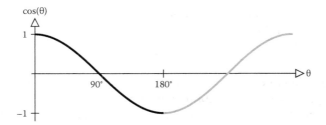

Figure 2.5.
Cosine function: its values at $\theta = 0°$, $\theta = 90°$, and $\theta = 180°$ are important to remember.

By examining a plot of the cosine function in Figure 2.5, some sense can be made of (2.20).

First we consider the special case of perpendicular vectors. Recall the dot product was zero, which makes $\cos(90°) = 0$, just as it should be.

If $\mathbf{v}$ has the same (or opposite) direction as $\mathbf{w}$, that is $\mathbf{v} = k\mathbf{w}$, then (2.20) becomes

$$\cos \theta = \frac{k\mathbf{w} \cdot \mathbf{w}}{\|k\mathbf{w}\| \|\mathbf{w}\|}.$$

Using (2.5), we have

$$\cos \theta = \frac{k\|\mathbf{w}\|^2}{|k| \|\mathbf{w}\| \|\mathbf{w}\|} = \pm 1.$$

Again, examining Figure 2.5, we see this corresponds to either $\theta = 0°$ or $\theta = 180°$, for vectors of the same or opposite direction, respectively.

The cosine values from (2.20) range between ± 1; this corresponds to angles between $0°$ and $180°$ (or 0 and π radians). Thus, the smaller angle between the two vectors is measured. This is clear from the derivation: the angle θ enclosed by completing the triangle defined by the two vectors must be less than $180°$. Three types of angles can be formed:

- *right*: $\cos(\theta) = 0 \rightarrow \mathbf{v} \cdot \mathbf{w} = 0$;

- *acute*: $\cos(\theta) > 0 \rightarrow \mathbf{v} \cdot \mathbf{w} > 0$;

- *obtuse*: $\cos(\theta) < 0 \rightarrow \mathbf{v} \cdot \mathbf{w} < 0$.

These are illustrated in counterclockwise order from twelve o'clock in Sketch 2.20.

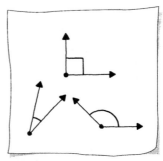

Sketch 2.20.
Three types of angles.

If the actual angle θ needs to be calculated, then the arccosine function has to be invoked: let

$$s = \frac{\mathbf{v} \cdot \mathbf{w}}{\|\mathbf{v}\|\|\mathbf{w}\|}$$

then $\theta = \mathrm{acos}(s)$ where acos is short for arccosine. One word of warning: in some math libraries, if $s > 1$ or $s < -1$ then an error occurs and a nonusable result (NaN—Not a Number) is returned.

Thus, if s is calculated, it is best to check that its value is within the appropriate range. It is not uncommon that an intended value of $s = 1.0$ is actually something like $s = 1.0000001$ due to *round-off*. Thus, the arccosine function should be used with caution. In many instances, as in comparing angles, the cosine of the angle is all you need! Additionally, computing the cosine or sine is 40 times more expensive than a multiplication, meaning that a cosine operation might take 200 cycles (operations) and a multiplication might take 5 cycles. Arccosine and arcsine are yet more expensive.

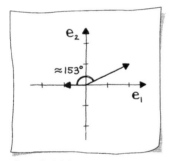

Sketch 2.21.
The angle between two vectors.

Example 2.5

Let's calculate the angle between the two vectors illustrated in Sketch 2.21, forming an obtuse angle:

$$\mathbf{v} = \begin{bmatrix} 2 \\ 1 \end{bmatrix} \quad \text{and} \quad \mathbf{w} = \begin{bmatrix} -1 \\ 0 \end{bmatrix}.$$

Calculate the length of each vector,

$$\|\mathbf{v}\| = \sqrt{2^2 + 1^2} = \sqrt{5}$$
$$\|\mathbf{w}\| = \sqrt{-1^2 + 0^2} = 1.$$

The cosine of the angle between the vectors is calculated using (2.20) as

$$\cos(\theta) = \frac{(2 \times -1) + (1 \times 0)}{\sqrt{5} \times 1} = \frac{-2}{\sqrt{5}} \approx -0.8944.$$

Then

$$\arccos(-0.8944) \approx 153.4°.$$

To convert an angle given in degrees to radians multiply by $\pi/180°$. (Recall that $\pi \approx 3.14159$ radians.) This means that

$$2.677 \text{ radians} \approx 153.4° \times \frac{\pi}{180°}.$$

2.8   Orthogonal Projections

Sketch 2.19 illustrates that the projection of the vector $\mathbf{w}$ onto $\mathbf{v}$ creates a footprint of length $b = ||\mathbf{w}|| \cos(\theta)$. This we derive from basic trigonometry: $\cos(\theta) = b/\text{hypotenuse}$. The *orthogonal projection* of $\mathbf{w}$ onto $\mathbf{v}$ is then the vector

$$\mathbf{u} = (||\mathbf{w}|| \cos(\theta)) \frac{\mathbf{v}}{||\mathbf{v}||} = \frac{\mathbf{v} \cdot \mathbf{w}}{||\mathbf{v}||^2} \mathbf{v}. \tag{2.21}$$

Sometimes this projection is expressed as

$$\mathbf{u} = \text{proj}_{\mathcal{V}_1} \mathbf{w},$$

where $\mathcal{V}_1$ is the set of all 2D vectors $k\mathbf{v}$ and it is referred to as a one-dimensional *subspace* of $\mathbb{R}^2$. Therefore, $\mathbf{u}$ is the *best approximation* to $\mathbf{w}$ in the subspace $\mathcal{V}_1$. This concept of closest or best approximation will be needed for several problems, such as finding the point at the end of the footprint in Section 3.7 and for least squares approximations in Section 12.7. We will revisit subspaces with more rigor in Chapter 14.

Using the orthogonal projection, it is easy to decompose the 2D vector $\mathbf{w}$ into a sum of two perpendicular vectors, namely $\mathbf{u}$ and $\mathbf{u}^\perp$ (a vector perpendicular to $\mathbf{u}$), such that

$$\mathbf{w} = \mathbf{u} + \mathbf{u}^\perp. \tag{2.22}$$

Another way to state this: we have *resolved* $\mathbf{w}$ into components with respect to two other vectors. Already having found the vector $\mathbf{u}$, we now set

$$\mathbf{u}^\perp = \mathbf{w} - \frac{\mathbf{v} \cdot \mathbf{w}}{||\mathbf{v}||^2} \mathbf{v}.$$

This can also be written as

$$\mathbf{u}^\perp = \mathbf{w} - \text{proj}_{\mathcal{V}_1} \mathbf{w},$$

and thus $\mathbf{u}^\perp$ is the component of $\mathbf{w}$ orthogonal to the space of $\mathbf{u}$.

See the Exercises of Chapter 8 for a 3D version of this decomposition. Orthogonal projections and vector decomposition are at the core of constructing the Gram-Schmidt orthonormal coordinate frame in Section 11.8 for 3D and in Section 14.4 for higher dimensions. An application that uses this frame is discussed in Section 20.7.

The ability to decompose a vector into its component parts is key to Fourier analysis, quantum mechanics, digital audio, and video recording.

2.9   Inequalities

Here are two important inequalities when dealing with vector lengths.

Let's start with the expression from (2.19), i.e.,

$$\mathbf{v} \cdot \mathbf{w} = \|\mathbf{v}\|\|\mathbf{w}\| \cos\theta.$$

Squaring both sides gives

$$(\mathbf{v} \cdot \mathbf{w})^2 = \|\mathbf{v}\|^2\|\mathbf{w}\|^2 \cos^2\theta.$$

Noting that $0 \le \cos^2\theta \le 1$, we conclude that

$$(\mathbf{v} \cdot \mathbf{w})^2 \le \|\mathbf{v}\|^2\|\mathbf{w}\|^2. \qquad (2.23)$$

This is the *Cauchy-Schwartz inequality*. Equality holds if and only if $\mathbf{v}$ and $\mathbf{w}$ are linearly dependent. This inequality is fundamental in the study of more general vector spaces, which are presented in Chapter 14.

Suppose we would like to find an inequality that describes the relationship between the length of two vectors $\mathbf{v}$ and $\mathbf{w}$ and the length of their sum $\mathbf{v} + \mathbf{w}$. In other words, how does the length of the third side of a triangle relate to the lengths of the other two? Let's begin with expanding $\|\mathbf{v} + \mathbf{w}\|^2$:

$$\begin{aligned}
\|\mathbf{v} + \mathbf{w}\|^2 &= (\mathbf{v} + \mathbf{w}) \cdot (\mathbf{v} + \mathbf{w}) \\
&= \mathbf{v} \cdot \mathbf{v} + 2\mathbf{v} \cdot \mathbf{w} + \mathbf{w} \cdot \mathbf{w} \\
&\le \mathbf{v} \cdot \mathbf{v} + 2|\mathbf{v} \cdot \mathbf{w}| + \mathbf{w} \cdot \mathbf{w} \\
&\le \mathbf{v} \cdot \mathbf{v} + 2\|\mathbf{v}\|\|\mathbf{w}\| + \mathbf{w} \cdot \mathbf{w} \\
&= \|\mathbf{v}\|^2 + 2\|\mathbf{v}\|\|\mathbf{w}\| + \|\mathbf{w}\|^2 \\
&= (\|\mathbf{v}\| + \|\mathbf{w}\|)^2.
\end{aligned} \qquad (2.24)$$

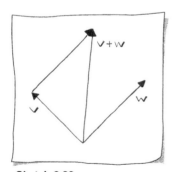

Taking square roots gives

$$\|\mathbf{v} + \mathbf{w}\| \le \|\mathbf{v}\| + \|\mathbf{w}\|,$$

Sketch 2.22.

The triangle inequality.

which is known as the *triangle inequality*. It states the intuitively obvious fact that the sum of any two edge lengths in a triangle is never smaller than the length of the third edge; see Sketch 2.22 for an illustration.

- point versus vector
- coordinates versus components
- $\mathbb{E}^2$ versus $\mathbb{R}^2$
- coordinate independent
- vector length
- unit vector
- zero divide tolerance
- Pythagorean theorem
- distance between two points
- parallelogram rule
- scaling
- ratio
- barycentric combination

- linear interpolation
- convex combination
- barycentric coordinates
- linearly dependent vectors
- linear combination
- basis for $\mathbb{R}^2$
- dot product
- Law of Cosines
- perpendicular vectors
- angle between vectors
- orthogonal projection
- vector decomposition
- Cauchy-Schwartz inequality
- triangle inequality

2.10 Exercises

1. Illustrate the parallelogram rule applied to the vectors
$$\mathbf{v} = \begin{bmatrix} -2 \\ 1 \end{bmatrix} \quad \text{and} \quad \mathbf{w} = \begin{bmatrix} 2 \\ 1 \end{bmatrix}.$$

2. The parallelogram rule states that adding or subtracting two vectors, $\mathbf{v}$ and $\mathbf{w}$, yields another vector. Why is it called the parallelogram rule?

3. Define your own $\mathbf{p}, \mathbf{q} \in \mathbb{E}^2$ and $\mathbf{v}, \mathbf{w} \in \mathbb{R}^2$. Determine which of the following expressions are geometrically meaningful. Illustrate those that are.

    (a) $\mathbf{p} + \mathbf{q}$ (b) $\frac{1}{2}\mathbf{p} + \frac{1}{2}\mathbf{q}$
    (c) $\mathbf{p} + \mathbf{v}$ (d) $3\mathbf{p} + \mathbf{v}$
    (e) $\mathbf{v} + \mathbf{w}$ (f) $2\mathbf{v} + \frac{1}{2}\mathbf{w}$
    (g) $\mathbf{v} - 2\mathbf{w}$ (h) $\frac{3}{2}\mathbf{p} - \frac{1}{2}\mathbf{q}$

4. Suppose we are given $\mathbf{p}, \mathbf{q} \in \mathbb{E}^2$ and $\mathbf{v}, \mathbf{w} \in \mathbb{R}^2$. Do the following operations result in a point or a vector?

    (a) $\mathbf{p} - \mathbf{q}$ (b) $\frac{1}{2}\mathbf{p} + \frac{1}{2}\mathbf{q}$
    (c) $\mathbf{p} + \mathbf{v}$ (d) $3\mathbf{v}$
    (e) $\mathbf{v} + \mathbf{w}$ (f) $\mathbf{p} + \frac{1}{2}\mathbf{w}$

5. What barycentric combination of the points $\mathbf{p}$ and $\mathbf{q}$ results in the midpoint of the line through these two points?

6. Illustrate a point with barycentric coordinates $(1/2, 1/4, 1/4)$ with respect to three other points.

7. Consider two points. Form the set of all convex combinations of these points. What is the geometry of this set?

8. Consider three noncollinear points. Form the set of all convex combinations of these points. What is the geometry of this set?

9. What is the length of the vector

$$\mathbf{v} = \begin{bmatrix} -4 \\ -3 \end{bmatrix}?$$

10. What is the magnitude of the vector

$$\mathbf{v} = \begin{bmatrix} 3 \\ -3 \end{bmatrix}?$$

11. If a vector $\mathbf{v}$ is length 10, then what is the length of the vector $-2\mathbf{v}$?

12. Find the distance between the points

$$\mathbf{p} = \begin{bmatrix} 3 \\ 3 \end{bmatrix} \quad \text{and} \quad \mathbf{q} = \begin{bmatrix} -2 \\ -3 \end{bmatrix}.$$

13. Find the distance between the points

$$\mathbf{p} = \begin{bmatrix} -3 \\ -3 \end{bmatrix} \quad \text{and} \quad \mathbf{q} = \begin{bmatrix} -2 \\ -3 \end{bmatrix}.$$

14. What is the length of the unit vector $\mathbf{u}$?

15. Normalize the vector $\mathbf{v} = \begin{bmatrix} -4 \\ -3 \end{bmatrix}.$

16. Normalize the vector $\mathbf{v} = \begin{bmatrix} 4 \\ 2 \end{bmatrix}.$

17. Given points

$$\mathbf{p} = \begin{bmatrix} 1 \\ 1 \end{bmatrix}, \quad \mathbf{q} = \begin{bmatrix} 7 \\ 7 \end{bmatrix}, \quad \mathbf{r} = \begin{bmatrix} 3 \\ 3 \end{bmatrix},$$

what are the barycentric coordinates of $\mathbf{r}$ with respect to $\mathbf{p}$ and $\mathbf{q}$?

18. Given points

$$\mathbf{p} = \begin{bmatrix} 1 \\ 1 \end{bmatrix}, \quad \mathbf{q} = \begin{bmatrix} 7 \\ 7 \end{bmatrix}, \quad \mathbf{r} = \begin{bmatrix} 5 \\ 5 \end{bmatrix},$$

what are the barycentric coordinates of $\mathbf{r}$ with respect to $\mathbf{p}$ and $\mathbf{q}$?

19. If $\mathbf{v} = 4\mathbf{w}$, are $\mathbf{v}$ and $\mathbf{w}$ linearly independent?

20. If $\mathbf{v} = 4\mathbf{w}$, what is the area of the parallelogram spanned by $\mathbf{v}$ and $\mathbf{w}$?

21. Do the vectors

$$\mathbf{v}_1 = \begin{bmatrix} 1 \\ 4 \end{bmatrix} \quad \text{and} \quad \mathbf{v}_2 = \begin{bmatrix} 3 \\ 0 \end{bmatrix}$$

    form a basis for $\mathbb{R}^2$?

22. What linear combination allows us to express $\mathbf{u}$ with respect to $\mathbf{v}_1$ and $\mathbf{v}_2$, where

$$\mathbf{u} = \begin{bmatrix} 6 \\ 4 \end{bmatrix}, \quad \mathbf{v}_1 = \begin{bmatrix} 1 \\ 0 \end{bmatrix}, \quad \mathbf{v}_2 = \begin{bmatrix} 4 \\ 4 \end{bmatrix}?$$

23. Show that the dot product has the following properties for vectors $\mathbf{u}, \mathbf{v}, \mathbf{w} \in \mathbb{R}^2$.

$$\mathbf{u} \cdot \mathbf{v} = \mathbf{v} \cdot \mathbf{u} \qquad \text{symmetric}$$

$$\mathbf{v} \cdot (s\mathbf{w}) = s(\mathbf{v} \cdot \mathbf{w}) \qquad \text{homogeneous}$$

$$(\mathbf{v} + \mathbf{w}) \cdot \mathbf{u} = \mathbf{v} \cdot \mathbf{u} + \mathbf{w} \cdot \mathbf{u} \qquad \text{distributive}$$

$$\mathbf{v} \cdot \mathbf{v} > 0 \quad \text{if} \quad \mathbf{v} \neq \mathbf{0} \quad \text{and} \quad \mathbf{v} \cdot \mathbf{v} = 0 \quad \text{if} \quad \mathbf{v} = \mathbf{0} \quad \text{positive}$$

24. What is $\mathbf{v} \cdot \mathbf{w}$ where

$$\mathbf{v} = \begin{bmatrix} 5 \\ 4 \end{bmatrix} \quad \text{and} \quad \mathbf{w} = \begin{bmatrix} 0 \\ 1 \end{bmatrix}?$$

    What is the scalar product of $\mathbf{w}$ and $\mathbf{v}$?

25. Compute the angle (in degrees) formed by the vectors

$$\mathbf{v} = \begin{bmatrix} 5 \\ 5 \end{bmatrix} \quad \text{and} \quad \mathbf{w} = \begin{bmatrix} 3 \\ -3 \end{bmatrix}.$$

26. Compute the cosine of the angle formed by the vectors

$$\mathbf{v} = \begin{bmatrix} 5 \\ 5 \end{bmatrix} \quad \text{and} \quad \mathbf{w} = \begin{bmatrix} 0 \\ 4 \end{bmatrix}.$$

    Is the angle less than or greater than $90°$?

27. Are the following angles acute, obtuse, or right?

$$\cos \theta_1 = -0.7 \quad \cos \theta_2 = 0 \quad \cos \theta_3 = 0.7$$

28. Given the vectors

$$\mathbf{v} = \begin{bmatrix} 1 \\ -1 \end{bmatrix} \quad \text{and} \quad \mathbf{w} = \begin{bmatrix} 3 \\ 2 \end{bmatrix},$$

    find the orthogonal projection $\mathbf{u}$ of $\mathbf{w}$ onto $\mathbf{v}$. Decompose $\mathbf{w}$ into components $\mathbf{u}$ and $\mathbf{u}^{\perp}$.

29. For
$$\mathbf{v} = \begin{bmatrix} 1/\sqrt{2} \\ 1/\sqrt{2} \end{bmatrix} \quad \text{and} \quad \mathbf{w} = \begin{bmatrix} 1 \\ 3 \end{bmatrix}$$
find
$$\mathbf{u} = \text{proj}_{\mathcal{V}_1} \mathbf{w},$$
where $\mathcal{V}_1$ is the set of all 2D vectors $k\mathbf{v}$, and find
$$\mathbf{u}^\perp = \mathbf{w} - \text{proj}_{\mathcal{V}_1} \mathbf{w}.$$

30. Given vectors $\mathbf{v}$ and $\mathbf{w}$, is it possible for $(\mathbf{v} \cdot \mathbf{w})^2$ to be greater than $\|\mathbf{v}\|^2 \|\mathbf{w}\|^2$?

31. Given vectors $\mathbf{v}$ and $\mathbf{w}$, under what conditions is $(\mathbf{v} \cdot \mathbf{w})^2$ equal to $\|\mathbf{v}\|^2 \|\mathbf{w}\|^2$? Give an example.

32. Given vectors $\mathbf{v}$ and $\mathbf{w}$, can the length of $\|\mathbf{v} + \mathbf{w}\|$ be longer than the length of $\|\mathbf{v}\| + \|\mathbf{w}\|$?

33. Given vectors $\mathbf{v}$ and $\mathbf{w}$, under what conditions is $\|\mathbf{v} + \mathbf{w}\| = \|\mathbf{v}\| + \|\mathbf{w}\|$? Give an example.

3

Lining Up: 2D Lines

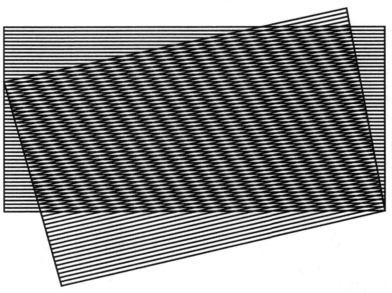

Figure 3.1.
Moiré patterns: overlaying two sets of lines at an angle results in an interesting pattern.

"Real" objects are three-dimensional, or 3D. So why should we consider 2D objects, such as the 2D lines in this chapter? Because they really are the building blocks for geometric constructions and play a key role in many applications. We'll look at various representations for lines, where each is suited for particular applications. Once we can represent a line, we can perform intersections and determine distances from a line.

Figure 3.1 shows how interesting playing with lines can be. Two sets of parallel lines are overlaid and the resulting interference pattern is called a Moiré pattern. Such patterns are used in optics for checking the properties of lenses.

3.1 Defining a Line

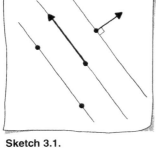

Sketch 3.1.
Elements to define a line.

As illustrated in Sketch 3.1, two elements of 2D geometry define a line:

- two points;

- a point and a vector parallel to the line;

- a point and a vector perpendicular to the line.

The unit vector that is perpendicular (or orthogonal) to a line is referred to as the *normal* to the line. Figure 3.2 shows two *families of lines*: one family of lines shares a common point and the other family of lines shares the same normal. Just as there are different ways to specify a line geometrically, there are different mathematical representations: parametric, implicit, and explicit. Each representation will be examined and the advantages of each will be explained. Additionally, we will explore how to convert from one form to another.

Figure 3.2.
Families of lines: one family shares a common point and the other shares a common normal.

3.2   Parametric Equation of a Line

The *parametric equation of a line* $\mathbf{l}(t)$ has the form

$$\mathbf{l}(t) = \mathbf{p} + t\mathbf{v}, \qquad (3.1)$$

where $\mathbf{p} \in \mathbb{E}^2$ and $\mathbf{v} \in \mathbb{R}^2$. The scalar value t is the *parameter*. (See Sketch 3.2.) Evaluating (3.1) for a specific parameter $t = \hat{t}$, generates a point on the line.

We encountered (3.1) in Section 2.5 in the context of barycentric coordinates. Interpreting $\mathbf{v}$ as a difference of points, $\mathbf{v} = \mathbf{q} - \mathbf{p}$, this equation was reformulated as

$$\mathbf{l}(t) = (1 - t)\mathbf{p} + t\mathbf{q}. \qquad (3.2)$$

A parametric line can be written in the form of either (3.1) or (3.2). The latter is typically referred to as *linear interpolation*.

One way to interpret the parameter t is as *time*; at time $t = 0$ we will be at point $\mathbf{p}$ and at time $t = 1$ we will be at point $\mathbf{q}$. Sketch 3.2 illustrates that as t varies between zero and one, $t \in [0, 1]$, points are generated on the line between $\mathbf{p}$ and $\mathbf{q}$. Recall from Section 2.5 that these values of t constitute a *convex combination*, which is a special case of a *barycentric combination*. If the parameter is a negative number, that is $t < 0$, the direction of $\mathbf{v}$ reverses, generating points on the line "behind" $\mathbf{p}$. The case $t > 1$ is similar: this scales $\mathbf{v}$ so that it is elongated, which generates points "past" $\mathbf{q}$. In the context of linear interpolation, when $t < 0$ or $t > 1$, it is called *extrapolation*.

The parametric form is very handy for computing points on a line. For example, to compute ten equally spaced points on the line segment between $\mathbf{p}$ and $\mathbf{q}$, simply define ten values of $t \in [0, 1]$ as

$$t = i/9, \qquad i = 0, \ldots, 9.$$

(Be sure this is a floating point calculation when programming!) Equally spaced parameter values correspond to equally spaced points.

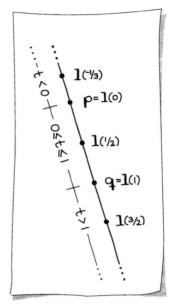

Sketch 3.2.
Parametric form of a line.

Example 3.1

Compute five points on the line defined by the points

$$\mathbf{p} = \begin{bmatrix} 1 \\ 2 \end{bmatrix} \quad \text{and} \quad \mathbf{q} = \begin{bmatrix} 6 \\ 4 \end{bmatrix}.$$

Define $\mathbf{v} = \mathbf{q} - \mathbf{p}$, then the line is defined as

$$\mathbf{l}(t) = \begin{bmatrix} 1 \\ 2 \end{bmatrix} + t \begin{bmatrix} 5 \\ 2 \end{bmatrix}.$$

Generate five t-values as

$$t = i/4, \qquad i = 0, \dots, 4.$$

Plug each t-value into the formulation for $\mathbf{l}(t)$:

$$i = 0, \quad t = 0, \qquad \mathbf{l}(0) = \begin{bmatrix} 1 \\ 2 \end{bmatrix};$$

$$i = 1, \quad t = 1/4, \quad \mathbf{l}(1/4) = \begin{bmatrix} 9/4 \\ 5/2 \end{bmatrix};$$

$$i = 2, \quad t = 2/4, \quad \mathbf{l}(2/4) = \begin{bmatrix} 7/2 \\ 3 \end{bmatrix};$$

$$i = 3, \quad t = 3/4, \quad \mathbf{l}(3/4) = \begin{bmatrix} 19/4 \\ 7/2 \end{bmatrix};$$

$$i = 4, \quad t = 1, \qquad \mathbf{l}(1) = \begin{bmatrix} 6 \\ 4 \end{bmatrix}.$$

Plot these values for yourself to verify them.

As you can see, the position of the point $\mathbf{p}$ and the direction and length of the vector $\mathbf{v}$ determine which points on the line are generated as we increment through $t \in [0, 1]$. This particular artifact of the parametric equation of a line is called the *parametrization*. The parametrization is related to the speed at which a point traverses the line. We may affect this speed by scaling $\mathbf{v}$: the larger the scale factor, the faster the point's motion!

3.3   Implicit Equation of a Line

Another way to represent the same line is to use the *implicit equation of a line*. For this representation, we start with a point $\mathbf{p}$, and as illustrated in Sketch 3.3, construct a vector $\mathbf{a}$ that is perpendicular to the line.

For any point $\mathbf{x}$ on the line, it holds that

$$\mathbf{a} \cdot (\mathbf{x} - \mathbf{p}) = 0. \tag{3.3}$$

This says that $\mathbf{a}$ and the vector $(\mathbf{x} - \mathbf{p})$ are perpendicular. If $\mathbf{a}$ has unit length, it is called the *normal* to the line, and then (3.3) is the *point normal form* of a line. Expanding this equation, we get

$$a_1 x_1 + a_2 x_2 + (-a_1 p_1 - a_2 p_2) = 0.$$

Commonly, this is written as

$$a x_1 + b x_2 + c = 0, \qquad (3.4)$$

where

$$a = a_1, \qquad (3.5)$$
$$b = a_2, \qquad (3.6)$$
$$c = -a_1 p_1 - a_2 p_2. \qquad (3.7)$$

Equation (3.4) is called the *implicit equation of the line*.

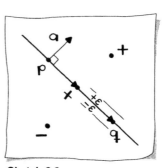

Sketch 3.3.
Implicit form of a line.

Example 3.2

Following Sketch 3.4, suppose we know two points,

$$\mathbf{p} = \begin{bmatrix} 2 \\ 2 \end{bmatrix} \quad \text{and} \quad \mathbf{q} = \begin{bmatrix} 6 \\ 4 \end{bmatrix},$$

on the line. To construct the coefficients a, b, and c in (3.4), first form the vector

$$\mathbf{v} = \mathbf{q} - \mathbf{p} = \begin{bmatrix} 4 \\ 2 \end{bmatrix}.$$

Now construct a vector $\mathbf{a}$ that is perpendicular to $\mathbf{v}$:

$$\mathbf{a} = \begin{bmatrix} -v_2 \\ v_1 \end{bmatrix} = \begin{bmatrix} -2 \\ 4 \end{bmatrix}. \qquad (3.8)$$

Note, equally as well, we could have chosen $\mathbf{a}$ to be

$$\begin{bmatrix} 2 \\ -4 \end{bmatrix}.$$

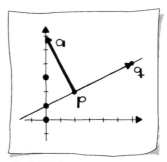

Sketch 3.4.
Implicit construction.

The coefficients a and b in (3.5) and (3.6) are now defined as $a = -2$ and $b = 4$. With $\mathbf{p}$ as defined above, solve for c as in (3.7). In this example,

$$c = 2 \times 2 - 4 \times 2 = -4.$$

The implicit equation of the line is complete:

$$-2x_1 + 4x_2 - 4 = 0.$$

The implicit form is very useful for deciding if an arbitrary point lies on the line. To test if a point $\mathbf{x}$ is on the line, just plug its coordinates into (3.4). If the value f of the left-hand side of this equation,

$$f = ax_1 + bx_2 + c,$$

is zero then the point is on the line.

A numerical caveat is needed here. Checking equality with floating point numbers should never be done. Instead, a tolerance ϵ around zero must be used. What is a meaningful tolerance in this situation? We'll see in Section 3.6 that

$$d = \frac{f}{\|\mathbf{a}\|} \tag{3.9}$$

reflects the true distance of $\mathbf{x}$ to the line. Now the tolerance has a physical meaning, which makes it much easier to specify. Sketch 3.3 illustrates the physical relationship of this tolerance to the line.

The sign of d indicates on which side of the line the point lies. This sign is dependent upon the definition of $\mathbf{a}$. (Remember, there were two possible orientations.) Positive d corresponds to the point on the side of the line to which $\mathbf{a}$ points.

Example 3.3

Let's continue with our example for the line

$$-2x_1 + 4x_2 - 4 = 0,$$

as illustrated in Sketch 3.4. We want to test if the point

$$\mathbf{x} = \begin{bmatrix} 0 \\ 1 \end{bmatrix}$$

lies on the line. First, calculate

$$\|\mathbf{a}\| = \sqrt{-2^2 + 4^2} = \sqrt{20}.$$

The distance is

$$d = (-2 \times 0 + 4 \times 1 - 4)/\sqrt{20} = 0/\sqrt{20} = 0,$$

which indicates the point is on the line.

Test the point
$$\mathbf{x} = \begin{bmatrix} 0 \\ 3 \end{bmatrix}.$$

For this point,
$$d = (-2 \times 0 + 4 \times 3 - 4)/\sqrt{20} = 8/\sqrt{20} \approx 1.79.$$

Checking Sketch 3.3, this is a positive number, indicating that it is on the same side of the line as the direction of $\mathbf{a}$. Check for yourself that d does indeed reflect the actual distance of this point to the line.

Test the point
$$\mathbf{x} = \begin{bmatrix} 0 \\ 0 \end{bmatrix}.$$

Calculating the distance for this point, we get
$$d = (-2 \times 0 + 4 \times 0 - 4)/\sqrt{20} = -4/\sqrt{20} \approx -0.894.$$

Checking Sketch 3.3, this is a negative number, indicating it is on the opposite side of the line as the direction of $\mathbf{a}$.

From Example 3.3, we see that if we want to know the distance of many points to the line, it is more economical to represent the implicit line equation with each coefficient divided by $\|\mathbf{a}\|$,
$$\frac{ax_1 + bx_2 + c}{\|\mathbf{a}\|} = 0,$$

which we know as the point normal form. The point normal form of the line from the example above is
$$-\frac{2}{\sqrt{20}}x_1 + \frac{4}{\sqrt{20}}x_2 - \frac{4}{\sqrt{20}} = 0.$$

Examining (3.4) you might notice that a *horizontal* line takes the form
$$bx_2 + c = 0.$$

This line intersects the $\mathbf{e}_2$-axis at $-c/b$. A *vertical* line takes the form
$$ax_1 + c = 0.$$

This line intersects the $\mathbf{e}_1$-axis at $-c/a$. Using the implicit form, these lines are in no need of special handling.

3.4   Explicit Equation of a Line

The *explicit equation of a line* is the third possible representation. The explicit form is closely related to the implicit form in (3.4). It expresses x_2 as a function of x_1: rearranging the implicit equation we have

$$x_2 = -\frac{a}{b}x_1 - \frac{c}{b}.$$

A more typical way of writing this is

$$x_2 = \hat{a}x_1 + \hat{b},$$

where $\hat{a} = -a/b$ and $\hat{b} = -c/b$.

The coefficients have geometric meaning: $\hat{a}$ is the *slope* of the line and $\hat{b}$ is the $\mathbf{e}_2$-intercept. Sketch 3.5 illustrates the geometry of the coefficients for the line

$$x_2 = 1/3x_1 + 1.$$

The slope measures the steepness of the line as a ratio of the change in x_2 to a change in x_1: "rise/run," or more precisely $\tan(\theta)$. The $\mathbf{e}_2$-intercept indicates that the line passes through $(0, \hat{b})$.

Immediately, a drawback of the explicit form is apparent. If the "run" is zero then the (vertical) line has infinite slope. This makes life very difficult when programming! When we study transformations (e.g., changing the orientation of some geometry) in Chapter 6, we will see that infinite slopes actually arise often.

The primary popularity of the explicit form comes from the study of calculus. Additionally, in computer graphics, this form is popular when *pixel* calculation is necessary. Examples are Bresenham's line drawing algorithm and scan line polygon fill algorithms (see [10]).

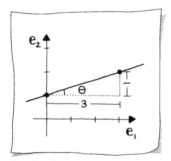

Sketch 3.5.
A line in explicit form.

3.5   Converting Between Parametric and Implicit Equations

As we have discussed, there are advantages to both the parametric and implicit representations of a line. Depending on the geometric algorithm, it may be convenient to use one form rather than the other. We'll ignore the explicit form, since as we said, it isn't very useful for general 2D geometry.

3.5.1   Parametric to Implicit

Given: The line l in parametric form,

$$l : l(t) = \mathbf{p} + t\mathbf{v}.$$

Find: The coefficients a, b, c that define the implicit equation of the line

$$l : ax_1 + bx_2 + c = 0.$$

Solution: First form a vector $\mathbf{a}$ that is perpendicular to the vector $\mathbf{v}$. Choose

$$\mathbf{a} = \begin{bmatrix} -v_2 \\ v_1 \end{bmatrix}.$$

This determines the coefficients a and b, as in (3.5) and (3.6), respectively. Simply let $a = a_1$ and $b = a_2$. Finally, solve for the coefficient c as in (3.7). Taking $\mathbf{p}$ from $l(t)$ and $\mathbf{a}$, form

$$c = -(a_1 p_1 + a_2 p_2).$$

We stepped through a numerical example of this in the derivation of the implicit form in Section 3.3, and it is illustrated in Sketch 3.4. In this example, $l(t)$ is given as

$$l(t) = \begin{bmatrix} 2 \\ 2 \end{bmatrix} + t \begin{bmatrix} 4 \\ 2 \end{bmatrix}.$$

3.5.2   Implicit to Parametric

Given: The line l in implicit form,

$$l : ax_1 + bx_2 + c = 0.$$

Find: The line l in parametric form,

$$l : l(t) = \mathbf{p} + t\mathbf{v}.$$

Solution: Recognize that we need one point on the line and a vector parallel to the line. The vector is easy: simply form a vector perpendicular to $\mathbf{a}$ of the implicit line. For example, we could set

$$\mathbf{v} = \begin{bmatrix} b \\ -a \end{bmatrix}.$$

Next, find a point on the line. Two candidate points are the intersections with the $\mathbf{e}_1$- or $\mathbf{e}_2$-axis,

$$\begin{bmatrix} -c/a \\ 0 \end{bmatrix} \quad \text{or} \quad \begin{bmatrix} 0 \\ -c/b \end{bmatrix},$$

respectively. For numerical stability, let's choose the intersection closest to the origin. Thus, we choose the former if $|a| > |b|$, and the latter otherwise.

Example 3.4

Revisit the numerical example from the implicit form derivation in Section 3.3; it is illustrated in Sketch 3.4. The implicit equation of the line is

$$-2x_1 + 4x_2 - 4 = 0.$$

We want to find a parametric equation of this line,

$$\mathbf{l} : \mathbf{l}(t) = \mathbf{p} + t\mathbf{v}.$$

First form

$$\mathbf{v} = \begin{bmatrix} 4 \\ 2 \end{bmatrix}.$$

Now determine which is greater in absolute value, a or b. Since $|-2| < |4|$, we choose

$$\mathbf{p} = \begin{bmatrix} 0 \\ 4/4 \end{bmatrix} = \begin{bmatrix} 0 \\ 1 \end{bmatrix}.$$

The parametric equation is

$$\mathbf{l}(t) = \begin{bmatrix} 0 \\ 1 \end{bmatrix} + t \begin{bmatrix} 4 \\ 2 \end{bmatrix}.$$

The implicit and parametric forms both allow an infinite number of representations for the same line. In fact, in the example we just finished, the loop

$$\text{parametric} \rightarrow \text{implicit} \rightarrow \text{parametric}$$

produced two different parametric forms. We started with

$$\mathbf{l}(t) = \begin{bmatrix} 2 \\ 2 \end{bmatrix} + t \begin{bmatrix} 4 \\ 2 \end{bmatrix},$$

and ended with

$$\mathbf{l}(t) = \begin{bmatrix} 0 \\ 1 \end{bmatrix} + t \begin{bmatrix} 4 \\ 2 \end{bmatrix}.$$

We could have just as easily generated the line

$$\mathbf{l}(t) = \begin{bmatrix} 0 \\ 1 \end{bmatrix} + t \begin{bmatrix} -4 \\ -2 \end{bmatrix},$$

if **v** was formed with the rule

$$\mathbf{v} = \begin{bmatrix} -b \\ a \end{bmatrix}.$$

Sketch 3.6 illustrates the first and third parametric representations of this line.

These three parametric forms represent the same line! However, the manner in which the lines will be *traced* will differ. This is referred to as the *parametrization* of the line. We already encountered this concept in Section 3.2.

Sketch 3.6.
Two parametric representations for the same line.

3.6   Distance of a Point to a Line

If you are given a point **r** and a line **l**, how far is that point from the line? For example, as in Figure 3.3 (left), suppose a robot will travel along the line and the points represent objects in the room. The robot needs a certain clearance as it moves, thus we must check that no point is closer than a given tolerance, thus it is necessary to check the distance of each point to the line. In Section 2.8 on orthogonal

Figure 3.3.
Left: robot path application where clearance around the robot's path must be measured. Right: measuring the perpendicular distance of each point to the line.

projections, we learned that the smallest distance $d(\mathbf{r}, \mathbf{l})$ of a point to a line is the *orthogonal or perpendicular distance*. This distance is illustrated in Figure 3.3 (right).

3.6.1 Starting with an Implicit Line

Suppose our problem is formulated as follows:

Given: A line $\mathbf{l}$ in implicit form defined by (3.3) or (3.4) and a point $\mathbf{r}$.

Find: $d(\mathbf{r}, \mathbf{l})$, or d for brevity.

Solution: The implicit line is given by coefficients a, b, and c, and thus also the vector

$$\mathbf{a} = \begin{bmatrix} a \\ b \end{bmatrix}.$$

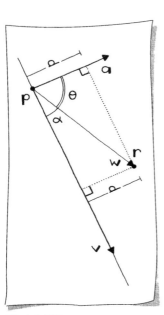

$$d = \frac{ar_1 + br_2 + c}{\|\mathbf{a}\|},$$

or in vector notation

$$d = \frac{\mathbf{a} \cdot (\mathbf{r} - \mathbf{p})}{\|\mathbf{a}\|},$$

where $\mathbf{p}$ is a point on the line.

Let's investigate why this is so. Recall that the implicit equation of a line was derived through use of the dot product

$$\mathbf{a} \cdot (\mathbf{x} - \mathbf{p}) = 0,$$

as in (3.3); a line is given by a point $\mathbf{p}$ and a vector $\mathbf{a}$ normal to the line. Any point $\mathbf{x}$ on the line will satisfy this equality.

As in Sketch 3.7, we now consider a point $\mathbf{r}$ that is clearly not on the line. As a result, the equality will not be satisfied; however, let's assign a value v to the left-hand side:

$$v = \mathbf{a} \cdot (\mathbf{r} - \mathbf{p}).$$

To simplify, define $\mathbf{w} = \mathbf{r} - \mathbf{p}$, as in Sketch 3.7. Recall the definition of the dot product in (2.19) as

$$\mathbf{v} \cdot \mathbf{w} = \|\mathbf{v}\| \|\mathbf{w}\| \cos\theta.$$

Thus, the expression for v becomes

$$v = \mathbf{a} \cdot \mathbf{w} = \|\mathbf{a}\| \|\mathbf{w}\| \cos(\theta). \qquad (3.10)$$

Sketch 3.7.

Distance point to line.

The right triangle in Sketch 3.7 allows for an expression for $\cos(\theta)$ as

$$\cos(\theta) = \frac{d}{\|\mathbf{w}\|}.$$

Substituting this into (3.10), we have

$$v = \|\mathbf{a}\|d.$$

This indicates that the actual distance of $\mathbf{r}$ to the line is

$$d = \frac{v}{\|\mathbf{a}\|} = \frac{\mathbf{a} \cdot (\mathbf{r} - \mathbf{p})}{\|\mathbf{a}\|} = \frac{ar_1 + br_2 + c}{\|\mathbf{a}\|}. \qquad (3.11)$$

If many points will be checked against a line, it is advantageous to store the line in *point normal form*. This means that $\|\mathbf{a}\| = 1$, eliminating the division in (3.11).

Example 3.5

Start with the line and the point

$$\mathbf{l} : 4x_1 + 2x_2 - 8 = 0; \qquad \mathbf{r} = \begin{bmatrix} 5 \\ 3 \end{bmatrix}.$$

Find the distance from $\mathbf{r}$ to the line. (Draw your own sketch for this example, similar to Sketch 3.7.)

First, calculate

$$\|\mathbf{a}\| = \sqrt{4^2 + 2^2} = 2\sqrt{5}.$$

Then the distance is

$$d(\mathbf{r}, \mathbf{l}) = \frac{4 \times 5 + 2 \times 3 - 8}{2\sqrt{5}} = \frac{9}{\sqrt{5}} \approx 4.02.$$

As another exercise, let's rewrite the line in point normal form with coefficients

$$\hat{a} = \frac{4}{2\sqrt{5}} = \frac{2}{\sqrt{5}},$$

$$\hat{b} = \frac{2}{2\sqrt{5}} = \frac{1}{\sqrt{5}},$$

$$\hat{c} = \frac{c}{\|\mathbf{a}\|} = \frac{-8}{2\sqrt{5}} = -\frac{4}{\sqrt{5}},$$

thus making the point normal form of the line

$$\frac{2}{\sqrt{5}}x_1 + \frac{1}{\sqrt{5}}x_2 - \frac{4}{\sqrt{5}} = 0.$$

3.6.2   Starting with a Parametric Line

Alternatively, suppose our problem is formulated as follows:

Given: A line **l** in parametric form, defined by a point **p** and a vector **v**, and a point **r**.

Find: $d(\mathbf{r}, \mathbf{l})$, or d for brevity. Again, this is illustrated in Sketch 3.7.

Solution: Form the vector $\mathbf{w} = \mathbf{r} - \mathbf{p}$. Use the relationship

$$d = \|\mathbf{w}\| \sin(\alpha).$$

Later in Section 8.2, we will see how to express $\sin(\alpha)$ directly in terms of **v** and **w**; for now, we express it in terms of the cosine:

$$\sin(\alpha) = \sqrt{1 - \cos(\alpha)^2},$$

and as before

$$\cos(\alpha) = \frac{\mathbf{v} \cdot \mathbf{w}}{\|\mathbf{v}\|\|\mathbf{w}\|}.$$

Thus, we have defined the distance d.

Example 3.6

We'll use the same line as in the previous example, but now it will be given in parametric form as

$$\mathbf{l}(t) = \begin{bmatrix} 0 \\ 4 \end{bmatrix} + t \begin{bmatrix} 2 \\ -4 \end{bmatrix}.$$

We'll also use the same point

$$\mathbf{r} = \begin{bmatrix} 5 \\ 3 \end{bmatrix}.$$

Add any new vectors for this example to the sketch you drew for the previous example.

First create the vector

$$\mathbf{w} = \begin{bmatrix} 5 \\ 3 \end{bmatrix} - \begin{bmatrix} 0 \\ 4 \end{bmatrix} = \begin{bmatrix} 5 \\ -1 \end{bmatrix}.$$

Next calculate $\|\mathbf{w}\| = \sqrt{26}$ and $\|\mathbf{v}\| = \sqrt{20}$. Compute

$$\cos(\alpha) = \frac{\begin{bmatrix} 2 \\ -4 \end{bmatrix} \cdot \begin{bmatrix} 5 \\ -1 \end{bmatrix}}{\sqrt{26}\sqrt{20}} \approx 0.614.$$

Thus, the distance to the line becomes

$$d(\mathbf{r}, \mathbf{l}) \approx \sqrt{26}\sqrt{1 - (0.614)^2} \approx 4.02,$$

which rightly produces the same result as the previous example.

3.7   The Foot of a Point

Section 3.6 detailed how to calculate the distance of a point from a line. A new question arises: which point on the line is closest to the point? This point will be called the *foot* of the given point.

If you are given a line in implicit form, it is best to convert it to parametric form for this problem. This illustrates how the implicit form is handy for *testing* if a point is on the line; however, it is not as handy for *finding* points on the line.

The problem at hand is thus:

Given: A line $\mathbf{l}$ in parametric form, defined by a point $\mathbf{p}$ and a vector $\mathbf{v}$, and another point $\mathbf{r}$.

Find: The point $\mathbf{q}$ on the line that is closest to $\mathbf{r}$. (See Sketch 3.8.)

Solution: The point $\mathbf{q}$ can be defined as

$$\mathbf{q} = \mathbf{p} + t\mathbf{v}, \tag{3.12}$$

so our problem is solved once we have found the scalar factor t. From Sketch 3.8, we see that

$$\cos(\theta) = \frac{\|t\mathbf{v}\|}{\|\mathbf{w}\|},$$

where $\mathbf{w} = \mathbf{r} - \mathbf{p}$. Using

$$\cos(\theta) = \frac{\mathbf{v} \cdot \mathbf{w}}{\|\mathbf{v}\|\|\mathbf{w}\|},$$

we find

$$t = \frac{\mathbf{v} \cdot \mathbf{w}}{\|\mathbf{v}\|^2}.$$

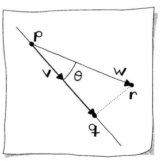

Sketch 3.8.
Closest point $\mathbf{q}$ on line to point $\mathbf{r}$.

Example 3.7

Given: The parametric line $\mathbf{l}$ defined as

$$\mathbf{l}(t) = \begin{bmatrix} 0 \\ 1 \end{bmatrix} + t \begin{bmatrix} 0 \\ 2 \end{bmatrix},$$

and point
$$\mathbf{r} = \begin{bmatrix} 3 \\ 4 \end{bmatrix}.$$

Find: The point $\mathbf{q}$ on l that is closest to $\mathbf{r}$. This example is easy enough to find the answer by simply drawing a sketch, but let's go through the steps.

Solution: Define the vector
$$\mathbf{w} = \mathbf{r} - \mathbf{p} = \begin{bmatrix} 3 \\ 4 \end{bmatrix} - \begin{bmatrix} 0 \\ 1 \end{bmatrix} = \begin{bmatrix} 3 \\ 3 \end{bmatrix}.$$

Compute $\mathbf{v} \cdot \mathbf{w} = 6$ and $\|\mathbf{v}\| = 2$. Thus, $t = 3/2$ and
$$\mathbf{q} = \begin{bmatrix} 0 \\ 4 \end{bmatrix}.$$

Try this example with $\mathbf{r} = \begin{bmatrix} 2 \\ -1 \end{bmatrix}.$

3.8   A Meeting Place: Computing Intersections

Finding a point in common between two lines is done many times over in a CAD or graphics package. Take for example Figure 3.4: the top part of the figure shows a great number of intersecting lines. In order to color some of the areas, as in the bottom part of the figure, it is necessary to know the intersection points. Intersection problems arise in many other applications. The first question to ask is what type of information do you want:

- Do you want to know merely whether the lines intersect?

- Do you want to know the point at which they intersect?

- Do you want a parameter value on one or both lines for the intersection point?

The particular question(s) you want to answer along with the line representation(s) will determine the best method for solving the intersection problem.

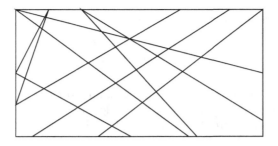

Figure 3.4.
Intersecting lines: the top figure may be drawn without knowing where the shown lines intersect. By finding line/line intersections (bottom), it is possible to color areas—creating an artistic image!

3.8.1 Parametric and Implicit

We then want to solve the following:

Given: Two lines l_1 and l_2:

$$l_1: \quad l_1(t) = \mathbf{p} + t\mathbf{v}$$
$$l_2: \quad ax_1 + bx_2 + c = 0.$$

Find: The intersection point $\mathbf{i}$. See Sketch 3.9 for an illustration.[1]

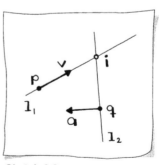

Sketch 3.9.
Parametric and implicit line intersection.

Solution: We will approach the problem by finding the specific parameter $\hat{t}$ with respect to l_1 of the intersection point.

This intersection point, when inserted into the equation of l_2, will cause the left-hand side to evaluate to zero:

$$a[p_1 + \hat{t}v_1] + b[p_2 + \hat{t}v_2] + c = 0.$$

---

[1]In Section 3.3, we studied the conversion from the geometric elements of a point $\mathbf{q}$ and perpendicular vector $\mathbf{a}$ to the implicit line coefficients.

This is one equation and one unknown! Just solve for $\hat{t}$,

$$\hat{t} = \frac{-c - ap_1 - bp_2}{av_1 + bv_2},\qquad(3.13)$$

then $\mathbf{i} = \mathbf{l}(\hat{t})$.

But wait—we must check if the denominator of (3.13) is zero before carrying out this calculation. Besides causing havoc numerically, what else does a zero denominator infer? The denominator

$$\text{denom} = av_1 + bv_2$$

can be rewritten as

$$\text{denom} = \mathbf{a} \cdot \mathbf{v}.$$

We know from (2.7) that a zero dot product implies that two vectors are perpendicular. Since $\mathbf{a}$ is perpendicular to the line $\mathbf{l}_2$ in implicit form, the lines are *parallel* if

$$\mathbf{a} \cdot \mathbf{v} = 0.$$

Of course, we always check for equality within a tolerance. A physically meaningful tolerance is best. Thus, it is better to check the quantity

$$\cos(\theta) = \frac{\mathbf{a} \cdot \mathbf{v}}{\|\mathbf{a}\|\|\mathbf{v}\|};\qquad(3.14)$$

the tolerance will be the cosine of an angle. It usually suffices to use a tolerance between $\cos(0.1°)$ and $\cos(0.5°)$. Angle tolerances are particularly nice to have because they are *dimension independent*. Note that we do not need to use the actual angle, just the cosine of the angle.

If the test in (3.14) indicates the lines are parallel, then we might want to determine if the lines are identical. By simply plugging in the coordinates of $\mathbf{p}$ into the equation of $\mathbf{l}_2$, and computing, we get

$$d = \frac{ap_1 + bp_2 + c}{\|\mathbf{a}\|}.$$

If d is equal to zero (within tolerance), then the lines are identical.

Example 3.8

Given: Two lines $\mathbf{l}_1$ and $\mathbf{l}_2$,

$$\mathbf{l}_1 : \quad \mathbf{l}_1(t) = \begin{bmatrix} 0 \\ 3 \end{bmatrix} + t \begin{bmatrix} -2 \\ -1 \end{bmatrix}$$
$$\mathbf{l}_2 : \quad 2x_1 + x_2 - 8 = 0.$$

Find: The intersection point **i**. Create your own sketch and try to predict what the answer should be.

Solution: Find the parameter $\hat{t}$ for l_1 as given in (3.13). First check the denominator:

$$\text{denom} = 2 \times (-2) + 1 \times (-1) = -5.$$

This is not zero, so we proceed to find

$$\hat{t} = \frac{8 - 2 \times 0 - 1 \times 3}{-5} = -1.$$

Plug this parameter value into l_1 to find the intersection point:

$$l_1(-1) = \begin{bmatrix} 0 \\ 3 \end{bmatrix} + -1 \begin{bmatrix} -2 \\ -1 \end{bmatrix} = \begin{bmatrix} 2 \\ 4 \end{bmatrix}.$$

3.8.2 Both Parametric

Another method for finding the intersection of two lines arises by using the parametric form for both, illustrated in Sketch 3.10.

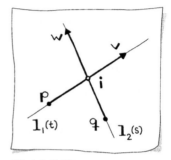

Sketch 3.10.
Intersection of lines in parametric form.

Given: Two lines in parametric form,

$$\begin{aligned} l_1 : \quad & l_1(t) = \mathbf{p} + t\mathbf{v} \\ l_2 : \quad & l_2(s) = \mathbf{q} + s\mathbf{w}. \end{aligned}$$

Note that we use two different parameters, t and s, here. This is because the lines are independent of each other.

Find: The intersection point **i**.

Solution: We need two parameter values $\hat{t}$ and $\hat{s}$ such that

$$\mathbf{p} + \hat{t}\mathbf{v} = \mathbf{q} + \hat{s}\mathbf{w}.$$

This may be rewritten as

$$\hat{t}\mathbf{v} - \hat{s}\mathbf{w} = \mathbf{q} - \mathbf{p}. \tag{3.15}$$

We have two equations (one for each coordinate) and two unknowns $\hat{t}$ and $\hat{s}$. To solve for the unknowns, we could formulate an expression for $\hat{t}$ using the first equation, and substitute this expression into the second equation. This then generates a solution for $\hat{s}$. Use this solution in the expression for $\hat{t}$, and solve for $\hat{t}$. (Equations like this are treated systematically in Chapter 5.) Once we have $\hat{t}$ and $\hat{s}$, the intersection point is found by inserting one of these values into its respective parametric line equation.

If the vectors $\mathbf{v}$ and $\mathbf{w}$ are linearly dependent, as discussed in Section 2.6, then it will not be possible to find a unique $\hat{t}$ and $\hat{s}$. The lines are parallel and possibly identical.

Example 3.9

Given: Two lines $\mathbf{l}_1$ and $\mathbf{l}_2$,

$$\mathbf{l}_1 : \quad \mathbf{l}_1(t) = \begin{bmatrix} 0 \\ 3 \end{bmatrix} + t \begin{bmatrix} -2 \\ -1 \end{bmatrix}$$

$$\mathbf{l}_2 : \quad \mathbf{l}_2(s) = \begin{bmatrix} 4 \\ 0 \end{bmatrix} + s \begin{bmatrix} -1 \\ 2 \end{bmatrix}.$$

Find: The intersection point $\mathbf{i}$. This means that we need to find $\hat{t}$ and $\hat{s}$ such that $\mathbf{l}_1(\hat{t}) = \mathbf{l}_2(\hat{s})$.[2] Again, create your own sketch and try to predict the answer.

Solution: Set up the two equations with two unknowns as in (3.15).

$$\hat{t} \begin{bmatrix} -2 \\ -1 \end{bmatrix} - \hat{s} \begin{bmatrix} -1 \\ 2 \end{bmatrix} = \begin{bmatrix} 4 \\ 0 \end{bmatrix} - \begin{bmatrix} 0 \\ 3 \end{bmatrix}.$$

Solve these equations, resulting in $\hat{t} = -1$ and $\hat{s} = 2$. Plug these values into the line equations to verify the same intersection point is produced for each:

$$\mathbf{l}_1(-1) = \begin{bmatrix} 0 \\ 3 \end{bmatrix} + -1 \begin{bmatrix} -2 \\ -1 \end{bmatrix} = \begin{bmatrix} 2 \\ 4 \end{bmatrix}$$

$$\mathbf{l}_2(2) = \begin{bmatrix} 4 \\ 0 \end{bmatrix} + 2 \begin{bmatrix} -1 \\ 2 \end{bmatrix} = \begin{bmatrix} 2 \\ 4 \end{bmatrix}.$$

---

[2]Really we only need $\hat{t}$ or $\hat{s}$ to find the intersection point.

3.8.3 Both Implicit

And yet a third method:

Given: Two lines in implicit form,

$$l_1 : \quad ax_1 + bx_2 + c = 0,$$
$$l_2 : \quad \bar{a}x_1 + \bar{b}x_2 + \bar{c} = 0.$$

As illustrated in Sketch 3.11, each line is geometrically given in terms of a point and a vector perpendicular to the line.

Find: The intersection point

$$\mathbf{i} = \hat{\mathbf{x}} = \begin{bmatrix} \hat{x}_1 \\ \hat{x}_2 \end{bmatrix}$$

that simultaneously satisfies l_1 and l_2.

Solution: We have two equations

$$a\hat{x}_1 + b\hat{x}_2 = -c, \tag{3.16}$$
$$\bar{a}\hat{x}_1 + \bar{b}\hat{x}_2 = -\bar{c} \tag{3.17}$$

with two unknowns, $\hat{x}_1$ and $\hat{x}_2$. Equations like this are solved in Chapter 5, but this is simple enough to solve without those methods.

If the lines are parallel then it will not be possible to find $\hat{\mathbf{x}}$. This means that $\mathbf{a}$ and $\bar{\mathbf{a}}$ are linearly dependent.

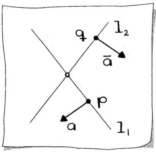

Sketch 3.11.

Intersection of two lines in implicit form.

Example 3.10

Given: Two lines l_1 and l_2,

$$l_1 : \quad x_1 - 2x_2 + 6 = 0$$
$$l_2 : \quad 2x_1 + x_2 - 8 = 0.$$

Find: The intersection point $\hat{\mathbf{x}}$ as above. Create your own sketch and try to predict the answer.

Solution: Reformulate the equations for l_1 and l_2 as in (3.16) and (3.17). You will find that

$$\hat{x} = \begin{bmatrix} 2 \\ 4 \end{bmatrix}.$$

Plug this point into the equations for l_1 and l_2 to verify.

- parametric form of a line
- linear interpolation
- point normal form
- implicit form of a line
- explicit form of a line
- equation of a line through two points
- equation of a line defined by a point and a vector parallel to the line

- equation of a line defined by a point and a vector perpendicular to the line
- distance of a point to a line
- line form conversions
- foot of a point
- intersection of lines

3.9 Exercises

1. Give the parametric form of the line $l(t)$ defined by points

$$\mathbf{p} = \begin{bmatrix} 0 \\ 1 \end{bmatrix} \quad \text{and} \quad \mathbf{q} = \begin{bmatrix} 4 \\ 2 \end{bmatrix},$$

such that $l(0) = \mathbf{p}$ and $l(1) = \mathbf{q}$.

2. Define the line $l(t)$ in the linear interpolation form that interpolates to points

$$\mathbf{p} = \begin{bmatrix} 2 \\ 3 \end{bmatrix} \quad \text{and} \quad \mathbf{q} = \begin{bmatrix} -1 \\ -1 \end{bmatrix},$$

such that $l(0) = \mathbf{p}$ and $l(1) = \mathbf{q}$.

3. For the line in Exercise 2, is the point $l(2)$ formed from a convex combination?

4. Using the line in Exercise 1, construct five equally spaced points on the line segment $l(t)$ where $t \in [0, 1]$. The first point should be $\mathbf{p}$ and the last point should be $\mathbf{q}$. What are the parameter values for these points?

5. Find the equation for a line in implicit form that passes through the points

$$\mathbf{p} = \begin{bmatrix} -2 \\ 0 \end{bmatrix} \quad \text{and} \quad \mathbf{q} = \begin{bmatrix} 0 \\ -1 \end{bmatrix}.$$

6. Test if the following points lie on the line defined in Exercise 5:

$$\mathbf{r}_0 = \begin{bmatrix} 0 \\ 0 \end{bmatrix}, \quad \mathbf{r}_1 = \begin{bmatrix} -4 \\ 1 \end{bmatrix}, \quad \mathbf{r}_2 = \begin{bmatrix} 5 \\ 1 \end{bmatrix}, \quad \mathbf{r}_3 = \begin{bmatrix} -3 \\ -1 \end{bmatrix}.$$

7. For the points in Exercise 6, if a point does not lie on the line, calculate the distance from the line.

8. What is the point normal form of the line in Exercise 5?

9. What is the implicit equation of the horizontal line through $x_2 = -1/2$? What is the implicit equation of the vertical line through $x_1 = 1/2$?

10. What is the explicit equation of the line $6x_1 + 3x_2 + 3 = 0$?

11. What is the slope and e_2-intercept of the line $x_2 = 4x_1 - 1$?

12. What is the explicit equation of the horizontal line through $x_2 = -1/2$? What is the explicit equation of the vertical line through $x_1 = 1/2$?

13. What is the explicit equation of the line with zero slope with e_2-intercept 3? What is the explicit equation of the line with slope 2 that passes through the origin?

14. What is the implicit equation of the line

$$\mathbf{l}(t) = \begin{bmatrix} 1 \\ 1 \end{bmatrix} + t \begin{bmatrix} -2 \\ 3 \end{bmatrix}?$$

15. What is the implicit equation of the line

$$\mathbf{l}(t) = \begin{bmatrix} 0 \\ 1 \end{bmatrix} + t \begin{bmatrix} 0 \\ 1 \end{bmatrix}?$$

16. What is the implicit equation of the line

$$\mathbf{l}(t) = (1 - t) \begin{bmatrix} -1 \\ 0 \end{bmatrix} + t \begin{bmatrix} 2 \\ 2 \end{bmatrix}?$$

17. What is the parametric form of the line $3x_1 + 2x_2 + 1 = 0$?

18. What is the parametric form of the line $5x_1 + 1 = 0$?

19. Given the line $\mathbf{l}(t) = \mathbf{p} + t\mathbf{v}$ and point $\mathbf{r}$ defined as

$$\mathbf{l}(t) = \begin{bmatrix} 0 \\ 0 \end{bmatrix} + t \begin{bmatrix} 1 \\ 1 \end{bmatrix} \quad \text{and} \quad \mathbf{r} = \begin{bmatrix} 5 \\ 0 \end{bmatrix},$$

   find the distance of the point to the line.

20. Given the line $\mathbf{l}(t) = \mathbf{p} + t\mathbf{v}$ and point $\mathbf{r}$ defined as

$$\mathbf{l}(t) = \begin{bmatrix} -2 \\ 0 \end{bmatrix} + t \begin{bmatrix} 2 \\ 2 \end{bmatrix} \quad \text{and} \quad \mathbf{r} = \begin{bmatrix} 0 \\ 0 \end{bmatrix},$$

   find the distance of the point to the line.

21. Find the foot of the point $\mathbf{r}$ in Exercise 19.

22. Find the foot of the point $\mathbf{r}$ in Exercise 20.

23. Given two lines: l_1 defined by points

$$\begin{bmatrix} 1 \\ 0 \end{bmatrix} \quad \text{and} \quad \begin{bmatrix} 0 \\ 3 \end{bmatrix},$$

and l_2 defined by points

$$\begin{bmatrix} -1 \\ 6 \end{bmatrix} \quad \text{and} \quad \begin{bmatrix} -4 \\ 1 \end{bmatrix},$$

find the intersection point using each of the three methods in Section 3.8.

24. Find the intersection of the lines

$$l_1 : \quad l_1(t) = \begin{bmatrix} 0 \\ -1 \end{bmatrix} + t \begin{bmatrix} 1 \\ 1 \end{bmatrix} \quad \text{and}$$
$$l_2 : \quad -x_1 + x_2 + 1 = 0.$$

25. Find the intersection of the lines

$$l_1 : \quad -x_1 + x_2 + 1 = 0 \quad \text{and}$$
$$l_2 : \quad l(t) = \begin{bmatrix} 2 \\ 2 \end{bmatrix} + t \begin{bmatrix} 2 \\ 2 \end{bmatrix}.$$

26. Find the closest point on the line $l(t) = \mathbf{p} + t\mathbf{v}$, where

$$l(t) = \begin{bmatrix} 2 \\ 2 \end{bmatrix} + t \begin{bmatrix} 2 \\ 2 \end{bmatrix}$$

to the point

$$\mathbf{r} = \begin{bmatrix} 0 \\ 2 \end{bmatrix}.$$

27. The line $l(t)$ passes through the points

$$\mathbf{p} = \begin{bmatrix} 0 \\ 1 \end{bmatrix} \quad \text{and} \quad \mathbf{q} = \begin{bmatrix} 4 \\ 2 \end{bmatrix}.$$

Now define a line $\mathbf{m}(t)$ that is perpendicular to l and that passes through the midpoint of $\mathbf{p}$ and $\mathbf{q}$.

4

Changing Shapes:
Linear Maps in 2D

Figure 4.1.
Linear maps in 2D: an interesting geometric figure constructed by applying 2D linear maps to a square.

Geometry always has two parts to it: one part is the description of the objects that can be generated; the other investigates how these objects can be changed (or transformed). Any object formed by several vectors may be mapped to an arbitrarily bizarre curved or distorted object—here, we are interested in those maps that map 2D vectors to 2D vectors and are "benign" in some well-defined sense. All these maps may be described using the tools of matrix operations, or linear maps. An interesting pattern is generated from a simple square in Figure 4.1 by such "benign" 2D linear maps—rotations and scalings.

Matrices were first introduced by H. Grassmann in 1844. They became the basis of *linear algebra*. Most of their properties can be studied by just considering the humble 2×2 case, which corresponds to 2D linear maps.

4.1   Skew Target Boxes

In Section 1.1, we saw how to map an object from a unit square to a rectangular target box. We will now look at the part of that mapping that is a linear map.

First, our unit square will be defined by vectors $\mathbf{e}_1$ and $\mathbf{e}_2$. Thus, a vector $\mathbf{v}$ in this $[\mathbf{e}_1, \mathbf{e}_2]$-system is defined as

$$\mathbf{v} = v_1\mathbf{e}_1 + v_2\mathbf{e}_2. \tag{4.1}$$

If we focus on mapping vectors to vectors, then we will limit the target box to having a lower-left corner at the origin. In Chapter 6 we will reintroduce the idea of a generally positioned target box. Instead of specifying two extreme points for a rectangular target box, we will describe a parallelogram target box by two vectors $\mathbf{a}_1, \mathbf{a}_2$, defining an $[\mathbf{a}_1, \mathbf{a}_2]$-system. A vector $\mathbf{v}$ is now mapped to a vector $\mathbf{v}'$ by

$$\mathbf{v}' = v_1\mathbf{a}_1 + v_2\mathbf{a}_2, \tag{4.2}$$

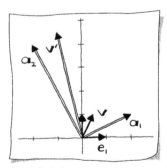

Sketch 4.1.

A skew target box defined by $\mathbf{a}_1$ and $\mathbf{a}_2$.

as illustrated by Sketch 4.1. This simply states that we duplicate the $[\mathbf{e}_1, \mathbf{e}_2]$-geometry in the $[\mathbf{a}_1, \mathbf{a}_2]$-system: The linear map transforms $\mathbf{e}_1, \mathbf{e}_2, \mathbf{v}$ to $\mathbf{a}_1, \mathbf{a}_2, \mathbf{v}'$, respectively. The components of $\mathbf{v}'$ are in the context of the $[\mathbf{e}_1, \mathbf{e}_2]$-system. However, the components of $\mathbf{v}'$ with respect to the $[\mathbf{a}_1, \mathbf{a}_2]$-system are the components of $\mathbf{v}$. Reviewing a definition from Section 2.6, we recall that (4.2) is called a *linear combination*.

Example 4.1

Let's look at an example of the action of the map from the linear combination in (4.2). Let the origin and

$$\mathbf{a}_1 = \begin{bmatrix} 2 \\ 1 \end{bmatrix}, \qquad \mathbf{a}_2 = \begin{bmatrix} -2 \\ 4 \end{bmatrix}$$

define a new $[\mathbf{a}_1, \mathbf{a}_2]$-coordinate system, and let

$$\mathbf{v} = \begin{bmatrix} 1/2 \\ 1 \end{bmatrix}$$

be a vector in the $[\mathbf{e}_1, \mathbf{e}_2]$-system. Applying the components of $\mathbf{v}$ in a linear combination of $\mathbf{a}_1$ and $\mathbf{a}_2$, results in

$$\mathbf{v}' = \frac{1}{2} \times \begin{bmatrix} 2 \\ 1 \end{bmatrix} + 1 \times \begin{bmatrix} -2 \\ 4 \end{bmatrix} = \begin{bmatrix} -1 \\ 9/2 \end{bmatrix}. \qquad (4.3)$$

Thus $\mathbf{v}'$ has components

$$\begin{bmatrix} 1/2 \\ 1 \end{bmatrix}$$

with respect to the $[\mathbf{a}_1, \mathbf{a}_2]$-system; with respect to the $[\mathbf{e}_1, \mathbf{e}_2]$-system, it has coordinates

$$\begin{bmatrix} -1 \\ 9/2 \end{bmatrix}.$$

See Sketch 4.1 for an illustration.

4.2   The Matrix Form

The components of a subscripted vector will be written with a double subscript as

$$\mathbf{a}_1 = \begin{bmatrix} a_{1,1} \\ a_{2,1} \end{bmatrix}.$$

The vector component index precedes the vector subscript.

The components for the vector $\mathbf{v}'$ in the $[\mathbf{e}_1, \mathbf{e}_2]$-system from Example 4.1 are expressed as

$$\begin{bmatrix} -1 \\ 9/2 \end{bmatrix} = \frac{1}{2} \times \begin{bmatrix} 2 \\ 1 \end{bmatrix} + 1 \times \begin{bmatrix} -2 \\ 4 \end{bmatrix}. \qquad (4.4)$$

This is strictly an equation between vectors. It invites a more concise notation using *matrix notation*:

$$\begin{bmatrix} -1 \\ 9/2 \end{bmatrix} = \begin{bmatrix} 2 & -2 \\ 1 & 4 \end{bmatrix} \begin{bmatrix} 1/2 \\ 1 \end{bmatrix}. \tag{4.5}$$

The 2×2 array in this equation is called a *matrix*. It has two columns, corresponding to the vectors $\mathbf{a}_1$ and $\mathbf{a}_2$. It also has two rows, namely the first row with entries $2, -2$ and the second one with $1, 4$.

In general, an equation like this one has the form

$$\mathbf{v}' = \begin{bmatrix} a_{1,1} & a_{1,2} \\ a_{2,1} & a_{2,2} \end{bmatrix} \begin{bmatrix} v_1 \\ v_2 \end{bmatrix}, \tag{4.6}$$

or,

$$\mathbf{v}' = A\mathbf{v}, \tag{4.7}$$

where A is the 2×2 matrix. The vector $\mathbf{v}'$ is called the *image* of $\mathbf{v}$, and thus $\mathbf{v}$ is the *preimage*. The *linear map* is described by the matrix A—we may think of A as being the map's coordinates. We will also refer to the linear map itself by A. Then $\mathbf{v}'$ is in the *range* of the map and $\mathbf{v}$ is in the *domain*.

The elements $a_{1,1}$ and $a_{2,2}$ form the *diagonal* of the matrix. The product $A\mathbf{v}$ has two components, each of which is obtained as a dot product between the corresponding row of the matrix and $\mathbf{v}$. In full generality, we have

$$A\mathbf{v} = \begin{bmatrix} v_1\mathbf{a}_1 + v_2\mathbf{a}_2 \end{bmatrix} = \begin{bmatrix} v_1 a_{1,1} + v_2 a_{1,2} \\ v_1 a_{2,1} + v_2 a_{2,2} \end{bmatrix}.$$

For example,

$$\begin{bmatrix} 0 & -1 \\ 2 & 4 \end{bmatrix} \begin{bmatrix} -1 \\ 4 \end{bmatrix} = \begin{bmatrix} -4 \\ 14 \end{bmatrix}.$$

In other words, $A\mathbf{v}$ is equivalent to forming the linear combination $\mathbf{v}$ of the columns of A. All such combinations, that is all such $\mathbf{v}'$, form the *column space* of A.

Another note on notation. Coordinate systems, such as the $[\mathbf{e}_1, \mathbf{e}_2]$-system, can be interpreted as a matrix with columns $\mathbf{e}_1$ and $\mathbf{e}_2$. Thus,

$$[\mathbf{e}_1, \mathbf{e}_2] \equiv \begin{bmatrix} 1 & 0 \\ 0 & 1 \end{bmatrix},$$

and this is called the 2×2 *identity matrix*.

There is a neat way to write the matrix-times-vector algebra that facilitates manual computation. As explained above, every entry in the resulting vector is a dot product of the input vector and a row of the matrix. Let's arrange this as follows:

$$
\begin{array}{cc|c}
 & & 2 \\
 & & 1/2 \\
\hline
2 & -2 & 3 \\
1 & 4 & 4
\end{array}
$$

Each entry of the resulting vector is now at the intersection of the corresponding matrix row and the input vector, which is written as a column. As you multiply and then add the terms in your dot product, this scheme guides you to the correct position in the result automatically! Here we multiplied a 2×2 matrix by a 2×1 vector. Note that the interior dimensions (both 2) must be identical and the outer dimensions, 2 and 1, indicate the resulting vector or matrix size, 2×1. Sometimes it is convenient to think of the vector $\mathbf{v}$ as a 2×1 matrix: it is a matrix with two rows and one column.

One fundamental matrix operation is *matrix addition*. Two matrices A and B may be added by adding corresponding elements:

$$
\begin{bmatrix} a_{1,1} & a_{1,2} \\ a_{2,1} & a_{2,2} \end{bmatrix} + \begin{bmatrix} b_{1,1} & b_{1,2} \\ b_{2,1} & b_{2,2} \end{bmatrix} = \begin{bmatrix} a_{1,1} + b_{1,1} & a_{1,2} + b_{1,2} \\ a_{2,1} + b_{2,1} & a_{2,2} + b_{2,2} \end{bmatrix}. \tag{4.8}
$$

Notice that the matrices must be of the same dimensions; this is not true for matrix multiplication, which we will demonstrate in Section 4.10.

Using matrix addition, we may write

$$
A\mathbf{v} + B\mathbf{v} = (A + B)\mathbf{v}.
$$

This works because of the very simple definition of matrix addition. This is also called the *distributive law*.

Forming the *transpose matrix* is another fundamental matrix operation. It is denoted by A^{T} and is formed by interchanging the rows and columns of A: the first row of A^{T} is A's first column, and the second row of A^{T} is A's second column. For example, if

$$
A = \begin{bmatrix} 1 & -2 \\ 3 & 5 \end{bmatrix}, \quad \text{then} \quad A^{\mathrm{T}} = \begin{bmatrix} 1 & 3 \\ -2 & 5 \end{bmatrix}.
$$

Since we may think of a vector $\mathbf{v}$ as a matrix, we should be able to find $\mathbf{v}$'s transpose. Not very hard: it is a vector with one row and two columns,

$$
\mathbf{v} = \begin{bmatrix} -1 \\ 4 \end{bmatrix} \quad \text{and} \quad \mathbf{v}^{\mathrm{T}} = \begin{bmatrix} -1 & 4 \end{bmatrix}.
$$

It is straightforward to confirm that

$$[A + B]^{\mathrm{T}} = A^{\mathrm{T}} + B^{\mathrm{T}}. \tag{4.9}$$

Two more identities are

$$A^{\mathrm{T}\,\mathrm{T}} = A \quad \text{and} \quad [cA]^{\mathrm{T}} = cA^{\mathrm{T}}. \tag{4.10}$$

A *symmetric matrix* is a special matrix that we will encounter many times. A matrix A is symmetric if $A = A^{\mathrm{T}}$, for example

$$\begin{bmatrix} 5 & 8 \\ 8 & 1 \end{bmatrix}.$$

There are no restrictions on the diagonal elements, but all other elements are equal to the element about the diagonal with reversed indices. For a 2×2 matrix, this means that $a_{2,1} = a_{1,2}$.

With matrix notation, we can now continue the discussion of independence from Section 2.6. The columns of a matrix define an $[\mathbf{a}_1, \mathbf{a}_2]$-system. If the vectors $\mathbf{a}_1$ and $\mathbf{a}_2$ are linearly independent then the matrix is said to have *full rank*, or for the 2×2 case, the matrix has *rank* 2. If $\mathbf{a}_1$ and $\mathbf{a}_2$ are linearly dependent then the matrix has rank 1. These two statements may be summarized as: the rank of a 2×2 matrix equals the number of linearly independent column (row) vectors. Matrices that do not have full rank are called *rank deficient* or *singular*. We will encounter an important example of a rank deficient matrix, a projection matrix, in Section 4.8. The only matrix with rank zero is the *zero matrix*, a matrix with all zero entries. The 2×2 zero matrix is

$$\begin{bmatrix} 0 & 0 \\ 0 & 0 \end{bmatrix}.$$

The importance of the rank equivalence of column and row vectors will come to light in later chapters when we deal with $n \times n$ matrices. For now, we can observe that this fact means that the ranks of A and A^{T} are equal.

4.3 Linear Spaces

2D linear maps act on vectors in 2D *linear spaces*, also known as 2D *vector spaces*. Recall from Section 2.1 that the set of all ordered pairs, or 2D vectors $\mathbf{v}$ is called $\mathbb{R}^2$. In Section 2.8, we encountered the concept of a *subspace* of $\mathbb{R}^2$ in finding the orthogonal projection of a vector **w** onto a vector **v**. (In Chapter 14, we will look at more general linear spaces.)

The standard operations in a linear space are addition and scalar multiplication, which are encapsulated for vectors in the linear combination in (4.2). This is called the *linearity property*. Additionally, linear maps, or matrices, are characterized by preservation of linear combinations. This statement can be encapsulated as follows

$$A(a\mathbf{u} + b\mathbf{v}) = aA\mathbf{u} + bA\mathbf{v}. \tag{4.11}$$

Let's break this statement down into the two basic elements: scalar multiplication and addition. For the sake of concreteness, we shall use the example

$$A = \begin{bmatrix} -1 & 1/2 \\ 0 & -1/2 \end{bmatrix}, \quad \mathbf{u} = \begin{bmatrix} 1 \\ 2 \end{bmatrix}, \quad \mathbf{v} = \begin{bmatrix} -1 \\ 4 \end{bmatrix}.$$

We may multiply all elements of a matrix by one factor; we then say that we have multiplied the matrix by that factor. Using our example, we may multiply the matrix A by a factor, say 2:

$$2 \times \begin{bmatrix} -1 & 1/2 \\ 0 & -1/2 \end{bmatrix} = \begin{bmatrix} -2 & 1 \\ 0 & -1 \end{bmatrix}.$$

When we say that matrices preserve multiplication by scalar factors we mean that if we scale a vector by a factor c, then its image will also be scaled by c:

$$A(c\mathbf{u}) = cA\mathbf{u}.$$

Example 4.2

Here are the computations that go along with Sketch 4.2:

$$\begin{bmatrix} -1 & 1/2 \\ 0 & -1/2 \end{bmatrix} \left(2 \times \begin{bmatrix} 1 \\ 2 \end{bmatrix} \right) = 2 \times \begin{bmatrix} -1 & 1/2 \\ 0 & -1/2 \end{bmatrix} \begin{bmatrix} 1 \\ 2 \end{bmatrix} = \begin{bmatrix} 0 \\ -2 \end{bmatrix}.$$

Matrices also preserve sums:

$$A(\mathbf{u} + \mathbf{v}) = A\mathbf{u} + A\mathbf{v}.$$

This is also called the *distributive law*. Sketch 4.3 illustrates this property (with a different set of $A, \mathbf{u}, \mathbf{v}$).

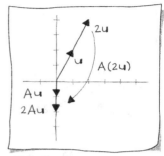

Sketch 4.2.
Matrices preserve scalings.

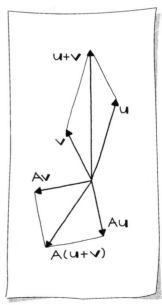

Sketch 4.3.
Matrices preserve sums.

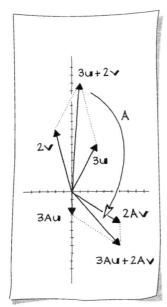

Sketch 4.4.

Matrices preserve linear
combinations.

Now an example to demonstrate that matrices preserve linear combinations, as expressed in (4.11).

Example 4.3

$$A(3\mathbf{u} + 2\mathbf{v}) = \begin{bmatrix} -1 & 1/2 \\ 0 & -1/2 \end{bmatrix} \left(3\begin{bmatrix} 1 \\ 2 \end{bmatrix} + 2\begin{bmatrix} -1 \\ 4 \end{bmatrix} \right)$$
$$= \begin{bmatrix} -1 & 1/2 \\ 0 & -1/2 \end{bmatrix} \begin{bmatrix} 1 \\ 14 \end{bmatrix} = \begin{bmatrix} 6 \\ -7 \end{bmatrix}.$$

$$3A\mathbf{u} + 2A\mathbf{v} = 3\begin{bmatrix} -1 & 1/2 \\ 0 & -1/2 \end{bmatrix}\begin{bmatrix} 1 \\ 2 \end{bmatrix} + 2\begin{bmatrix} -1 & 1/2 \\ 0 & -1/2 \end{bmatrix}\begin{bmatrix} -1 \\ 4 \end{bmatrix}$$
$$= \begin{bmatrix} 0 \\ -3 \end{bmatrix} + \begin{bmatrix} 6 \\ -4 \end{bmatrix} = \begin{bmatrix} 6 \\ -7 \end{bmatrix}.$$

Sketch 4.4 illustrates this example.

Preservation of linear combinations is a key property of matrices—we will make substantial use of it throughout this book.

4.4   Scalings

Consider the linear map given by

$$\mathbf{v}' = \begin{bmatrix} 1/2 & 0 \\ 0 & 1/2 \end{bmatrix} \mathbf{v} = \begin{bmatrix} v_1/2 \\ v_2/2 \end{bmatrix}. \tag{4.12}$$

This map will shorten $\mathbf{v}$ since $\mathbf{v}' = 1/2\mathbf{v}$. Its effect is illustrated in Figure 4.2. That figure—and more to follow—has two parts. The left part is a Phoenix whose feathers form rays that correspond to a sampling of unit vectors. The right part shows what happens if we map the Phoenix, and in turn the unit vectors, using the matrix from (4.12). In this figure we have drawn the $\mathbf{e}_1$ and $\mathbf{e}_2$ vectors,

$$\mathbf{e}_1 = \begin{bmatrix} 1 \\ 0 \end{bmatrix} \quad \text{and} \quad \mathbf{e}_2 = \begin{bmatrix} 0 \\ 1 \end{bmatrix},$$

but in future figures we will not. Notice the positioning of these vectors relative to the Phoenix. Now in the right half, $\mathbf{e}_1$ and $\mathbf{e}_2$ have

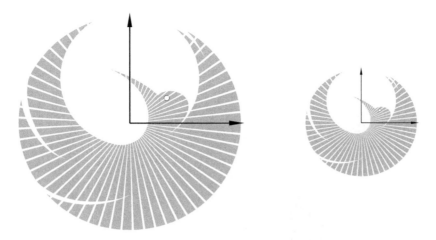

Figure 4.2.
Scaling: a uniform scaling.

been mapped to the vectors $\mathbf{a}_1$ and $\mathbf{a}_2$. These are the column vectors of the matrix in (4.12),

$$\mathbf{a}_1 = \begin{bmatrix} 1/2 \\ 0 \end{bmatrix} \quad \text{and} \quad \mathbf{a}_2 = \begin{bmatrix} 0 \\ 1/2 \end{bmatrix}.$$

The Phoenix's shape provides a sense of orientation. In this example, the linear map did not change orientation, but more complicated maps will.

Next, consider

$$\mathbf{v}' = \begin{bmatrix} 2 & 0 \\ 0 & 2 \end{bmatrix} \mathbf{v}.$$

Now, $\mathbf{v}$ will be "enlarged."

In general, a scaling is defined by the operation

$$\mathbf{v}' = \begin{bmatrix} s_{1,1} & 0 \\ 0 & s_{2,2} \end{bmatrix} \mathbf{v}, \tag{4.13}$$

thus allowing for nonuniform scalings in the $\mathbf{e}_1$- and $\mathbf{e}_2$-direction. Figure 4.3 gives an example for $s_{1,1} = 1/2$ and $s_{2,2} = 2$.

A scaling affects the *area* of the object that is scaled. If we scale an object by $s_{1,1}$ in the $\mathbf{e}_1$-direction, then its area will be changed by a factor $s_{1,1}$. Similarly, it will change by a factor of $s_{2,2}$ when we

Figure 4.3.
Scaling: a nonuniform scaling.

apply that scaling to the $\mathbf{e}_2$-direction. The total effect is thus a factor of $s_{1,1}s_{2,2}$.

You can see this from Figure 4.2 by mentally constructing the square spanned by $\mathbf{e}_1$ and $\mathbf{e}_2$ and comparing its area to the rectangle spanned by the image vectors. It is also interesting to note that, in Figure 4.3, the scaling factors result in no change of area, although a distortion did occur.

The distortion of the circular Phoenix that we see in Figure 4.3 is actually well-defined—it is an ellipse! In fact, all 2×2 matrices will map circles to ellipses. (In higher dimensions, we will speak of ellipsoids.) We will refer to this ellipse that characterizes the action of the matrix as the *action ellipse*.[1] In Figure 4.2, the action ellipse is a scaled circle, which is a special case of an ellipse. In Chapter 16, we will relate the shape of the ellipse to the linear map.

---

[1] We will study ellipses in Chapter 19. An ellipse is symmetric about two axes that intersect at the center of the ellipse. The longer axis is called the *major axis* and the shorter axis is called the *minor axis*. The semi-major and semi-minor axes are one-half their respective axes.

Figure 4.4.
Reflections: a reflection about the $\mathbf{e}_1$-axis.

4.5 Reflections

Consider the scaling

$$\mathbf{v}' = \begin{bmatrix} 1 & 0 \\ 0 & -1 \end{bmatrix} \mathbf{v}. \tag{4.14}$$

We may rewrite this as

$$\begin{bmatrix} v_1' \\ v_2' \end{bmatrix} = \begin{bmatrix} v_1 \\ -v_2 \end{bmatrix}.$$

The effect of this map is apparently a change in sign of the second component of $\mathbf{v}$, as shown in Figure 4.4. Geometrically, this means that the input vector $\mathbf{v}$ is reflected about the $\mathbf{e}_1$-axis, or the line $x_1 = 0$.

Obviously, reflections like the one in Figure 4.4 are just a special case of scalings—previously we simply had not given much thought to negative scaling factors. However, a reflection takes a more general form, and it results in the mirror image of the vectors. Mathematically, a reflection maps each vector about a line through the origin.

The most common reflections are those about the coordinate axes, with one such example illustrated in Figure 4.4, and about the lines $x_1 = x_2$ and $x_1 = -x_2$. The reflection about the line $x_1 = x_2$ is

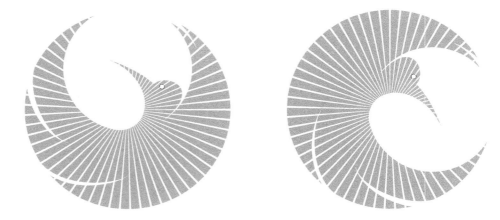

Figure 4.5.
Reflections: a reflection about the line $x_1 = x_2$.

achieved by the matrix

$$\mathbf{v}' = \begin{bmatrix} 0 & 1 \\ 1 & 0 \end{bmatrix} \mathbf{v} = \begin{bmatrix} v_2 \\ v_1 \end{bmatrix}.$$

Its effect is shown in Figure 4.5; that is, the components of the input vector are interchanged.

By inspection of the figures in this section, it appears that reflections do not change areas. But be careful—they do change the *sign* of the area due to a change in orientation. If we rotate $\mathbf{e}_1$ into $\mathbf{e}_2$, we move in a counterclockwise direction. Now, rotate

$$\mathbf{a}_1 = \begin{bmatrix} 0 \\ 1 \end{bmatrix} \quad \text{into} \quad \mathbf{a}_2 = \begin{bmatrix} 1 \\ 0 \end{bmatrix},$$

and notice that we move in a clockwise direction. This change in orientation is reflected in the sign of the area. We will examine this in detail in Section 4.9.

The matrix

$$\mathbf{v}' = \begin{bmatrix} -1 & 0 \\ 0 & -1 \end{bmatrix} \mathbf{v}, \qquad (4.15)$$

as seen in Figure 4.6, appears to be a reflection, but it is really a rotation of 180°. (Rotations are covered in Section 4.6.) If we rotate $\mathbf{a}_1$ into $\mathbf{a}_2$ we move in a counterclockwise direction, confirming that this is not a reflection.

Notice that all reflections result in an action ellipse that is a circle.

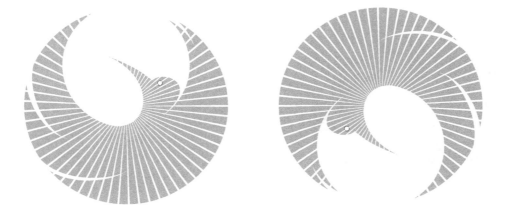

Figure 4.6.
Reflections: a reflection about both axes is also a rotation of $180°$.

4.6 Rotations

The notion of rotating a vector around the origin is intuitively clear, but a corresponding matrix takes a few moments to construct. To keep it easy at the beginning, let us rotate the unit vector

$$\mathbf{e}_1 = \begin{bmatrix} 1 \\ 0 \end{bmatrix}$$

by α degrees counterclockwise, resulting in a new (rotated) vector

$$\mathbf{e}_1' = \begin{bmatrix} \cos\alpha \\ \sin\alpha \end{bmatrix}.$$

Notice that $\cos^2\alpha + \sin^2\alpha = 1$, thus this is a rotation. Consult Sketch 4.5 to convince yourself of this fact.

Thus, we need to find a matrix R that achieves

$$\begin{bmatrix} \cos\alpha \\ \sin\alpha \end{bmatrix} = \begin{bmatrix} r_{1,1} & r_{1,2} \\ r_{2,1} & r_{2,2} \end{bmatrix} \begin{bmatrix} 1 \\ 0 \end{bmatrix}.$$

Additionally, we know that $\mathbf{e}_2$ will rotate to

$$\mathbf{e}_2' = \begin{bmatrix} -\sin\alpha \\ \cos\alpha \end{bmatrix}.$$

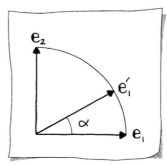

Sketch 4.5.
Rotating a unit vector.

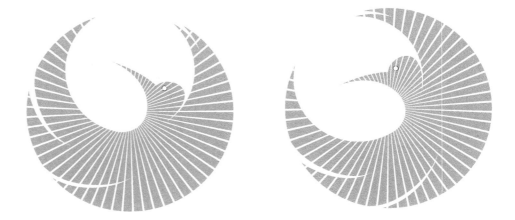

Figure 4.7.
Rotations: a rotation by $45°$.

This leads to the correct *rotation matrix*; it is given by

$$R = \begin{bmatrix} \cos\alpha & -\sin\alpha \\ \sin\alpha & \cos\alpha \end{bmatrix}. \tag{4.16}$$

But let's verify that we have already found the solution to the general rotation problem.

Let $\mathbf{v}$ be an arbitrary vector. We claim that the matrix R from (4.16) will rotate it by α degrees to a new vector $\mathbf{v}'$. If this is so, then we must have

$$\mathbf{v} \cdot \mathbf{v}' = \|\mathbf{v}\|^2 \cos\alpha$$

according to the rules of dot products (see Section 2.7). Here, we made use of the fact that a rotation does not change the length of a vector, i.e., $\|\mathbf{v}\| = \|\mathbf{v}'\|$ and hence $\|\mathbf{v}\| \cdot \|\mathbf{v}'\| = \|\mathbf{v}\|^2$.

Since

$$\mathbf{v}' = \begin{bmatrix} v_1 \cos\alpha - v_2 \sin\alpha \\ v_1 \sin\alpha + v_2 \cos\alpha \end{bmatrix},$$

the dot product $\mathbf{v} \cdot \mathbf{v}'$ is given by

$$\begin{aligned} \mathbf{v} \cdot \mathbf{v}' &= v_1^2 \cos\alpha - v_1 v_2 \sin\alpha + v_1 v_2 \sin\alpha + v_2^2 \cos\alpha \\ &= (v_1^2 + v_2^2) \cos\alpha \\ &= \|\mathbf{v}\|^2 \cos\alpha, \end{aligned}$$

and all is shown! See Figure 4.7 for an illustration. There, $\alpha = 45°$, and the rotation matrix is thus given by

$$R = \begin{bmatrix} \sqrt{2}/2 & -\sqrt{2}/2 \\ \sqrt{2}/2 & \sqrt{2}/2 \end{bmatrix}.$$

Rotations are in a special class of transformations; these are called *rigid body motions.* (See Section 5.9 for more details on these special matrices.) The action ellipse of a rotation is a circle. Finally, it should come without saying that rotations do not change areas.

4.7 Shears

What map takes a rectangle to a parallelogram? Pictorially, one such map is shown in Sketch 4.6.

In this example, we have a map:

$$\mathbf{v} = \begin{bmatrix} 0 \\ 1 \end{bmatrix} \longrightarrow \mathbf{v}' = \begin{bmatrix} d_1 \\ 1 \end{bmatrix}.$$

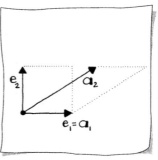

Sketch 4.6.
A special shear.

In matrix form, this is realized by

$$\begin{bmatrix} d_1 \\ 1 \end{bmatrix} = \begin{bmatrix} 1 & d_1 \\ 0 & 1 \end{bmatrix} \begin{bmatrix} 0 \\ 1 \end{bmatrix}. \tag{4.17}$$

Verify! The 2×2 matrix in this equation is called a *shear matrix.* It is the kind of matrix that is used when you generate italic fonts from standard ones.

A shear matrix may be applied to arbitrary vectors. If $\mathbf{v}$ is an input vector, then a shear maps it to $\mathbf{v}'$:

$$\mathbf{v}' = \begin{bmatrix} 1 & d_1 \\ 0 & 1 \end{bmatrix} \begin{bmatrix} v_1 \\ v_2 \end{bmatrix} = \begin{bmatrix} v_1 + v_2 d_1 \\ v_2 \end{bmatrix},$$

as illustrated in Figure 4.8. Clearly, the circular Phoenix is mapped to an elliptical one.

We have so far restricted ourselves to shears along the $\mathbf{e}_1$-axis; we may also shear along the $\mathbf{e}_2$-axis. Then we would have

$$\mathbf{v}' = \begin{bmatrix} 1 & 0 \\ d_2 & 1 \end{bmatrix} \begin{bmatrix} v_1 \\ v_2 \end{bmatrix} = \begin{bmatrix} v_1 \\ v_1 d_2 + v_2 \end{bmatrix},$$

as illustrated in Figure 4.9.

Since it will be needed later, we look at the following. What is the shear that achieves

$$\mathbf{v} = \begin{bmatrix} v_1 \\ v_2 \end{bmatrix} \longrightarrow \mathbf{v}' = \begin{bmatrix} v_1 \\ 0 \end{bmatrix}?$$

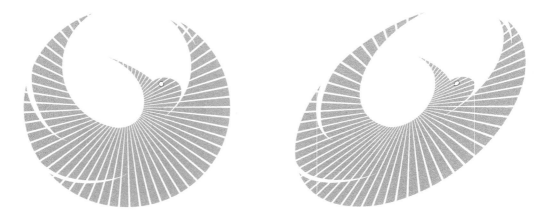

Figure 4.8.
Shears: shearing parallel to the $\mathbf{e}_1$-axis.

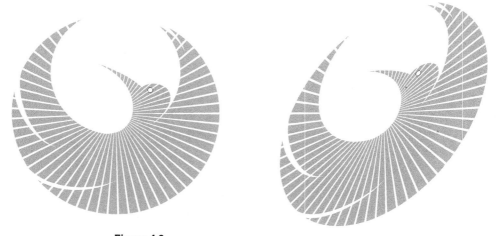

Figure 4.9.
Shears: shearing parallel to the $\mathbf{e}_2$-axis.

It is obviously a shear parallel to the $\mathbf{e}_2$-axis and is given by the map

$$\mathbf{v}' = \begin{bmatrix} v_1 \\ 0 \end{bmatrix} = \begin{bmatrix} 1 & 0 \\ -v_2/v_1 & 1 \end{bmatrix} \begin{bmatrix} v_1 \\ v_2 \end{bmatrix}. \tag{4.18}$$

Shears do not change areas. In Sketch 4.6, we see that the rectangle and its image, a parallelogram, have the same area: both have the same base and the same height.

4.8 Projections

Projections—parallel projections, for our purposes—act like sunlight casting shadows. Parallel projections are characterized by the fact that all vectors are projected in a parallel direction. In 2D, all vectors are projected onto a line. If the angle of incidence with the line is ninety degrees then it is an *orthogonal projection*, otherwise it is an *oblique projection*. In linear algebra, orthogonal projections are very important, as we have already seen in Section 2.8, they give us a best approximation in a particular subspace. Oblique projections are important to applications in fields such as computer graphics and architecture. On the other hand, in a *perspective projection*, the projection direction is not constant. These are not linear maps, however; they are introduced in Section 10.5.

Let's look at a simple 2D orthogonal projection. Take any vector **v** and "flatten it out" onto the $\mathbf{e}_1$-axis. This simply means: set the v_2-coordinate of the vector to zero. For example, if we project the vector

$$\mathbf{v} = \begin{bmatrix} 3 \\ 1 \end{bmatrix}$$

onto the $\mathbf{e}_1$-axis, it becomes

$$\mathbf{v}' = \begin{bmatrix} 3 \\ 0 \end{bmatrix},$$

as shown in Sketch 4.7.

What matrix achieves this map? That's easy:

$$\begin{bmatrix} 3 \\ 0 \end{bmatrix} = \begin{bmatrix} 1 & 0 \\ 0 & 0 \end{bmatrix} \begin{bmatrix} 3 \\ 1 \end{bmatrix}.$$

This matrix will not only project the vector

$$\begin{bmatrix} 3 \\ 1 \end{bmatrix}$$

onto the $\mathbf{e}_1$-axis, but in fact *every vector*! This is so since

$$\begin{bmatrix} 1 & 0 \\ 0 & 0 \end{bmatrix} \begin{bmatrix} v_1 \\ v_2 \end{bmatrix} = \begin{bmatrix} v_1 \\ 0 \end{bmatrix}.$$

While this is a somewhat trivial example of a projection, we see that this projection does indeed feature a main property of a projection: it *reduces dimensionality*. In this example, every vector from 2D

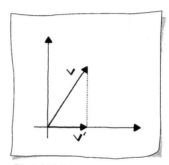

Sketch 4.7.
An orthogonal, parallel projection.

Figure 4.10.
Projections: all vectors are "flattened out" onto the $\mathbf{e}_1$-axis.

space is mapped into 1D space, namely onto the $\mathbf{e}_1$-axis. Figure 4.10 illustrates this property and that the action ellipse of a projection is a straight line segment that is covered twice.

To construct a 2D orthogonal projection matrix, first choose a unit vector $\mathbf{u}$ to define a line onto which to project. The matrix is defined by $\mathbf{a}_1$ and $\mathbf{a}_2$, or in other words, the projections of $\mathbf{e}_1$ and $\mathbf{e}_2$, respectively onto $\mathbf{u}$. From (2.21), we have

$$\mathbf{a}_1 = \frac{\mathbf{u} \cdot \mathbf{e}_1}{\|\mathbf{u}\|^2} \mathbf{u} = u_1 \mathbf{u},$$

$$\mathbf{a}_2 = \frac{\mathbf{u} \cdot \mathbf{e}_2}{\|\mathbf{u}\|^2} \mathbf{u} = u_2 \mathbf{u},$$

thus

$$A = \begin{bmatrix} u_1 \mathbf{u} & u_2 \mathbf{u} \end{bmatrix} \tag{4.19}$$

$$= \mathbf{u}\mathbf{u}^T. \tag{4.20}$$

Forming a matrix as in (4.20), from the product of a vector and its transpose, results in a *dyadic matrix*. Clearly the columns of A are linearly dependent and thus the matrix has rank one. This map reduces dimensionality, and as far as areas are concerned, projections take a lean approach: whatever an area was before application of the map, it is zero afterward.

Figure 4.11 shows the effect of (4.20) on the $\mathbf{e}_1$ and $\mathbf{e}_2$ axes. On the left side, the vector $\mathbf{u} = [\cos 30° \ \sin 30°]^{\mathrm{T}}$ and thin lines show the projection of $\mathbf{e}_1$ (black) and $\mathbf{e}_2$ (dark gray) onto $\mathbf{u}$. On the right side, many $\mathbf{u}$ vectors are illustrated: every $10°$, forming 36 arrows or $\mathbf{u}_i$ for $i = 1, 36$. The black circle of arrows is formed by the projection of $\mathbf{e}_1$ onto each $\mathbf{u}_i$. The gray circle of arrows is formed by the projection of $\mathbf{e}_2$ onto each $\mathbf{u}_i$.

In addition to reducing dimensionality, a projection matrix is also *idempotent*: $A = AA$. Geometrically, this means that once a vector has been projected onto a line, application of the same projection will leave the result unchanged.

Example 4.4

Let the projection line be defined by

$$\mathbf{u} = \begin{bmatrix} 1/\sqrt{2} \\ 1/\sqrt{2} \end{bmatrix}.$$

Then (4.20) defines the projection matrix to be

$$A = \begin{bmatrix} 0.5 & 0.5 \\ 0.5 & 0.5 \end{bmatrix}.$$

This $\mathbf{u}$ vector, corresponding to a $45°$ rotation of $\mathbf{e}_1$, is absent from the right part of Figure 4.11, so find where it belongs. In this case, the projection of $\mathbf{e}_1$ and $\mathbf{e}_2$ are identical.

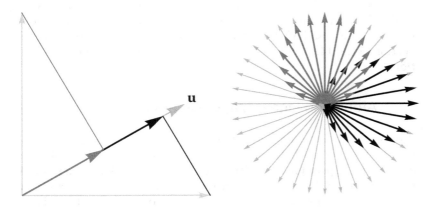

Figure 4.11.
Projections: $\mathbf{e}_1$ and $\mathbf{e}_2$ vectors orthogonally projected onto $\mathbf{u}$ results in $\mathbf{a}_1$ (black) and $\mathbf{a}_2$ (dark gray), respectively. Left: vector $\mathbf{u} = [\cos 30° \ \sin 30°]^{\mathsf{T}}$. Right: vectors $\mathbf{u}_i$ for $i = 1, 36$ are at $10°$ increments.

Try sketching the projection of a few vectors yourself to get a feel for how this projection works. In particular, try

$$\mathbf{v} = \begin{bmatrix} 1 \\ 2 \end{bmatrix}.$$

Sketch $\mathbf{v}' = A\mathbf{v}$. Now compute $\mathbf{v}'' = A\mathbf{v}'$. Surprised?

Let's revisit the orthogonal projection discussion from Section 2.8 by examining the action of A in (4.20) on a vector $\mathbf{x}$,

$$A\mathbf{x} = \mathbf{u}\mathbf{u}^\mathrm{T}\mathbf{x} = (\mathbf{u} \cdot \mathbf{x})\mathbf{u}.$$

We see this is the same result as (2.21). Suppose the projection of $\mathbf{x}$ onto $\mathbf{u}$ is $\mathbf{y}$, then $\mathbf{x} = \mathbf{y} + \mathbf{y}^\perp$, and we then have

$$A\mathbf{x} = \mathbf{u}\mathbf{u}^\mathrm{T}\mathbf{y} + \mathbf{u}\mathbf{u}^\mathrm{T}\mathbf{y}^\perp,$$

and since $\mathbf{u}^\mathrm{T}\mathbf{y} = \|\mathbf{y}\|$ and $\mathbf{u}^\mathrm{T}\mathbf{y}^\perp = 0$,

$$A\mathbf{x} = \|\mathbf{y}\|\mathbf{u}.$$

Projections will be revisited many times, and some examples include: homogeneous linear systems in Section 5.8, 3D projections in Sections 9.7 and 10.4, creating orthonormal coordinate frames in Sections 11.8, 14.4, and 20.7, and least squares approximation in Section 12.7.

4.9   Areas and Linear Maps: Determinants

As you might have noticed, we discussed one particular aspect of linear maps for each type: how areas are changed. We will now discuss this aspect for an arbitrary 2D linear map. Such a map takes the two vectors $[\mathbf{e}_1, \mathbf{e}_2]$ to the two vectors $[\mathbf{a}_1, \mathbf{a}_2]$. The area of the square spanned by $[\mathbf{e}_1, \mathbf{e}_2]$ is 1, that is area$(\mathbf{e}_1, \mathbf{e}_2) = 1$. If we knew the area of the parallelogram spanned by $[\mathbf{a}_1, \mathbf{a}_2]$, then we could say how the linear map affects areas.

How do we find the area P of a parallelogram spanned by two vectors $\mathbf{a}_1$ and $\mathbf{a}_2$? Referring to Sketch 4.8, let us first determine the

Sketch 4.8.
Area formed by $\mathbf{a}_1$ and $\mathbf{a}_2$.

area T of the triangle formed by $\mathbf{a}_1$ and $\mathbf{a}_2$. We see that

$$T = a_{1,1}a_{2,2} - T_1 - T_2 - T_3.$$

We then observe that

$$T_1 = \frac{1}{2}a_{1,1}a_{2,1}, \tag{4.21}$$

$$T_2 = \frac{1}{2}(a_{1,1} - a_{1,2})(a_{2,2} - a_{2,1}), \tag{4.22}$$

$$T_3 = \frac{1}{2}a_{1,2}a_{2,2}. \tag{4.23}$$

Working out the algebra, we arrive at

$$T = \frac{1}{2}a_{1,1}a_{2,2} - \frac{1}{2}a_{1,2}a_{2,1}.$$

Our aim was not really T, but the parallelogram area P. Clearly (see Sketch 4.9),

$$P = 2T,$$

and we have our desired area.

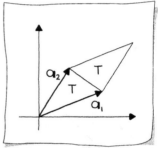

Sketch 4.9.
Parallelogram and triangles.

It is customary to use the term *determinant* for the (signed) area of the parallelogram spanned by $[\mathbf{a}_1, \mathbf{a}_2]$. Since the two vectors $\mathbf{a}_1$ and $\mathbf{a}_2$ form the columns of the matrix A, we also speak of the determinant of the matrix A, and denote it by $\det A$ or $|A|$:

$$|A| = \begin{vmatrix} a_{1,1} & a_{1,2} \\ a_{2,1} & a_{2,2} \end{vmatrix} = a_{1,1}a_{2,2} - a_{1,2}a_{2,1}. \tag{4.24}$$

Since A maps a square with area one onto a parallelogram with area $|A|$, the determinant of a matrix characterizes it as follows:

- If $|A| = 1$, then the linear map does not change areas.

- If $0 \le |A| < 1$, then the linear map shrinks areas.

- If $|A| = 0$, then the matrix is rank deficient.

- If $|A| > 1$, then the linear map expands areas.

- If $|A| < 0$, then the linear map changes the orientation of objects. (We'll look at this closer after Example 4.5.) Areas may still contract or expand depending on the magnitude of the determinant.

Example 4.5

We will look at a few examples. Let

$$A = \begin{bmatrix} 1 & 5 \\ 0 & 1 \end{bmatrix},$$

then $|A| = (1)(1) - (5)(0) = 1$. Since A represents a *shear*, we see again that those maps do not change areas.

For another example, let

$$A = \begin{bmatrix} 1 & 0 \\ 0 & -1 \end{bmatrix},$$

then $|A| = (1)(-1) - (0)(0) = -1$. This matrix corresponds to a *reflection*, and it leaves areas unchanged, except for a *sign change*.

Finally, let

$$A = \begin{bmatrix} 0.5 & 0.5 \\ 0.5 & 0.5 \end{bmatrix},$$

then $|A| = (0.5)(0.5) - (0.5)(0.5) = 0$. This matrix corresponds to the *projection* from Section 4.8. In that example, we saw that projections collapse any object onto a straight line, i.e., to an object with zero area.

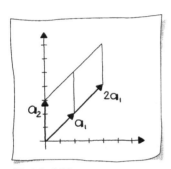

Sketch 4.10.

Resulting area after scaling one column of *A*.

There are some rules for working with determinants: If $A = [\mathbf{a}_1, \mathbf{a}_2]$, then

$$|c\mathbf{a}_1, \mathbf{a}_2| = c|\mathbf{a}_1, \mathbf{a}_2| = c|A|.$$

In other words, if one of the columns of A is scaled by a factor c, then A's determinant is also scaled by c. Verify that this is true from the definition of the determinant of A. Sketch 4.10 illustrates this for the example, $c = 2$. If *both* columns of A are scaled by c, then the determinant is scaled by c^2:

$$|c\mathbf{a}_1, c\mathbf{a}_2| = c^2|\mathbf{a}_1, \mathbf{a}_2| = c^2|A|.$$

Sketch 4.11 illustrates this for the example $c = 1/2$.

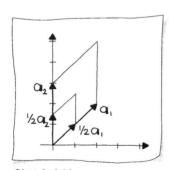

Sketch 4.11.

Resulting area after scaling both columns of *A*.

If $|A|$ is positive and c is negative, then replacing $\mathbf{a}_1$ by $c\mathbf{a}_1$ will cause $c|A|$, the area formed by $c\mathbf{a}_1$ and $\mathbf{a}_2$, to become negative. The notion of a negative area is very useful computationally. Two 2D vectors whose determinant is positive are called *right-handed*. The standard example is the pair of vectors $\mathbf{e}_1$ and $\mathbf{e}_2$. Two 2D vectors whose determinant is negative are called *left-handed*.[2] Sketch 4.12 shows a pair of right-handed vectors (top) and a pair of left-handed ones (bottom). Our definition of positive and negative area is not totally arbitrary: the triangle formed by vectors $\mathbf{a}_1$ and $\mathbf{a}_2$ has area $1/2 \times \sin(\alpha)\|\mathbf{a}_1\|\|\mathbf{a}_2\|$. Here, the angle α indicates how much we have to rotate $\mathbf{a}_1$ in order to line up with $\mathbf{a}_2$. If we interchange the two vectors, the sign of α and hence the sign of $\sin(\alpha)$ also changes!

There is also an area sign change when we *interchange* the columns of A:

$$|\mathbf{a}_1, \mathbf{a}_2| = -|\mathbf{a}_2, \mathbf{a}_1|. \tag{4.25}$$

This fact is easily verified using the definition of a determinant:

$$|\mathbf{a}_2, \mathbf{a}_1| = a_{1,2}a_{2,1} - a_{2,2}a_{1,1}.$$

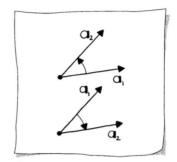

Sketch 4.12.

Right-handed and left-handed vectors.

4.10 Composing Linear Maps

Suppose you have mapped a vector $\mathbf{v}$ to $\mathbf{v}'$ using a matrix A. Next, you want to map $\mathbf{v}'$ to $\mathbf{v}''$ using a matrix B. We start out with

$$\mathbf{v}' = \begin{bmatrix} a_{1,1} & a_{1,2} \\ a_{2,1} & a_{2,2} \end{bmatrix} \begin{bmatrix} v_1 \\ v_2 \end{bmatrix} = \begin{bmatrix} a_{1,1}v_1 + a_{1,2}v_2 \\ a_{2,1}v_1 + a_{2,2}v_2 \end{bmatrix}.$$

Next, we have

$$\mathbf{v}'' = \begin{bmatrix} b_{1,1} & b_{1,2} \\ b_{2,1} & b_{2,2} \end{bmatrix} \begin{bmatrix} a_{1,1}v_1 + a_{1,2}v_2 \\ a_{2,1}v_1 + a_{2,2}v_2 \end{bmatrix}$$

$$= \begin{bmatrix} b_{1,1}(a_{1,1}v_1 + a_{1,2}v_2) + b_{1,2}(a_{2,1}v_1 + a_{2,2}v_2) \\ b_{2,1}(a_{1,1}v_1 + a_{1,2}v_2) + b_{2,2}(a_{2,1}v_1 + a_{2,2}v_2) \end{bmatrix}.$$

Collecting the terms in v_1 and v_2, we get

$$\mathbf{v}'' = \begin{bmatrix} b_{1,1}a_{1,1} + b_{1,2}a_{2,1} & b_{1,1}a_{1,2} + b_{1,2}a_{2,2} \\ b_{2,1}a_{1,1} + b_{2,2}a_{2,1} & b_{2,1}a_{1,2} + b_{2,2}a_{2,2} \end{bmatrix} \begin{bmatrix} v_1 \\ v_2 \end{bmatrix}.$$

The matrix that we have created here, let's call it C, is called the *product matrix* of B and A:

$$BA = C.$$

[2]The reason for this terminology will become apparent when we revisit these definitions for the 3D case (see Section 8.2).

In more detail,

$$\begin{bmatrix} b_{1,1} & b_{1,2} \\ b_{2,1} & b_{2,2} \end{bmatrix} \begin{bmatrix} a_{1,1} & a_{1,2} \\ a_{2,1} & a_{2,2} \end{bmatrix} = \begin{bmatrix} b_{1,1}a_{1,1} + b_{1,2}a_{2,1} & b_{1,1}a_{1,2} + b_{1,2}a_{2,2} \\ b_{2,1}a_{1,1} + b_{2,2}a_{2,1} & b_{2,1}a_{1,2} + b_{2,2}a_{2,2} \end{bmatrix}.$$
$$(4.26)$$

This looks messy, but a simple rule puts order into chaos: the element $c_{i,j}$ is computed as the dot product of B's ith row and A's jth column. We can use this product to describe the *composite map*:

$$\mathbf{v}'' = B\mathbf{v}' = B[A\mathbf{v}] = BA\mathbf{v}.$$

Example 4.6

Let

$$\mathbf{v} = \begin{bmatrix} 2 \\ -1 \end{bmatrix}, \quad A = \begin{bmatrix} -1 & 2 \\ 0 & 3 \end{bmatrix}, \quad B = \begin{bmatrix} 0 & -2 \\ -3 & 1 \end{bmatrix}.$$

Then

$$\mathbf{v}' = \begin{bmatrix} -1 & 2 \\ 0 & 3 \end{bmatrix} \begin{bmatrix} 2 \\ -1 \end{bmatrix} = \begin{bmatrix} -4 \\ -3 \end{bmatrix}$$

and

$$\mathbf{v}'' = \begin{bmatrix} 0 & -2 \\ -3 & 1 \end{bmatrix} \begin{bmatrix} -4 \\ -3 \end{bmatrix} = \begin{bmatrix} 6 \\ 9 \end{bmatrix}.$$

We can also compute $\mathbf{v}''$ using the matrix product BA:

$$C = BA = \begin{bmatrix} 0 & -2 \\ -3 & 1 \end{bmatrix} \begin{bmatrix} -1 & 2 \\ 0 & 3 \end{bmatrix} = \begin{bmatrix} 0 & -6 \\ 3 & -3 \end{bmatrix}.$$

Verify for yourself that $\mathbf{v}'' = C\mathbf{v}$.

Analogous to the matrix/vector product from Section 4.2, there is a neat way to arrange two matrices when forming their product for manual computation (yes, that is still encountered!). Using the matrices of Example 4.6, and highlighting the computation of $c_{2,1}$, we write

$$\begin{array}{cc|cc} & & -1 & 2 \\ & & \mathbf{0} & 3 \\ \hline 0 & -2 & \\ \mathbf{-3} & \mathbf{1} & \mathbf{3} \end{array}$$

Figure 4.12.
Linear map composition is order dependent. Top: rotate by $-120°$, then reflect about the (rotated) $\mathbf{e}_1$-axis. Bottom: reflect, then rotate.

You see how $c_{2,1}$ is at the intersection of column one of the "top" matrix and row two of the "left" matrix.

The complete multiplication scheme is then arranged like this

$$
\begin{array}{cc|cc}
 & & -1 & 2 \\
 & & 0 & 3 \\
\hline
0 & -2 & 0 & -6 \\
-3 & 1 & 3 & -3 \\
\end{array}
$$

While we use the term "product" for BA, it is very important to realize that this kind of product differs significantly from products of real numbers: it is not *commutative*. That is, in general

$$AB \neq BA.$$

Matrix products correspond to linear map compositions—since the products are not commutative, it follows that it matters in which order we carry out linear maps. *Linear map composition is order dependent.* Figure 4.12 gives an example.

Example 4.7

Let us take two very simple matrices and demonstrate that the product is not commutative. This example is illustrated in Figure 4.12. A rotates by $-120°$, and B reflects about the $\mathbf{e}_1$-axis:

$$A = \begin{bmatrix} -0.5 & 0.866 \\ -0.866 & -0.5 \end{bmatrix}, \qquad B = \begin{bmatrix} 1 & 0 \\ 0 & -1 \end{bmatrix}.$$

We first form AB (reflect and then rotate),

$$AB = \begin{bmatrix} -0.5 & 0.866 \\ -0.866 & -0.5 \end{bmatrix} \begin{bmatrix} 1 & 0 \\ 0 & -1 \end{bmatrix} = \begin{bmatrix} -0.5 & -0.866 \\ -0.866 & 0.5 \end{bmatrix}.$$

Next, we form BA (rotate and then reflect),

$$BA = \begin{bmatrix} 1 & 0 \\ 0 & -1 \end{bmatrix} \begin{bmatrix} -0.5 & 0.866 \\ -0.866 & -0.5 \end{bmatrix} = \begin{bmatrix} -0.5 & 0.866 \\ 0.866 & 0.5 \end{bmatrix}.$$

Clearly, these are not the same!

Of course, *some* maps *do* commute; for example, the 2D rotations. It does not matter if we rotate by α first and then by β or the other way around. In either case, we have rotated by $\alpha + \beta$. In terms of matrices,

$$\begin{bmatrix} \cos\alpha & -\sin\alpha \\ \sin\alpha & \cos\alpha \end{bmatrix} \begin{bmatrix} \cos\beta & -\sin\beta \\ \sin\beta & \cos\beta \end{bmatrix} =$$

$$\begin{bmatrix} \cos\alpha\cos\beta - \sin\alpha\sin\beta & -\cos\alpha\sin\beta - \sin\alpha\cos\beta \\ \sin\alpha\cos\beta + \cos\alpha\sin\beta & -\sin\alpha\sin\beta + \cos\alpha\cos\beta \end{bmatrix}.$$

Check for yourself that the other alternative gives the same result! By referring to a trigonometry reference, we see that this product matrix can be written as

$$\begin{bmatrix} \cos(\alpha + \beta) & -\sin(\alpha + \beta) \\ \sin(\alpha + \beta) & \cos(\alpha + \beta) \end{bmatrix},$$

which corresponds to a rotation by $\alpha+\beta$. As we will see in Section 9.9, rotations in 3D do not commute.

What about the rank of a composite map?

$$\text{rank}(AB) \leq \min\{\text{rank}(A), \text{rank}(B)\}. \tag{4.27}$$

This says that matrix multiplication does not increase rank. You should try a few rank 1 and 2 matrices to convince yourself of this fact.

One more example on composing linear maps, which we have seen in Section 4.8, is that projections are idempotent. If A is a projection matrix, then this means

$$A\mathbf{v} = AA\mathbf{v}$$

for any vector $\mathbf{v}$. Written out using only matrices, this becomes

$$A = AA \quad \text{or} \quad A = A^2. \tag{4.28}$$

Verify this property for the projection matrix

$$\begin{bmatrix} 0.5 & 0.5 \\ 0.5 & 0.5 \end{bmatrix}.$$

Excluding the identity matrix, only rank deficient matrices are idempotent.

4.11   More on Matrix Multiplication

Matrix multiplication is not limited to the product of 2×2 matrices. In fact, when we multiply a matrix by a vector, we follow the rules of matrix multiplication! If $\mathbf{v}' = A\mathbf{v}$, then the first component of $\mathbf{v}'$ is the dot product of A's first row and $\mathbf{v}$; the second component of $\mathbf{v}'$ is the dot product of A's second row and $\mathbf{v}$.

In Section 4.2, we introduced the transpose A^{T} of a matrix A and a vector $\mathbf{v}$. Usually, we write $\mathbf{u} \cdot \mathbf{v}$ for the dot product of $\mathbf{u}$ and $\mathbf{v}$, but sometimes considering the vectors as matrices is useful as well; that is, we can form the product as

$$\mathbf{u}^{\mathrm{T}}\mathbf{v} = \mathbf{u} \cdot \mathbf{v}. \tag{4.29}$$

For examples in this section, let

$$\mathbf{u} = \begin{bmatrix} 3 \\ 4 \end{bmatrix} \quad \text{and} \quad \mathbf{v} = \begin{bmatrix} -3 \\ 6 \end{bmatrix}.$$

Then it is straightforward to show that the left- and right-hand sides of (4.29) are equal to 15.

If we have a product $\mathbf{u}^{\mathrm{T}}\mathbf{v}$, what is $[\mathbf{u}^{\mathrm{T}}\mathbf{v}]^{\mathrm{T}}$? This has an easy answer:

$$(\mathbf{u}^{\mathrm{T}}\mathbf{v})^{\mathrm{T}} = \mathbf{v}^{\mathrm{T}}\mathbf{u},$$

as an example will clarify.

Example 4.8

$$[\mathbf{u}^{\mathrm{T}}\mathbf{v}]^{\mathrm{T}} = \left(\begin{bmatrix} 3 & 4 \end{bmatrix} \begin{bmatrix} -3 \\ 6 \end{bmatrix}\right)^{\mathrm{T}} = [15]^{\mathrm{T}} = 15$$

$$\mathbf{v}^{\mathrm{T}}\mathbf{u} = \begin{bmatrix} -3 & 6 \end{bmatrix} \begin{bmatrix} 3 \\ 4 \end{bmatrix} = [15] = 15.$$

The results are the same.

We saw that addition of matrices is straightforward under transposition; matrix multiplication is not that straightforward. We have

$$(AB)^{\mathrm{T}} = B^{\mathrm{T}}A^{\mathrm{T}}. \qquad (4.30)$$

To see why this is true, recall that each element of a product matrix is obtained as a dot product. In the matrix products below, we show the calculation of one element of the left- and right-hand sides of (4.30) to demonstrate that the dot products for this one element are identical. Let transpose matrix elements be referred to as $b_{i,j}^{\mathrm{T}}$.

$$(AB)^{\mathrm{T}} = \left(\begin{bmatrix} a_{1,1} & a_{1,2} \\ \mathbf{a_{2,1}} & \mathbf{a_{2,2}} \end{bmatrix} \begin{bmatrix} \mathbf{b_{1,1}} & b_{1,2} \\ \mathbf{b_{2,1}} & b_{2,2} \end{bmatrix}\right)^{\mathrm{T}} = \begin{bmatrix} c_{1,1} & \mathbf{c_{1,2}} \\ c_{2,1} & c_{2,2} \end{bmatrix},$$

$$B^{\mathrm{T}}A^{\mathrm{T}} = \begin{bmatrix} \mathbf{b_{1,1}^{\mathrm{T}}} & \mathbf{b_{1,2}^{\mathrm{T}}} \\ b_{2,1}^{\mathrm{T}} & b_{2,2}^{\mathrm{T}} \end{bmatrix} \begin{bmatrix} a_{1,1}^{\mathrm{T}} & \mathbf{a_{1,2}^{\mathrm{T}}} \\ a_{2,1}^{\mathrm{T}} & \mathbf{a_{2,2}^{\mathrm{T}}} \end{bmatrix} = \begin{bmatrix} c_{1,1} & \mathbf{c_{1,2}} \\ c_{2,1} & c_{2,2} \end{bmatrix}.$$

Since $b_{i,j} = b_{j,i}^{\mathrm{T}}$, we see the identical dot product is calculated to form $c_{1,2}$.

What is the *determinant* of a product matrix? If $C = AB$ denotes a matrix product, then

$$|AB| = |A||B|, \qquad (4.31)$$

which tells us that B scales objects by $|B|$, A scales objects by $|A|$, and the composition of the maps scales by the product of the individual scales.

Example 4.9

As a simple example, take two scalings

$$A = \begin{bmatrix} 1/2 & 0 \\ 0 & 1/2 \end{bmatrix}, \quad B = \begin{bmatrix} 4 & 0 \\ 0 & 4 \end{bmatrix}.$$

We have $|A| = 1/4$ and $|B| = 16$. Thus, A scales down, and B scales up, but the effect of B's scaling is greater than that of A's. The product

$$AB = \begin{bmatrix} 2 & 0 \\ 0 & 2 \end{bmatrix}$$

thus scales up: $|AB| = |A||B| = 4$.

Just as for real numbers, we can define *exponents* for matrices:

$$A^r = \underbrace{A \cdot \ldots \cdot A}_{r \text{ times}}.$$

Here are some rules.

$$\boxed{\begin{aligned} A^{r+s} &= A^r A^s \\ A^{rs} &= (A^r)^s \\ A^0 &= I \end{aligned}}$$

For now, assume r and s are positive integers. See Sections 5.9 and 9.10 for a discussion of A^{-1}, the *inverse matrix*.

4.12 Matrix Arithmetic Rules

We encountered some of these rules throughout this chapter in terms of matrix and vector multiplication: a vector is simply a special matrix. The focus of this chapter is on 2×2 and 2×1 matrices, but these rules apply to matrices of any size. In the rules that follow, let a, b be scalars.

Importantly, however, the matrix sizes must be compatible for the operations to be performed. Specifically, matrix addition requires the matrices to have the same dimensions and matrix multiplication requires the "inside" dimensions to be equal: suppose A's dimensions are $m \times r$ and B's are $r \times n$, then the product $C = AB$ is permissible since the dimension r is shared, and the resulting matrix C will have the "outer" dimensions, reading left to right: $m \times n$.

Commutative Law for Addition
$$A + B = B + A$$

Associative Law for Addition
$$A + (B + C) = (A + B) + C$$

No Commutative Law for Multiplication
$$AB \neq BA$$

Associative Law for Multiplication
$$A(BC) = (AB)C$$

Distributive Law
$$A(B + C) = AB + AC$$

Distributive Law
$$(B + C)A = BA + CA$$

Rules involving scalars:

$$a(B + C) = aB + aC$$
$$(a + b)C = aC + bC$$
$$(ab)C = a(bC)$$
$$a(BC) = (aB)C = B(aC)$$

Rules involving the transpose:

$$(A + B)^{\mathrm{T}} = A^{\mathrm{T}} + B^{\mathrm{T}}$$
$$(bA)^{\mathrm{T}} = bA^{\mathrm{T}}$$
$$(AB)^{\mathrm{T}} = B^{\mathrm{T}}A^{\mathrm{T}}$$
$$A^{\mathrm{T}^{\mathrm{T}}} = A$$

Chapter 9 will introduce 3×3 matrices and Chapter 12 will introduce $n \times n$ matrices.

- linear combination
- matrix form
- preimage and image
- domain and range
- column space
- identity matrix
- matrix addition
- distributive law
- transpose matrix
- symmetric matrix
- rank of a matrix
- rank deficient
- singular matrix
- linear space or vector space
- subspace
- linearity property
- scalings

- action ellipse
- reflections
- rotations
- rigid body motions
- shears
- projections
- parallel projection
- oblique projection
- dyadic matrix
- idempotent map
- determinant
- signed area
- matrix multiplication
- composite map
- noncommutative property of matrix multiplication
- transpose of a product or sum of matrices
- rules of matrix arithmetic

4.13 Exercises

For the following exercises, let

$$A = \begin{bmatrix} 0 & -1 \\ 1 & 0 \end{bmatrix}, \quad B = \begin{bmatrix} 1 & -1 \\ -1 & 1/2 \end{bmatrix}, \quad \mathbf{v} = \begin{bmatrix} 2 \\ 3 \end{bmatrix}.$$

1. What linear combination of

$$\mathbf{c}_1 = \begin{bmatrix} 1 \\ 0 \end{bmatrix} \quad \text{and} \quad \mathbf{c}_2 = \begin{bmatrix} 0 \\ 2 \end{bmatrix}$$

   results in

$$\mathbf{w} = \begin{bmatrix} 2 \\ 1 \end{bmatrix}?$$

   Write the result in matrix form.

2. Suppose $\mathbf{w}' = w_1 \mathbf{c}_1 + w_2 \mathbf{c}_2$. Express this in matrix form.

3. Is the vector $\mathbf{v}$ in the column space of A?

4. Construct a matrix C such that the vector

$$\mathbf{w} = \begin{bmatrix} 0 \\ 1 \end{bmatrix}$$

   is *not* in its column space.

5. Are either A or B symmetric matrices?

6. What is the transpose of A, B, and $\mathbf{v}$?

7. What is the transpose of the 2×2 identity matrix I?

8. Compute $A^{\mathrm{T}} + B^{\mathrm{T}}$ and $[A + B]^{\mathrm{T}}$.

9. What is the rank of B?

10. What is the rank of the matrix $C = [\mathbf{c} \ 3\mathbf{c}]$?

11. What is the rank of the zero matrix?

12. For the matrix A, vectors $\mathbf{v}$ and $\mathbf{u} = [1 \ 0]^{\mathrm{T}}$, and scalars $a = 4$ (applied to $\mathbf{u}$) and $b = 2$ (applied to $\mathbf{v}$), demonstrate the linearity property of linear maps.

13. Describe geometrically the effect of A and B. (You may do this analytically or by using software to illustrate the action of the matrices.)

14. Compute $A\mathbf{v}$ and $B\mathbf{v}$.

15. Construct the matrix S that maps the vector $\mathbf{w}$ to $3\mathbf{w}$.

16. What scaling matrix will result in an action ellipse with major axis twice the length of the minor axis?

17. Construct the matrix that reflects about the $\mathbf{e}_2$-axis.

18. What is the shear matrix that maps $\mathbf{v}$ onto the $\mathbf{e}_2$-axis?

19. What type of linear map is the matrix

$$\begin{bmatrix} -1 & 0 \\ 0 & -1 \end{bmatrix}?$$

20. Are either A or B a rigid body motion?

21. Is the matrix A idempotent?

22. Construct an orthogonal projection onto

$$\mathbf{u} = \begin{bmatrix} 2/\sqrt{5} \\ 1/\sqrt{5} \end{bmatrix}.$$

23. For an arbitrary unit vector $\mathbf{u}$, show that $P = \mathbf{u}\mathbf{u}^{\mathrm{T}}$ is idempotent.

24. Suppose we apply each of the following linear maps to the vertices of a unit square: scaling, reflection, rotation, shear, projection. For each map, state if there is a change in area and the reason.

25. What is the determinant of A?

26. What is the determinant of B?

27. What is the determinant of $4B$?

28. Compute $A + B$. Show that $A\mathbf{v} + B\mathbf{v} = (A + B)\mathbf{v}$.

29. Compute $AB\mathbf{v}$ and $BA\mathbf{v}$.

30. Compute $B^{\mathrm{T}}A$.

31. What is A^2?

32. Let M and N be 2×2 matrices and each is rank one. What can you say about the rank of $M + N$?

33. Let two square matrices M and N each have rank one. What can you say about the rank of MN?

34. Find matrices C and D, both having rank greater than zero, such that the product CD has rank zero.

2×2 Linear Systems

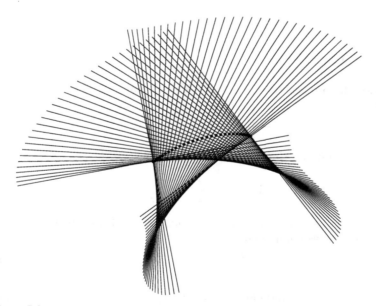

Figure 5.1.
Intersection of lines: two families of lines are shown; the intersections of corresponding line pairs are marked by black boxes. For each intersection, a 2×2 linear system has to be solved.

Just about anybody can solve two equations in two unknowns by somehow manipulating the equations. In this chapter, we will develop a systematic way for finding the solution, simply by checking the underlying geometry. This approach will later enable us to solve much larger systems of equations in Chapters 12 and 13. Figure 5.1 illustrates many instances of intersecting two lines: a problem that can be formulated as a 2 × 2 linear system.

5.1 Skew Target Boxes Revisited

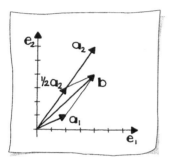

Sketch 5.1.

Geometry of a 2 × 2 system.

In our standard $[\mathbf{e}_1, \mathbf{e}_2]$-coordinate system, suppose we are given two vectors $\mathbf{a}_1$ and $\mathbf{a}_2$. In Section 4.1, we showed how these vectors define a skew target box with its lower-left corner at the origin. As illustrated in Sketch 5.1, suppose we also are given a vector $\mathbf{b}$ with respect to the $[\mathbf{e}_1, \mathbf{e}_2]$-system. Now the question arises, what are the components of $\mathbf{b}$ with respect to the $[\mathbf{a}_1, \mathbf{a}_2]$-system? In other words, we want to find a vector $\mathbf{u}$ with components u_1 and u_2, satisfying

$$u_1\mathbf{a}_1 + u_2\mathbf{a}_2 = \mathbf{b}. \tag{5.1}$$

Example 5.1

Before we proceed further, let's look at an example. Following Sketch 5.1, let

$$\mathbf{a}_1 = \begin{bmatrix} 2 \\ 1 \end{bmatrix}, \quad \mathbf{a}_2 = \begin{bmatrix} 4 \\ 6 \end{bmatrix}, \quad \mathbf{b} = \begin{bmatrix} 4 \\ 4 \end{bmatrix}.$$

Upon examining the sketch, we see that

$$1 \times \begin{bmatrix} 2 \\ 1 \end{bmatrix} + \frac{1}{2} \times \begin{bmatrix} 4 \\ 6 \end{bmatrix} = \begin{bmatrix} 4 \\ 4 \end{bmatrix}.$$

In the $[\mathbf{a}_1, \mathbf{a}_2]$-system, $\mathbf{b}$ has components $(1, 1/2)$. In the $[\mathbf{e}_1, \mathbf{e}_2]$-system, it has components $(4, 4)$.

What we have here are really two equations in the two unknowns u_1 and u_2, which we see by expanding the vector equations into

$$\begin{aligned} 2u_1 + 4u_2 &= 4 \\ u_1 + 6u_2 &= 4. \end{aligned} \tag{5.2}$$

And as we saw in Example 5.1, these two equations in two unknowns have the solution $u_1 = 1$ and $u_2 = 1/2$, as is seen by inserting these values for u_1 and u_2 into the equations.

Being able to solve two simultaneous sets of equations allows us to switch back and forth between different coordinate systems. The rest of this chapter is dedicated to a detailed discussion of how to solve these equations.

5.2   The Matrix Form

The two equations in (5.2) are also called a *linear system*. It can be written more compactly if we use matrix notation:

$$\begin{bmatrix} 2 & 4 \\ 1 & 6 \end{bmatrix} \begin{bmatrix} u_1 \\ u_2 \end{bmatrix} = \begin{bmatrix} 4 \\ 4 \end{bmatrix}. \tag{5.3}$$

In general, a 2×2 linear system looks like this:

$$\begin{bmatrix} a_{1,1} & a_{1,2} \\ a_{2,1} & a_{2,2} \end{bmatrix} \begin{bmatrix} u_1 \\ u_2 \end{bmatrix} = \begin{bmatrix} b_1 \\ b_2 \end{bmatrix}. \tag{5.4}$$

Equation (5.4) is shorthand notation for the equations

$$a_{1,1}u_1 + a_{1,2}u_2 = b_1, \tag{5.5}$$
$$a_{2,1}u_1 + a_{2,2}u_2 = b_2. \tag{5.6}$$

We sometimes write it even shorter, using a matrix A:

$$A\mathbf{u} = \mathbf{b}, \tag{5.7}$$

where

$$A = \begin{bmatrix} a_{1,1} & a_{1,2} \\ a_{2,1} & a_{2,2} \end{bmatrix}, \quad \mathbf{u} = \begin{bmatrix} u_1 \\ u_2 \end{bmatrix}, \quad \mathbf{b} = \begin{bmatrix} b_1 \\ b_2 \end{bmatrix}.$$

Both $\mathbf{u}$ and $\mathbf{b}$ represent vectors, not points! (See Sketch 5.1 for an illustration.) The vector $\mathbf{u}$ is called the *solution* of the linear system.

While the savings of this notation is not completely obvious in the 2×2 case, it will save a lot of work for more complicated cases with more equations and unknowns.

The columns of the matrix A correspond to the vectors $\mathbf{a}_1$ and $\mathbf{a}_2$. We could then rewrite our linear system as (5.1). Geometrically, we are trying to express the given vector $\mathbf{b}$ as a linear combination of the given vectors $\mathbf{a}_1$ and $\mathbf{a}_2$; we need to determine the factors u_1 and u_2. If we are able to find at least one solution, then the linear system is called *consistent*, otherwise it is called inconsistent. Three possibilities for our *solution space* exist.

1. There is exactly one solution vector $\mathbf{u}$. In this case, $|A| \neq 0$, thus the matrix has full rank and is nonsingular.

2. There is no solution, or in other words, the system is inconsistent. (See Section 5.6 for a geometric description.)

3. There are infinitely many solutions. (See Sections 5.7 and 5.8 for examples.)

5.3   A Direct Approach: Cramer's Rule

Sketch 5.2 offers a direct solution to our linear system. By inspecting the areas of the parallelograms in the sketch, we see that

$$u_1 = \frac{\text{area}(\mathbf{b}, \mathbf{a}_2)}{\text{area}(\mathbf{a}_1, \mathbf{a}_2)},$$

$$u_2 = \frac{\text{area}(\mathbf{a}_1, \mathbf{b})}{\text{area}(\mathbf{a}_1, \mathbf{a}_2)}.$$

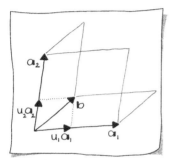

Sketch 5.2.

Cramer's rule.

An easy way to see how these ratios of areas correspond to u_1 and u_2 is to shear the parallelograms formed by $\mathbf{b}, \mathbf{a}_2$ and $\mathbf{b}, \mathbf{a}_1$ onto the $\mathbf{a}_1$ and $\mathbf{a}_2$ axes, respectively. (Shears preserve areas.) The area of a parallelogram is given by the determinant of the two vectors spanning it. Recall from Section 4.9 that this is a signed area. This method of solving for the solution of a linear system is called *Cramer's rule*.

Example 5.2

Applying Cramer's rule to the linear system in (5.3), we get

$$u_1 = \frac{\begin{vmatrix} 4 & 4 \\ 4 & 6 \end{vmatrix}}{\begin{vmatrix} 2 & 4 \\ 1 & 6 \end{vmatrix}} = \frac{8}{8},$$

$$u_2 = \frac{\begin{vmatrix} 2 & 4 \\ 1 & 4 \end{vmatrix}}{\begin{vmatrix} 2 & 4 \\ 1 & 6 \end{vmatrix}} = \frac{4}{8}.$$

Examining the determinant in the numerator, notice that $\mathbf{b}$ replaces $\mathbf{a}_1$ in the solution for u_1 and then $\mathbf{b}$ replaces $\mathbf{a}_2$ in the solution for u_2.

Notice that if the area spanned by $\mathbf{a}_1$ and $\mathbf{a}_2$ is zero, that is, the vectors are multiples of each other, then Cramer's rule will not result in a solution. (See Sections 5.6–5.8 for more information on this situation.)

Cramer's rule is primarily of theoretical importance. For larger systems, Cramer's rule is both expensive and numerically unstable. Hence, we now study a more effective method.

5.4   Gauss Elimination

Let's consider a special 2×2 linear system:

$$\begin{bmatrix} a_{1,1} & a_{1,2} \\ 0 & a_{2,2} \end{bmatrix} \mathbf{u} = \mathbf{b}. \qquad (5.8)$$

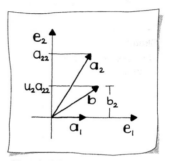

Sketch 5.3.
A special linear system.

This situation is shown in Sketch 5.3. This matrix is called *upper triangular* because all elements below the diagonal are zero, forming a triangle of numbers above the diagonal.

We can solve this system without much work. Examining Equation (5.8), we see it is possible to solve for

$$u_2 = b_2/a_{2,2}.$$

With u_2 in hand, we can solve the first equation from (5.5) for

$$u_1 = \frac{1}{a_{1,1}}(b_1 - u_2 a_{1,2}).$$

This technique of solving the equations from the bottom up is called *back substitution.*

Notice that the process of back substitution requires divisions. Therefore, if the diagonal elements, $a_{1,1}$ or $a_{2,2}$, equal zero then the algorithm will fail. This type of failure indicates that the columns of A are not linearly independent. (See Sections 5.6–5.8 for more information on this situation.) Because of the central role that the diagonal elements play in Gauss elimination, they are called *pivots.*

In general, we will not be so lucky to encounter an upper triangular system as in (5.8). But any linear system in which A is nonsingular may be *transformed* to this simple form, as we shall see by reexamining the system in (5.3). We write it as

$$u_1 \begin{bmatrix} 2 \\ 1 \end{bmatrix} + u_2 \begin{bmatrix} 4 \\ 6 \end{bmatrix} = \begin{bmatrix} 4 \\ 4 \end{bmatrix}.$$

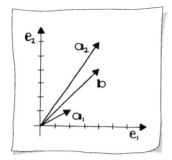

Sketch 5.4.
The geometry of a linear system.

This situation is shown in Sketch 5.4. Clearly, $\mathbf{a}_1$ is not on the $\mathbf{e}_1$-axis as we would like, but we can apply a stepwise procedure so that it will become just that. This systematic, stepwise procedure is called *forward elimination*. The process of forward elimination followed by back substitution is called *Gauss elimination.*[1]

Recall one key fact from Chapter 4: *linear maps do not change linear combinations.* That means if we apply the same linear map to all vectors in our system, then the factors u_1 and u_2 won't change. If the map is given by a matrix S, then

$$S\left[u_1 \begin{bmatrix} 2 \\ 1 \end{bmatrix} + u_2 \begin{bmatrix} 4 \\ 6 \end{bmatrix} \right] = S \begin{bmatrix} 4 \\ 4 \end{bmatrix}.$$

In order to get $\mathbf{a}_1$ to line up with the $\mathbf{e}_1$-axis, we will employ a *shear* parallel to the $\mathbf{e}_2$-axis, such that

$$\begin{bmatrix} 2 \\ 1 \end{bmatrix} \text{ is mapped to } \begin{bmatrix} 2 \\ 0 \end{bmatrix}.$$

That shear (see Section 4.7) is given by the matrix

$$S_1 = \begin{bmatrix} 1 & 0 \\ -1/2 & 1 \end{bmatrix}.$$

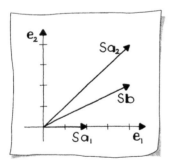

Sketch 5.5.
Shearing the vectors in a linear system.

We apply S_1 to all vectors involved in our system:

$$\begin{bmatrix} 1 & 0 \\ -1/2 & 1 \end{bmatrix} \begin{bmatrix} 2 \\ 1 \end{bmatrix} = \begin{bmatrix} 2 \\ 0 \end{bmatrix}, \qquad \begin{bmatrix} 1 & 0 \\ -1/2 & 1 \end{bmatrix} \begin{bmatrix} 4 \\ 6 \end{bmatrix} = \begin{bmatrix} 4 \\ 4 \end{bmatrix},$$

$$\begin{bmatrix} 1 & 0 \\ -1/2 & 1 \end{bmatrix} \begin{bmatrix} 4 \\ 4 \end{bmatrix} = \begin{bmatrix} 4 \\ 2 \end{bmatrix}.$$

The effect of this map is shown in Sketch 5.5.

Our transformed system now reads

$$\begin{bmatrix} 2 & 4 \\ 0 & 4 \end{bmatrix} \begin{bmatrix} u_1 \\ u_2 \end{bmatrix} = \begin{bmatrix} 4 \\ 2 \end{bmatrix}.$$

Now we can employ *back substitution* to find

$$u_2 = 2/4 = 1/2,$$
$$u_1 = \frac{1}{2} \left(4 - 4 \times \frac{1}{2} \right) = 1.$$

---

[1]Gauss elimination and forward elimination are often used interchangeably.

For 2×2 linear systems there is only one matrix entry to zero in the forward elimination procedure. We will restate the procedure in a more algorithmic way in Chapter 12 when there is more work to do.

Example 5.3

We will look at one more example of forward elimination and back substitution. Let a linear system be given by

$$\begin{bmatrix} -1 & 4 \\ 2 & 2 \end{bmatrix} \begin{bmatrix} u_1 \\ u_2 \end{bmatrix} = \begin{bmatrix} 0 \\ 2 \end{bmatrix}.$$

The shear that takes $\mathbf{a}_1$ to the $\mathbf{e}_1$-axis is given by

$$S_1 = \begin{bmatrix} 1 & 0 \\ 2 & 1 \end{bmatrix},$$

and it transforms the system to

$$\begin{bmatrix} -1 & 4 \\ 0 & 10 \end{bmatrix} \begin{bmatrix} u_1 \\ u_2 \end{bmatrix} = \begin{bmatrix} 0 \\ 2 \end{bmatrix}.$$

Draw your own sketch to understand the geometry.

Using back substitution, the solution is now easily found as $u_1 = 8/10$ and $u_2 = 2/10$.

5.5 Pivoting

Consider the system

$$\begin{bmatrix} 0 & 1 \\ 1 & 0 \end{bmatrix} \begin{bmatrix} u_1 \\ u_2 \end{bmatrix} = \begin{bmatrix} 1 \\ 1 \end{bmatrix},$$

illustrated in Sketch 5.6.

Our standard approach, shearing $\mathbf{a}_1$ onto the $\mathbf{e}_1$-axis, will not work here; there is no shear that takes

$$\begin{bmatrix} 0 \\ 1 \end{bmatrix}$$

onto the $\mathbf{e}_1$-axis. However, there is no problem if we simply exchange the two equations! Then we have

$$\begin{bmatrix} 1 & 0 \\ 0 & 1 \end{bmatrix} \begin{bmatrix} u_1 \\ u_2 \end{bmatrix} = \begin{bmatrix} 1 \\ 1 \end{bmatrix},$$

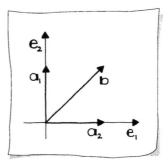

Sketch 5.6.
A linear system that needs pivoting.

and thus $u_1 = u_2 = 1$. So we cannot blindly apply a shear to $\mathbf{a}_1$; we must first check that one exists. If it does not—i.e., if $a_{1,1} = 0$—exchange the equations.

As a rule of thumb, if a method fails because some number equals zero, then it will work poorly if that number is small. It is thus advisable to exchange the two equations anytime we have $|a_{1,1}| < |a_{2,1}|$. The absolute value is used here since we are interested in the magnitude of the involved numbers, not their sign. The process of exchanging equations (rows) so that the pivot is the largest in absolute value is called *row pivoting* or *partial pivoting*, and it is used to improve numerical stability. In Section 5.8, a special type of linear system is introduced that sometimes needs another type pivoting. However, since row pivoting is the most common, we'll refer to this simply as "pivoting."

Example 5.4

Let's study an example taken from [11]:

$$\begin{bmatrix} 0.0001 & 1 \\ 1 & 1 \end{bmatrix} \begin{bmatrix} u_1 \\ u_2 \end{bmatrix} = \begin{bmatrix} 1 \\ 2 \end{bmatrix}.$$

If we shear $\mathbf{a}_1$ onto the $\mathbf{e}_1$-axis, thus applying one forward elimination step, the new system reads

$$\begin{bmatrix} 0.0001 & 1 \\ 0 & -9999 \end{bmatrix} \begin{bmatrix} u_1 \\ u_2 \end{bmatrix} = \begin{bmatrix} 1 \\ -9998 \end{bmatrix}.$$

Performing back substitution, we find the solution is

$$\mathbf{u}_t = \begin{bmatrix} 1.0001 \\ 0.9998\bar{9} \end{bmatrix},$$

which we will call the "true" solution. Note the magnitude of changes in $\mathbf{a}_2$ and $\mathbf{b}$ relative to $\mathbf{a}_1$. This is the type of behavior that causes numerical problems. It can often be dealt with by using a larger number of digits.

Suppose we have a machine that stores only three digits, although it calculates with six digits. Due to round-off, the system above would be stored as

$$\begin{bmatrix} 0.0001 & 1 \\ 0 & -10000 \end{bmatrix} \begin{bmatrix} u_1 \\ u_2 \end{bmatrix} = \begin{bmatrix} 1 \\ -10000 \end{bmatrix},$$

which would result in a "round-off" solution of

$$\mathbf{u}_r = \begin{bmatrix} 0 \\ 1 \end{bmatrix},$$

which is not very close to the true solution $\mathbf{u}_t$, as

$$\|\mathbf{u}_t - \mathbf{u}_r\| = 1.0001.$$

Pivoting is a tool to damper the effects of round-off. Now employ pivoting by exchanging the rows, yielding the system

$$\begin{bmatrix} 1 & 1 \\ 0.0001 & 1 \end{bmatrix} \begin{bmatrix} u_1 \\ u_2 \end{bmatrix} = \begin{bmatrix} 2 \\ 1 \end{bmatrix}.$$

Shear $\mathbf{a}_1$ onto the $\mathbf{e}_1$-axis, and the new system reads

$$\begin{bmatrix} 1 & 1 \\ 0 & 0.9999 \end{bmatrix} \begin{bmatrix} u_1 \\ u_2 \end{bmatrix} = \begin{bmatrix} 2 \\ 0.9998 \end{bmatrix},$$

which results in the "pivoting solution"

$$\mathbf{u}_p = \begin{bmatrix} 1 \\ 1 \end{bmatrix}.$$

Notice that the vectors of the linear systems are all within the same range. Even with the three-digit machine, this system will allow us to compute a result that is closer to the true solution because the effects of round-off have been minimized. Now the error is

$$\|\mathbf{u}_t - \mathbf{u}_p\| = 0.00014.$$

Numerical strategies are the primary topic of numerical analysis, but they cannot be ignored in the study of linear algebra. Because this is an important real-world topic, we will revisit it. In Section 12.2 we will present Gauss elimination with pivoting integrated into the algorithm. In Section 13.4 we will introduce the condition number of a matrix, which is a measure for closeness to being singular. Chapters 13 and 16 will introduce other methods for solving linear systems that are better to use when numerical issues are a concern.

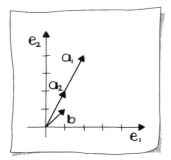

Sketch 5.7.
An unsolvable linear system.

5.6 Unsolvable Systems

Consider the situation shown in Sketch 5.7. The two vectors $\mathbf{a}_1$ and $\mathbf{a}_2$ are multiples of each other. In other words, they are *linearly dependent*.

The corresponding linear system is

$$\begin{bmatrix} 2 & 1 \\ 4 & 2 \end{bmatrix} \begin{bmatrix} u_1 \\ u_2 \end{bmatrix} = \begin{bmatrix} 1 \\ 1 \end{bmatrix}.$$

It is obvious from the sketch that we have a problem here, but let's just blindly apply forward elimination; apply a shear such that $\mathbf{a}_1$ is mapped to the $\mathbf{e}_1$-axis. The resulting system is

$$\begin{bmatrix} 2 & 1 \\ 0 & 0 \end{bmatrix} \begin{bmatrix} u_1 \\ u_2 \end{bmatrix} = \begin{bmatrix} 1 \\ -1 \end{bmatrix}.$$

But the last equation reads $0 = -1$, and now we really are in trouble! This means that our system is *inconsistent*, and therefore does not have a solution.

It is possible however, to find an *approximate* solution. This is done in the context of least squares methods, see Section 12.7.

5.7 Underdetermined Systems

Consider the system

$$\begin{bmatrix} 2 & 1 \\ 4 & 2 \end{bmatrix} \begin{bmatrix} u_1 \\ u_2 \end{bmatrix} = \begin{bmatrix} 3 \\ 6 \end{bmatrix},$$

shown in Sketch 5.8.

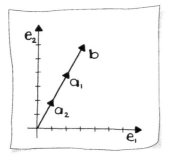

Sketch 5.8.
An underdetermined linear system.

We shear $\mathbf{a}_1$ onto the $\mathbf{e}_1$-axis, and obtain

$$\begin{bmatrix} 2 & 1 \\ 0 & 0 \end{bmatrix} \begin{bmatrix} u_1 \\ u_2 \end{bmatrix} = \begin{bmatrix} 3 \\ 0 \end{bmatrix}.$$

Now the last equation reads $0 = 0$—true, but a bit trivial! In reality, our system is just one equation written down twice in slightly different forms. This is also clear from the sketch: $\mathbf{b}$ may be written as a multiple of either $\mathbf{a}_1$ or $\mathbf{a}_2$, thus the system is underdetermined. This type of system is *consistent* because at least one solution exists. We can find a solution by setting $u_2 = 1$, and then back substitution results in $u_1 = 1$.

5.8   Homogeneous Systems

A system of the form
$$A\mathbf{u} = \mathbf{0}, \tag{5.9}$$
i.e., one where the right-hand side consists of the zero vector, is called *homogeneous*. One obvious solution is the zero vector itself; this is called the *trivial solution* and is usually of little interest. If it has a solution $\mathbf{u}$ that is not the zero vector, then clearly all multiples $c\mathbf{u}$ are also solutions: we multiply both sides of the equations by a common factor c. In other words, the system has an *infinite number of solutions*.

Not all homogeneous systems do have a nontrivial solution, however. Equation (5.9) may be read as follows: What vector $\mathbf{u}$, when mapped by A, has the zero vector as its image? The only 2×2 maps capable of achieving this have rank 1. They are characterized by the fact that their two columns $\mathbf{a}_1$ and $\mathbf{a}_2$ are parallel, or linearly dependent. If the system has only the trivial solution, then A is invertible.

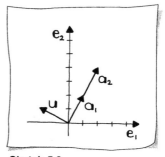

Sketch 5.9.
Homogeneous system with nontrivial solution.

Example 5.5

An example, illustrated in Sketch 5.9, should help. Let our homogeneous system be
$$\begin{bmatrix} 1 & 2 \\ 2 & 4 \end{bmatrix} \mathbf{u} = \begin{bmatrix} 0 \\ 0 \end{bmatrix}.$$

Clearly, $\mathbf{a}_2 = 2\mathbf{a}_1$; the matrix A maps all vectors onto the line defined by $\mathbf{a}_1$ and the origin. In this example, any vector $\mathbf{u}$ that is perpendicular to $\mathbf{a}_1$ will be projected to the zero vector:
$$A[c\mathbf{u}] = \mathbf{0}.$$

After one step of forward elimination, we have
$$\begin{bmatrix} 1 & 2 \\ 0 & 0 \end{bmatrix} \mathbf{u} = \begin{bmatrix} 0 \\ 0 \end{bmatrix}.$$

Any u_2 solves the last equation. So let's pick $u_2 = 1$. Back substitution then gives $u_1 = -2$, therefore
$$\mathbf{u} = \begin{bmatrix} -2 \\ 1 \end{bmatrix}$$
is a solution to the system; so is any multiple of it. Also check that $\mathbf{a}_1 \cdot \mathbf{u} = 0$, so they are in fact perpendicular.

All vectors $\mathbf{u}$ that satisfy a homogeneous system make up the *kernel* or *null space* of the matrix.

Example 5.6

We now consider an example of a homogeneous system that has only the trivial solution:

$$\begin{bmatrix} 1 & 2 \\ 2 & 1 \end{bmatrix} \mathbf{u} = \begin{bmatrix} 0 \\ 0 \end{bmatrix}.$$

The two columns of A are linearly independent; therefore, A does not reduce dimensionality. Then it cannot map any nonzero vector $\mathbf{u}$ to the zero vector!

This is clear after one step of forward elimination,

$$\begin{bmatrix} 1 & 2 \\ 0 & -3 \end{bmatrix} \mathbf{u} = \begin{bmatrix} 0 \\ 0 \end{bmatrix},$$

and back substitution results in

$$\mathbf{u} = \begin{bmatrix} 0 \\ 0 \end{bmatrix}.$$

In general, we may state that a homogeneous system has nontrivial solutions (and therefore, infinitely many solutions) only if the columns of the matrix are linearly dependent.

In some situations, row pivoting will not prepare the linear system for back substitution, necessitating *column pivoting*. When columns are exchanged, the corresponding unknowns must be exchanged as well.

Example 5.7

This next linear system might seem like a silly one to pose; however, systems of this type do arise in Section 7.3 in the context of finding eigenvectors:

$$\begin{bmatrix} 0 & 1/2 \\ 0 & 0 \end{bmatrix} \mathbf{u} = \mathbf{0}.$$

In order to apply back substitution to this system, column pivoting is necessary, thus the system becomes

$$\begin{bmatrix} 1/2 & 0 \\ 0 & 0 \end{bmatrix} \begin{bmatrix} u_2 \\ u_1 \end{bmatrix} = \mathbf{0}.$$

Now we set $u_1 = 1$ and proceed with back substitution to find that $u_2 = 0$. All vectors of the form

$$\mathbf{u} = c \begin{bmatrix} 1 \\ 0 \end{bmatrix}$$

are solutions.

5.9   Undoing Maps: Inverse Matrices

In this section, we will see how to *undo* a linear map. Reconsider the linear system

$$A\mathbf{u} = \mathbf{b}.$$

The matrix A maps $\mathbf{u}$ to $\mathbf{b}$. Now that we know $\mathbf{u}$, what is the matrix B that maps $\mathbf{b}$ back to $\mathbf{u}$,

$$\mathbf{u} = B\mathbf{b}? \tag{5.10}$$

Defining B—the inverse map—is the purpose of this section.

In solving the original linear system, we applied shears to the column vectors of A and to $\mathbf{b}$. After the first shear, we had

$$S_1 A\mathbf{u} = S_1 \mathbf{b}.$$

This demonstrated how shears can be used to zero elements of the matrix. Let's return to the example linear system in (5.3). After applying S_1 the system became

$$\begin{bmatrix} 2 & 4 \\ 0 & 4 \end{bmatrix} \begin{bmatrix} u_1 \\ u_2 \end{bmatrix} = \begin{bmatrix} 4 \\ 2 \end{bmatrix}.$$

Let's use another shear to zero the upper right element. Geometrically, this corresponds to constructing a shear that will map the new $\mathbf{a}_2$ to the $\mathbf{e}_2$-axis. It is given by the matrix

$$S_2 = \begin{bmatrix} 1 & -1 \\ 0 & 1 \end{bmatrix}.$$

Applying it to all vectors gives the new system

$$\begin{bmatrix} 2 & 0 \\ 0 & 4 \end{bmatrix} \begin{bmatrix} u_1 \\ u_2 \end{bmatrix} = \begin{bmatrix} 2 \\ 2 \end{bmatrix}.$$

After the second shear, our linear system has been changed to

$$S_2 S_1 A \mathbf{u} = S_2 S_1 \mathbf{b}.$$

Next, apply a nonuniform scaling S_3 in the $\mathbf{e}_1$ and $\mathbf{e}_2$ directions that will map the latest $\mathbf{a}_1$ and $\mathbf{a}_2$ onto the vectors $\mathbf{e}_1$ and $\mathbf{e}_2$. For our current example,

$$S_3 = \begin{bmatrix} 1/2 & 0 \\ 0 & 1/4 \end{bmatrix}.$$

The new system becomes

$$\begin{bmatrix} 1 & 0 \\ 0 & 1 \end{bmatrix} \begin{bmatrix} u_1 \\ u_2 \end{bmatrix} = \begin{bmatrix} 1 \\ 1/2 \end{bmatrix},$$

which corresponds to

$$S_3 S_2 S_1 A \mathbf{u} = S_3 S_2 S_1 \mathbf{b}.$$

This is a very special system. First of all, to solve for $\mathbf{u}$ is now trivial because A has been transformed into the *unit matrix* or *identity matrix* I,

$$I = \begin{bmatrix} 1 & 0 \\ 0 & 1 \end{bmatrix}. \tag{5.11}$$

This process of transforming A until it becomes the identity is theoretically equivalent to the back substitution process of Section 5.4. However, back substitution uses fewer operations and thus is the method of choice for solving linear systems.

Yet we have now found the matrix B in (5.10)! The two shears and scaling transformed A into the identity matrix I:

$$S_3 S_2 S_1 A = I; \tag{5.12}$$

thus, the solution of the system is

$$\mathbf{u} = S_3 S_2 S_1 \mathbf{b}. \tag{5.13}$$

This leads to the *inverse matrix* A^{-1} of a matrix A:

$$A^{-1} = S_3 S_2 S_1. \tag{5.14}$$

The matrix A^{-1} *undoes* the effect of the matrix A: the vector $\mathbf{u}$ was mapped to $\mathbf{b}$ by A, and $\mathbf{b}$ is mapped back to $\mathbf{u}$ by A^{-1}. Thus, we can now write (5.13) as

$$\mathbf{u} = A^{-1}\mathbf{b}.$$

If this transformation result can be achieved, then A is called *invertible*. At the end of this section and in Sections 5.6 and 5.7, we discuss cases in which A^{-1} does not exist.

If we combine (5.12) and (5.14), we immediately get

$$A^{-1}A = I. \tag{5.15}$$

This makes intuitive sense, since the actions of a map and its inverse should cancel out, i.e., not change anything—that is what I does! Figures 5.2 and 5.3 illustrate this. Then by the definition of the inverse,

$$AA^{-1} = I.$$

If A^{-1} exits, then it is unique.

The inverse of the identity is the identity

$$I^{-1} = I.$$

The inverse of a scaling is given by:

$$\begin{bmatrix} s & 0 \\ 0 & t \end{bmatrix}^{-1} = \begin{bmatrix} 1/s & 0 \\ 0 & 1/t \end{bmatrix}.$$

Multiply this out to convince yourself.

Figure 5.2 shows the effects of a matrix and its inverse for the scaling

$$\begin{bmatrix} 1 & 0 \\ 0 & 0.5 \end{bmatrix}.$$

Figure 5.3 shows the effects of a matrix and its inverse for the shear

$$\begin{bmatrix} 1 & 1 \\ 0 & 1 \end{bmatrix}.$$

We consider the inverse of a rotation as follows: if R_α rotates by α degrees counterclockwise, then $R_{-\alpha}$ rotates by α degrees clockwise, or

$$R_{-\alpha} = R_\alpha^{-1} = R_\alpha^{\mathrm{T}},$$

as we can see from the definition of a rotation matrix (4.16).

Figure 5.2.

Inverse matrices: illustrating scaling and its inverse, and that $AA^{-1} = A^{-1}A = I$. Top: the original Phoenix, the result of applying a scale, then the result of the inverse scale. Bottom: the original Phoenix, the result of applying the inverse scale, then the result of the original scale.

Figure 5.3.

Inverse matrices: illustrating a shear and its inverse, and that $AA^{-1} = A^{-1}A = I$. Top: the original Phoenix, the result of applying a shear, then the result of the inverse shear. Bottom: the original Phoenix, the result of applying the inverse shear, then the result of the original shear.

The rotation matrix is an example of an *orthogonal matrix*. An orthogonal matrix A is characterized by the fact that

$$A^{-1} = A^{\mathrm{T}}.$$

The column vectors $\mathbf{a}_1$ and $\mathbf{a}_2$ of an orthogonal matrix satisfy $\|\mathbf{a}_1\| = 1$, $\|\mathbf{a}_2\| = 1$ and $\mathbf{a}_1 \cdot \mathbf{a}_2 = 0$. In words, the column vectors are *orthonormal*. The row vectors are orthonormal as well. Those transformations that are described by orthogonal matrices are called *rigid body motions*. The determinant of an orthogonal matrix is ± 1.

We add without proof two fairly obvious identities that involve the inverse:

$$A^{-1^{-1}} = A, \tag{5.16}$$

which should be obvious from Figures 5.2 and 5.3, and

$$A^{-1^{\mathrm{T}}} = A^{\mathrm{T}^{-1}}. \tag{5.17}$$

Figure 5.4 illustrates this for

$$A = \begin{bmatrix} 1 & 0 \\ 1 & 0.5 \end{bmatrix}.$$

Figure 5.4.

Inverse matrices: the top illustrates I, A^{-1}, $A^{-1^{\mathrm{T}}}$ and the bottom illustrates I, A^{T}, $A^{\mathrm{T}^{-1}}$.

Given a matrix A, how do we compute its inverse? Let us start with

$$AA^{-1} = I. \qquad (5.18)$$

If we denote the two (unknown) columns of A^{-1} by $\overline{\mathbf{a}}_1$ and $\overline{\mathbf{a}}_2$, and those of I by $\mathbf{e}_1$ and $\mathbf{e}_2$, then (5.18) may be written as

$$A \begin{bmatrix} \overline{\mathbf{a}}_1 & \overline{\mathbf{a}}_2 \end{bmatrix} = \begin{bmatrix} \mathbf{e}_1 & \mathbf{e}_2 \end{bmatrix}.$$

This is really short for two linear systems

$$A\overline{\mathbf{a}}_1 = \mathbf{e}_1 \quad \text{and} \quad A\overline{\mathbf{a}}_2 = \mathbf{e}_2.$$

Both systems have the same matrix A and can thus be solved *simultaneously*. All we have to do is to apply the familiar shears and scale—those that transform A to I—to both $\mathbf{e}_1$ and $\mathbf{e}_2$.

Example 5.8

Let's revisit Example 5.3 with

$$A = \begin{bmatrix} -1 & 4 \\ 2 & 2 \end{bmatrix}.$$

Our two simultaneous systems are:

$$\begin{bmatrix} -1 & 4 \\ 2 & 2 \end{bmatrix} \begin{bmatrix} \overline{\mathbf{a}}_1 & \overline{\mathbf{a}}_2 \end{bmatrix} = \begin{bmatrix} 1 & 0 \\ 0 & 1 \end{bmatrix}.$$

The first shear takes this to

$$\begin{bmatrix} -1 & 4 \\ 0 & 10 \end{bmatrix} \begin{bmatrix} \overline{\mathbf{a}}_1 & \overline{\mathbf{a}}_2 \end{bmatrix} = \begin{bmatrix} 1 & 0 \\ 2 & 1 \end{bmatrix}.$$

The second shear yields

$$\begin{bmatrix} -1 & 0 \\ 0 & 10 \end{bmatrix} \begin{bmatrix} \overline{\mathbf{a}}_1 & \overline{\mathbf{a}}_2 \end{bmatrix} = \begin{bmatrix} 2/10 & -4/10 \\ 2 & 1 \end{bmatrix}.$$

Finally the scaling produces

$$\begin{bmatrix} 1 & 0 \\ 0 & 1 \end{bmatrix} \begin{bmatrix} \overline{\mathbf{a}}_1 & \overline{\mathbf{a}}_2 \end{bmatrix} = \begin{bmatrix} -2/10 & 4/10 \\ 2/10 & 1/10 \end{bmatrix}.$$

Thus the inverse matrix

$$A^{-1} = \begin{bmatrix} -2/10 & 4/10 \\ 2/10 & 1/10 \end{bmatrix}.$$

It can be the case that a matrix A does not have an inverse. For example, the matrix

$$\begin{bmatrix} 2 & 1 \\ 4 & 2 \end{bmatrix}$$

is *not invertible* because the columns are linearly dependent (and therefore the determinant is zero). A noninvertible matrix is also referred to as *singular*. If we try to compute the inverse by setting up two simultaneous systems,

$$\begin{bmatrix} 2 & 1 \\ 4 & 2 \end{bmatrix} \begin{bmatrix} \overline{\mathbf{a}}_1 & \overline{\mathbf{a}}_2 \end{bmatrix} = \begin{bmatrix} 1 & 0 \\ 0 & 1 \end{bmatrix},$$

then the first shear produces

$$\begin{bmatrix} 2 & 1 \\ 0 & 0 \end{bmatrix} \begin{bmatrix} \overline{\mathbf{a}}_1 & \overline{\mathbf{a}}_2 \end{bmatrix} = \begin{bmatrix} 1 & 0 \\ -2 & 1 \end{bmatrix}.$$

At this point it is clear that we will not be able to construct linear maps to achieve the identity matrix on the left side of the equation. Examples of singular matrices were introduced in Sections 5.6–5.8.

5.10   Defining a Map

Matrices map vectors to vectors. If we know the result of such a map, namely that two vectors $\mathbf{v}_1$ and $\mathbf{v}_2$ were mapped to $\mathbf{v}_1'$ and $\mathbf{v}_2'$, can we find the matrix that did it?

Suppose some matrix A was responsible for the map. We would then have the two equations

$$A\mathbf{v}_1 = \mathbf{v}_1' \quad \text{and} \quad A\mathbf{v}_2 = \mathbf{v}_2'.$$

Combining them, we can write

$$A[\mathbf{v}_1, \mathbf{v}_2] = [\mathbf{v}_1', \mathbf{v}_2'],$$

or, even shorter,

$$AV = V'. \tag{5.19}$$

To define A, we simply find V^{-1}, then

$$A = V'V^{-1}.$$

Of course $\mathbf{v}_1$ and $\mathbf{v}_2$ must be linearly independent for V^{-1} to exist. If the $\mathbf{v}_i$ and $\mathbf{v}_i'$ are each linearly independent, then A represents a *change of basis*.

Example 5.9

Let's find the linear map A that maps the basis V formed by vectors

$$\mathbf{v}_1 = \begin{bmatrix} 1 \\ 1 \end{bmatrix} \quad \text{and} \quad \mathbf{v}_2 = \begin{bmatrix} -1 \\ 1 \end{bmatrix}$$

and the basis V' formed by vectors

$$\mathbf{v}_1' = \begin{bmatrix} -1 \\ -1 \end{bmatrix} \quad \text{and} \quad \mathbf{v}_2' = \begin{bmatrix} 1 \\ -1 \end{bmatrix}$$

as illustrated in Sketch 5.10.

First, we find V^{-1} following the steps in Example 5.8, resulting in

$$V^{-1} = \begin{bmatrix} 1/2 & 1/2 \\ -1/2 & 1/2 \end{bmatrix}.$$

The change of basis linear map is

$$A = V'V^{-1} = \begin{bmatrix} -1 & -1 \\ 1 & 1 \end{bmatrix} \begin{bmatrix} 1/2 & 1/2 \\ -1/2 & 1/2 \end{bmatrix} = \begin{bmatrix} -1 & 0 \\ 0 & -1 \end{bmatrix}.$$

Check that $\mathbf{v}_i$ is mapped to $\mathbf{v}_i'$. If we have any vector $\mathbf{v}$ in the V basis, this map will return the coordinates of its corresponding vector $\mathbf{v}'$ in V'. Sketch 5.10 illustrates that $\mathbf{v} = [0 \ \ 1]^{\mathrm{T}}$ is mapped to $\mathbf{v}' = [0 \ \ -1]^{\mathrm{T}}$.

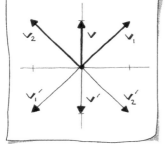

Sketch 5.10.

Change of basis.

In Section 6.5 we'll revisit this topic with an application.

5.11   A Dual View

Let's take a moment to recognize a dual view of linear systems. A coordinate system or linear combination approach (5.1) represents what we might call the "column view." If instead we focus on the row equations (5.5–5.6) we take a "row view." A great example of this can be found by revisiting two line intersection scenarios in Examples 3.9 and 3.10. In the former, we are intersecting two lines in parametric form, and the problem statement takes the column view by asking what linear combination of the column vectors results in the right-hand side. In the latter, we are intersecting two lines in implicit form,

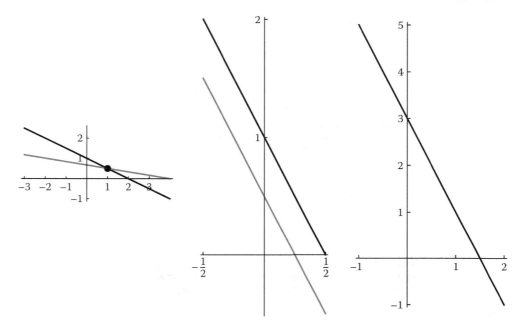

Figure 5.5.

Linear system classification: Three linear systems interpreted as line intersection problems. Left to right: unique solution, inconsistent, underdetermined.

and the problem statement takes the row view by asking what u_1 and u_2 satisfy both line equations. Depending on the problem at hand, we can choose the view that best suits our given information.

We took a column view in our approach to presenting 2×2 linear systems, but equally valid would be a row view. Let's look at the key examples from this chapter as if they were posed as implicit line intersection problems. Figure 5.5 illustrates linear systems from Example 5.1 (unique solution), the example in Section 5.6 (inconsistent linear system), and the example in Section 5.7 (underdetermined system). Importantly, the column and row views of the systems result in the same classification of the solution sets.

Figure 5.6 illustrates two types of homogeneous systems from examples in Section 5.8. Since the right-hand side of each line equation is zero, the lines will pass through the origin. This guarantees the trivial solution for both intersection problems. The system with nontrivial solutions is depicted on the right as two identical lines.

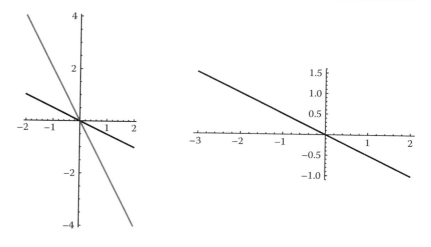

Figure 5.6.

Homogeneous linear system classification: Two homogeneous linear systems interpreted as line intersection problems. Left to right: trivial solution only, nontrivial solutions.

- linear system
- solution spaces
- consistent linear system
- Cramer's rule
- upper triangular
- Gauss elimination
- forward elimination
- back substitution
- linear combination
- inverse matrix
- orthogonal matrix
- orthonormal
- rigid body motion

- inconsistent system of equations
- underdetermined system of equations
- homogeneous system
- kernel
- null space
- row pivoting
- column pivoting
- complete pivoting
- change of basis
- column and row views of linear systems

5.12 Exercises

1. Using the matrix form, write down the linear system to express

$$\begin{bmatrix} 6 \\ 3 \end{bmatrix}$$

in terms of the local coordinate system defined by the origin,

$$\mathbf{a}_1 = \begin{bmatrix} 2 \\ -3 \end{bmatrix}, \quad \text{and} \quad \mathbf{a}_2 = \begin{bmatrix} 6 \\ 0 \end{bmatrix}.$$

2. Is the following linear system consistent? Why?

$$\begin{bmatrix} 1 & 2 \\ 0 & 0 \end{bmatrix} \mathbf{u} = \begin{bmatrix} 0 \\ 4 \end{bmatrix}.$$

3. What are the three possibilities for the solution space of a linear system $A\mathbf{u} = \mathbf{b}$?

4. Use Cramer's rule to solve the system in Exercise 1.

5. Use Cramer's rule to solve the system

$$\begin{bmatrix} 2 & 1 \\ 0 & 1 \end{bmatrix} \begin{bmatrix} x_1 \\ x_2 \end{bmatrix} = \begin{bmatrix} 8 \\ 2 \end{bmatrix}.$$

6. Give an example of an upper triangular matrix.

7. Use Gauss elimination to solve the system in Exercise 1.

8. Use Gauss elimination to solve the system

$$\begin{bmatrix} 4 & -2 \\ 2 & 1 \end{bmatrix} \begin{bmatrix} x_1 \\ x_2 \end{bmatrix} = \begin{bmatrix} -2 \\ 1 \end{bmatrix}.$$

9. Use Gauss elimination with pivoting to solve the system

$$\begin{bmatrix} 0 & 4 \\ 2 & 2 \end{bmatrix} \begin{bmatrix} x_1 \\ x_2 \end{bmatrix} = \begin{bmatrix} 8 \\ 6 \end{bmatrix}.$$

10. Resolve the system in Exercise 1 with Gauss elimination with pivoting.

11. Give an example by means of a sketch of an unsolvable system. Do the same for an underdetermined system.

12. Under what conditions can a nontrivial solution to a homogeneous system be found?

13. Does the following homogeneous system have a nontrivial solution?

$$\begin{bmatrix} 2 & 2 \\ 0 & 4 \end{bmatrix} \begin{bmatrix} x_1 \\ x_2 \end{bmatrix} = \begin{bmatrix} 0 \\ 0 \end{bmatrix}.$$

14. What is the kernel of the matrix

$$C = \begin{bmatrix} 2 & 6 \\ 4 & 12 \end{bmatrix}?$$

15. What is the null space of the matrix

$$C = \begin{bmatrix} 2 & 4 \\ 1 & 2 \end{bmatrix}?$$

16. Find the inverse of the matrix in Exercise 1.

17. What is the inverse of the matrix

$$\begin{bmatrix} 10 & 0 \\ 0 & 0.5 \end{bmatrix}?$$

18. What is the inverse of the matrix

$$\begin{bmatrix} \cos 30° & -\sin 30° \\ \sin 30° & \cos 30° \end{bmatrix}?$$

19. What type of matrix has the property that $A^{-1} = A^T$? Give an example.

20. What is the inverse of

$$\begin{bmatrix} 1 & 1 \\ 0 & 0 \end{bmatrix}?$$

21. Define the matrix A that maps

$$\begin{bmatrix} 1 \\ 0 \end{bmatrix} \rightarrow \begin{bmatrix} 1 \\ 0 \end{bmatrix} \quad \text{and} \quad \begin{bmatrix} 1 \\ 1 \end{bmatrix} \rightarrow \begin{bmatrix} 1 \\ -1 \end{bmatrix}.$$

22. Define the matrix A that maps

$$\begin{bmatrix} 2 \\ 0 \end{bmatrix} \rightarrow \begin{bmatrix} 1 \\ 1 \end{bmatrix} \quad \text{and} \quad \begin{bmatrix} 0 \\ 4 \end{bmatrix} \rightarrow \begin{bmatrix} -1 \\ 1 \end{bmatrix}.$$

6

Moving Things Around:
Affine Maps in 2D

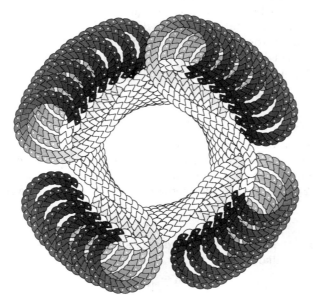

Figure 6.1.

Moving things around: affine maps in 2D applied to an old and familiar video game character.

Imagine playing a video game. As you press a button, figures and objects on the screen start moving around; they shift their positions,

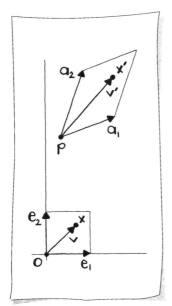

Sketch 6.1.
A skew target box.

they rotate, they zoom in or out. As you see this kind of motion, the game software must carry out quite a few transformations. In Figure 6.1, they have been applied to a familiar face in gaming. These computations are implementations of *affine maps*, the subject of this chapter.

6.1 Coordinate Transformations

In Section 4.1 the focus was on constructing a linear map that takes a vector $\mathbf{v}$ in the $[\mathbf{e}_1, \mathbf{e}_2]$-system,

$$\mathbf{v} = v_1 \mathbf{e}_1 + v_2 \mathbf{e}_2,$$

to a vector $\mathbf{v}'$ in the $[\mathbf{a}_1, \mathbf{a}_2]$-system

$$\mathbf{v}' = v_1 \mathbf{a}_1 + v_2 \mathbf{a}_2.$$

Recall that this latter system describes a parallelogram (skew) target box with lower-left corner at the origin. In this chapter, we want to construct this skew target box anywhere, and we want to map points rather than vectors, as illustrated by Sketch 6.1.

Now we will describe a skew target box by a point $\mathbf{p}$ and two vectors $\mathbf{a}_1, \mathbf{a}_2$. A point $\mathbf{x}$ is mapped to a point $\mathbf{x}'$ by

$$\mathbf{x}' = \mathbf{p} + x_1 \mathbf{a}_1 + x_2 \mathbf{a}_2, \qquad (6.1)$$
$$= \mathbf{p} + A\mathbf{x}, \qquad (6.2)$$

as illustrated by Sketch 6.1. This simply states that we duplicate the $[\mathbf{e}_1, \mathbf{e}_2]$-geometry in the $[\mathbf{a}_1, \mathbf{a}_2]$-system: $\mathbf{x}'$ has the same coordinates in the new system as $\mathbf{x}$ did in the old one. Technically, the linear map A in (6.2) is applied to the vector $\mathbf{x} - \mathbf{o}$, so it should be written as

$$\mathbf{x}' = \mathbf{p} + A(\mathbf{x} - \mathbf{o}), \qquad (6.3)$$

where $\mathbf{o}$ is the origin of $\mathbf{x}$'s coordinate system. In most cases, we will have the familiar

$$\mathbf{o} = \begin{bmatrix} 0 \\ 0 \end{bmatrix},$$

and then we will simply drop the "$-\mathbf{o}$" part, as in (6.2).

Affine maps are the basic tools to move and orient objects. All are of the form given in (6.3) and thus have two parts: a *translation*, given by $\mathbf{p}$, and a *linear map*, given by A.

Let's try representing the coordinate transformation of Section 1.1 as an affine map. The point $\mathbf{u}$ lives in the $[\mathbf{e}_1, \mathbf{e}_2]$-system, and we wish to find $\mathbf{x}$ in the $[\mathbf{a}_1, \mathbf{a}_2]$-system. Recall that the extents of the target box defined $\Delta_1 = \max_1 - \min_1$ and $\Delta_2 = \max_2 - \min_2$, so we set

$$\mathbf{p} = \begin{bmatrix} \min_1 \\ \min_2 \end{bmatrix}, \qquad \mathbf{a}_1 = \begin{bmatrix} \Delta_1 \\ 0 \end{bmatrix}, \qquad \mathbf{a}_2 = \begin{bmatrix} 0 \\ \Delta_2 \end{bmatrix}.$$

The affine map is defined as

$$\begin{bmatrix} x_1 \\ x_2 \end{bmatrix} = \begin{bmatrix} \min_1 \\ \min_2 \end{bmatrix} + \begin{bmatrix} \Delta_1 & 0 \\ 0 & \Delta_2 \end{bmatrix} \begin{bmatrix} u_1 \\ u_2 \end{bmatrix},$$

and we recover (1.3) and (1.4).

Of course we are not restricted to target boxes that are parallel to the $[\mathbf{e}_1, \mathbf{e}_2]$-coordinate axes. Let's look at such an example.

Example 6.1

Let

$$\mathbf{p} = \begin{bmatrix} 2 \\ 2 \end{bmatrix}, \qquad \mathbf{a}_1 = \begin{bmatrix} 2 \\ 1 \end{bmatrix}, \qquad \mathbf{a}_2 = \begin{bmatrix} -2 \\ 4 \end{bmatrix}$$

define a new coordinate system, and let

$$\mathbf{x} = \begin{bmatrix} 2 \\ 1/2 \end{bmatrix}$$

be a point in the $[\mathbf{e}_1, \mathbf{e}_2]$-system. In the new coordinate system, the $[\mathbf{a}_1, \mathbf{a}_2]$-system, the coordinates of $\mathbf{x}$ define a new point $\mathbf{x}'$. What is this point with respect to the $[\mathbf{e}_1, \mathbf{e}_2]$-system? The solution:

$$\mathbf{x}' = \begin{bmatrix} 2 \\ 2 \end{bmatrix} + \begin{bmatrix} 2 & -2 \\ 1 & 4 \end{bmatrix} \begin{bmatrix} 2 \\ 1/2 \end{bmatrix} = \begin{bmatrix} 5 \\ 6 \end{bmatrix}. \tag{6.4}$$

Thus, $\mathbf{x}'$ has coordinates

$$\begin{bmatrix} 2 \\ 1/2 \end{bmatrix}$$

with respect to the $[\mathbf{a}_1, \mathbf{a}_2]$-system; with respect to the $[\mathbf{e}_1, \mathbf{e}_2]$-system, it has coordinates

$$\begin{bmatrix} 5 \\ 6 \end{bmatrix}.$$

(See Sketch 6.2 for an illustration.)

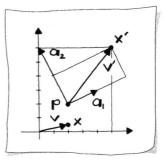

Sketch 6.2.
Mapping a point and a vector.

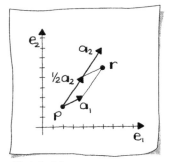

Sketch 6.3.
A new coordinate system.

And now an example of a skew target box. Let's revisit Example 5.1 from Section 5.1, and add an affine aspect to it by translating our target box.

Example 6.2

Sketch 6.3 illustrates the given geometry,

$$\mathbf{p} = \begin{bmatrix} 2 \\ 2 \end{bmatrix}, \quad \mathbf{a}_1 = \begin{bmatrix} 2 \\ 1 \end{bmatrix}, \quad \mathbf{a}_2 = \begin{bmatrix} 4 \\ 6 \end{bmatrix},$$

and the point

$$\mathbf{r} = \begin{bmatrix} 6 \\ 6 \end{bmatrix}$$

with respect to the $[\mathbf{e}_1, \mathbf{e}_2]$-system. We may ask, what are the coordinates of $\mathbf{r}$ with respect to the $[\mathbf{a}_1, \mathbf{a}_2]$-system? This was the topic of Chapter 5; we simply set up the linear system, $A\mathbf{u} = (\mathbf{r} - \mathbf{p})$, or

$$\begin{bmatrix} 2 & 4 \\ 1 & 6 \end{bmatrix} \begin{bmatrix} u_1 \\ u_2 \end{bmatrix} = \begin{bmatrix} 4 \\ 4 \end{bmatrix}.$$

Using Cramer's rule or Gauss elimination from Chapter 5, we find that $\mathbf{u} = \begin{bmatrix} 1 \\ 1/2 \end{bmatrix}$.

6.2   Affine and Linear Maps

A map of the form $\mathbf{v}' = A\mathbf{v}$ is called a *linear map* because it preserves linear combinations of vectors. This idea is expressed in (4.11) and illustrated in Sketch 4.4. A very fundamental property of linear maps has to do with *ratios*, which are defined in Section 2.5. What happens to the ratio of three collinear points when we map them by an affine map? The answer to this question is fairly fundamental to all of geometry, and it is: nothing. In other words, affine maps leave ratios unchanged, or *invariant*. To see this, let

$$\mathbf{p}_2 = (1 - t)\mathbf{p}_1 + t\mathbf{p}_3$$

and let an affine map be defined by

$$\mathbf{x}' = A\mathbf{x} + \mathbf{p}.$$

We now have

$$\mathbf{p}_2' = A((1-t)\mathbf{p}_1 + t\mathbf{p}_3) + \mathbf{p}$$
$$= (1-t)A\mathbf{p}_1 + tA\mathbf{p}_3 + [(1-t)+t]\mathbf{p}$$
$$= (1-t)[A\mathbf{p}_1 + \mathbf{p}] + t[A\mathbf{p}_3 + \mathbf{p}]$$
$$= (1-t)\mathbf{p}_1' + t\mathbf{p}_3'.$$

The step from the first to the second equation may seem a bit contrived; yet it is the one that makes crucial use of the fact that we are combining points using *barycentric combinations*: $(1-t)+t = 1$.

The last equation shows that the linear $(1-t), t$ relationship among three points is not changed by affine maps—meaning that their ratio is invariant, as is illustrated in Sketch 6.4. In particular, the midpoint of two points will be mapped to the midpoint of the image points.

The other basic property of affine maps is this: they map parallel lines to parallel lines. If two lines do not intersect before they are mapped, then they will not intersect afterward either. Conversely, two lines that intersect before the map will also do so afterward. Figure 6.2 shows how two families of parallel lines are mapped to two families of parallel lines. The two families intersect before and after the affine map. The map uses the matrix

$$A = \begin{bmatrix} 1 & 2 \\ 2 & 1 \end{bmatrix}.$$

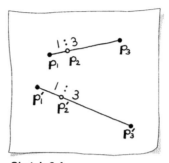

Sketch 6.4.

Ratios are invariant under affine maps.

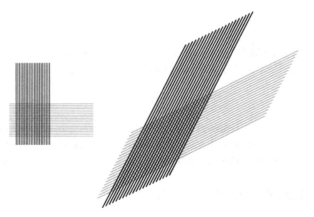

Figure 6.2.

Affine maps: parallel lines are mapped to parallel lines.

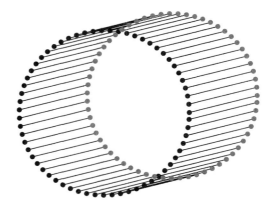

Figure 6.3.
Translations: points on a circle are translated by a fixed amount, and a line connects corresponding points.

6.3 Translations

If an object is moved without changing its orientation, then it is *translated*. See Figure 6.3 in which points on a circle have been translated by a fixed amount.

How is this action covered by the general affine map in (6.3)? Recall the *identity matrix* from Section 5.9, which has no effect whatsoever on any vector: we always have

$$I\mathbf{x} = \mathbf{x},$$

which you should be able to verify without effort.

A translation is thus written in the context of (6.3) as

$$\mathbf{x}' = \mathbf{p} + I\mathbf{x}.$$

One property of translations is that they do not change areas; the two circles in Figure 6.3 have the same area. A translation causes a *rigid body motion*. Recall that rotations are also of this type.

6.4 More General Affine Maps

In this section, we present two very common geometric problems that demand affine maps. It is one thing to say "every affine map is of the form $A\mathbf{x} + \mathbf{p}$," but it is not always clear what A and $\mathbf{p}$ should be for a given problem. Sometimes a more constructive approach is called for, as is the case with Problem 2 in this section.

Problem 1: Let $\mathbf{r}$ be some point around which you would like to *rotate* some other point $\mathbf{x}$ by α degrees, as shown in Sketch 6.5. Let $\mathbf{x}'$ be the rotated point.

Rotations have been defined only around the origin, not around arbitrary points. Hence, we translate our given geometry (the two points $\mathbf{r}$ and $\mathbf{x}$) such that $\mathbf{r}$ moves to the origin. This is easy:

$$\bar{\mathbf{r}} = \mathbf{r} - \mathbf{r} = \mathbf{0}, \quad \bar{\mathbf{x}} = \mathbf{x} - \mathbf{r}.$$

Now we rotate the vector $\bar{\mathbf{x}}$ around the origin by α degrees:

$$\bar{\bar{\mathbf{x}}} = A\bar{\mathbf{x}}.$$

The matrix A would be taken directly from (4.16). Finally, we translate $\bar{\bar{\mathbf{x}}}$ back to the center $\mathbf{r}$ of rotation:

$$\mathbf{x}' = A\bar{\mathbf{x}} + \mathbf{r}.$$

Let's reformulate this in terms of the given information. This is achieved by replacing $\bar{\mathbf{x}}$ by its definition:

$$\mathbf{x}' = A(\mathbf{x} - \mathbf{r}) + \mathbf{r}. \tag{6.5}$$

Example 6.3

Let

$$\mathbf{r} = \begin{bmatrix} 2 \\ 1 \end{bmatrix}, \quad \mathbf{x} = \begin{bmatrix} 3 \\ 0 \end{bmatrix},$$

and $\alpha = 90°$. We obtain

$$\mathbf{x}' = \begin{bmatrix} 0 & -1 \\ 1 & 0 \end{bmatrix} \begin{bmatrix} 1 \\ -1 \end{bmatrix} + \begin{bmatrix} 2 \\ 1 \end{bmatrix} = \begin{bmatrix} 3 \\ 2 \end{bmatrix}.$$

See Sketch 6.6 for an illustration.

Problem 2: Let $\mathbf{l}$ be a line and $\mathbf{x}$ be a point. You want to *reflect* $\mathbf{x}$ across $\mathbf{l}$, with result $\mathbf{x}'$, as shown in Sketch 6.7. This problem could be solved using the following affine maps. Find the intersection $\mathbf{r}$ of $\mathbf{l}$ with the $\mathbf{e}_1$-axis. Find the cosine of the angle between $\mathbf{l}$ and $\mathbf{e}_1$. Rotate $\mathbf{x}$

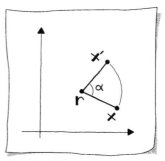

Sketch 6.5.
Rotating a point about another point.

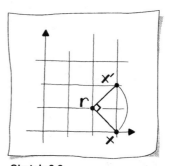

Sketch 6.6.
Rotate $\mathbf{x}$ 90° around $\mathbf{r}$.

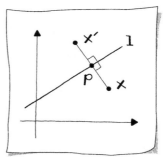

Sketch 6.7.
Reflect a point across a line.

around $\mathbf{r}$ such that $\mathbf{l}$ is mapped to the $\mathbf{e}_1$-axis. Reflect the rotated $\mathbf{x}$ across the $\mathbf{e}_1$-axis, and finally undo the rotation. Complicated!

It is much easier to employ the "foot of a point" algorithm that finds the closest point $\mathbf{p}$ on a line $\mathbf{l}$ to a point $\mathbf{x}$, which was developed in Section 3.7. Then $\mathbf{p}$ must be the midpoint of $\mathbf{x}$ and $\mathbf{x}'$:

$$\mathbf{p} = \frac{1}{2}\mathbf{x} + \frac{1}{2}\mathbf{x}',$$

from which we conclude

$$\mathbf{x}' = 2\mathbf{p} - \mathbf{x}. \tag{6.6}$$

While this does not have the standard affine map form, it is equivalent to it, yet computationally much less complex.

6.5 Mapping Triangles to Triangles

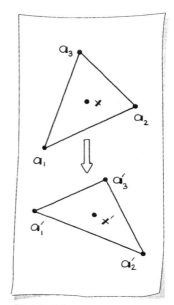

Sketch 6.8.

Two triangles define an affine map.

Affine maps may be viewed as combinations of linear maps and translations. Another flavor of affine maps is described in this section; it draws from concepts in Chapter 5.

This other flavor arises like this: given a (source) triangle T with vertices $\mathbf{a}_1, \mathbf{a}_2, \mathbf{a}_3$, and a (target) triangle T' with vertices $\mathbf{a}_1', \mathbf{a}_2', \mathbf{a}_3'$, what affine map takes T to T'? More precisely, if $\mathbf{x}$ is a point inside T, it will be mapped to a point $\mathbf{x}'$ inside T': how do we find $\mathbf{x}'$? For starters, see Sketch 6.8.

Our desired affine map will be of the form

$$\mathbf{x}' = A[\mathbf{x} - \mathbf{a}_1] + \mathbf{a}_1',$$

thus we need to find the matrix A. (We have chosen $\mathbf{a}_1$ and $\mathbf{a}_1'$ arbitrarily as the origins in the two coordinate systems.) We define

$$\mathbf{v}_2 = \mathbf{a}_2 - \mathbf{a}_1, \qquad \mathbf{v}_3 = \mathbf{a}_3 - \mathbf{a}_1,$$

and

$$\mathbf{v}_2' = \mathbf{a}_2' - \mathbf{a}_1', \qquad \mathbf{v}_3' = \mathbf{a}_3' - \mathbf{a}_1'.$$

We know

$$A\mathbf{v}_2 = \mathbf{v}_2',$$
$$A\mathbf{v}_3 = \mathbf{v}_3'.$$

These two vector equations may be combined into one matrix equation:

$$A \begin{bmatrix} \mathbf{v}_2 & \mathbf{v}_3 \end{bmatrix} = \begin{bmatrix} \mathbf{v}_2' & \mathbf{v}_3' \end{bmatrix},$$

which we abbreviate as
$$AV = V'.$$
We multiply both sides of this equation by V's inverse V^{-1} and obtain A as
$$A = V'V^{-1}.$$
This is the matrix we derived in Section 5.10, "Defining a Map."

Example 6.4

Triangle T is defined by the vertices
$$\mathbf{a}_1 = \begin{bmatrix} 0 \\ 1 \end{bmatrix}, \quad \mathbf{a}_2 = \begin{bmatrix} -1 \\ -1 \end{bmatrix}, \quad \mathbf{a}_3 = \begin{bmatrix} 1 \\ -1 \end{bmatrix},$$
and triangle T' is defined by the vertices
$$\mathbf{a}_1' = \begin{bmatrix} 0 \\ 1 \end{bmatrix}, \quad \mathbf{a}_2' = \begin{bmatrix} 1 \\ 3 \end{bmatrix}, \quad \mathbf{a}_3' = \begin{bmatrix} -1 \\ 3 \end{bmatrix}.$$
The matrices V and V' are then defined as
$$V = \begin{bmatrix} -1 & 1 \\ -2 & -2 \end{bmatrix}, \quad V' = \begin{bmatrix} 1 & -1 \\ 2 & 2 \end{bmatrix}.$$
The inverse of the matrix V is
$$V^{-1} = \begin{bmatrix} -1/2 & -1/4 \\ 1/2 & -1/4 \end{bmatrix},$$
thus the linear map A is defined as
$$A = \begin{bmatrix} -1 & 0 \\ 0 & -1 \end{bmatrix}.$$
Do you recognize the map?

Let's try a sample point
$$\mathbf{x} = \begin{bmatrix} 0 \\ -1/3 \end{bmatrix}$$
in T. This point is mapped to
$$\mathbf{x}' = \begin{bmatrix} -1 & 0 \\ 0 & -1 \end{bmatrix} \left[\begin{bmatrix} 0 \\ -1/3 \end{bmatrix} - \begin{bmatrix} 0 \\ 1 \end{bmatrix} \right] + \begin{bmatrix} 0 \\ 1 \end{bmatrix} = \begin{bmatrix} 0 \\ 7/3 \end{bmatrix}$$
in T'.

Note that V's inverse V^{-1} might not exist; this is the case when $\mathbf{v}_2$ and $\mathbf{v}_3$ are linearly dependent and thus $|V| = 0$.

6.6   Composing Affine Maps

Linear maps are an important theoretical tool, but ultimately we are interested in affine maps; they map objects that are defined by points to other such objects.

If an affine map is given by

$$\mathbf{x}' = A(\mathbf{x} - \mathbf{o}) + \mathbf{p},$$

nothing keeps us from applying it twice, resulting in $\mathbf{x}''$:

$$\mathbf{x}'' = A(\mathbf{x}' - \mathbf{o}) + \mathbf{p}.$$

This may be repeated several times—for interesting choices of A and $\mathbf{p}$, interesting images will result.

Example 6.5

For Figure 6.4 (left), the affine map is defined by

$$A = \begin{bmatrix} \cos 45° & -\sin 45° \\ \sin 45° & \cos 45° \end{bmatrix} \begin{bmatrix} 0.5 & 0.0 \\ 0 & 0.5 \end{bmatrix} \quad \text{and} \quad \mathbf{p} = \begin{bmatrix} 0 \\ 0 \end{bmatrix}. \tag{6.7}$$

In Figure 6.4 (right), a translation was added:

$$\mathbf{p} = \begin{bmatrix} 0.2 \\ 0 \end{bmatrix}. \tag{6.8}$$

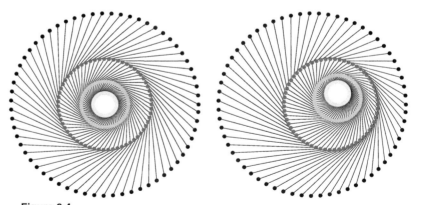

Figure 6.4.
Composing affine maps: The affine maps from (6.7) and (6.8) are applied iteratively, resulting in the left and right images, respectively. The starting object is a set of points on the unit circle centered at the origin. Successive iterations are lighter gray. The same linear map is used for both images; however, a translation has been added to create the right image.

For both images, the linear map is a composition of a scale and a rotation. In the left image, each successive iteration is applied to geometry that is centered about the origin. In the right image, the translation steps away from the origin, thus the rotation action moves the geometry in the $\mathbf{e}_2$-direction even though the translation is strictly in $\mathbf{e}_1$.

Example 6.6

For Figure 6.5 (left), the affine map is

$$A = \begin{bmatrix} \cos 90° & -\sin 90° \\ \sin 90° & \cos 90° \end{bmatrix} \begin{bmatrix} 0.5 & 0.0 \\ 0 & 0.5 \end{bmatrix} \begin{bmatrix} 1 & 1 \\ 0 & 1 \end{bmatrix} \quad \text{and} \quad \mathbf{p} = \begin{bmatrix} 0 \\ 0 \end{bmatrix}. \quad (6.9)$$

In Figure 6.5 (right), a translation was added:

$$\mathbf{p} = \begin{bmatrix} 2 \\ 0 \end{bmatrix}. \quad (6.10)$$

For both images, the linear map is a composition of a shear, scale, and rotation. In the left image, each successive iteration is applied to geometry that is centered about the origin. In the right image,

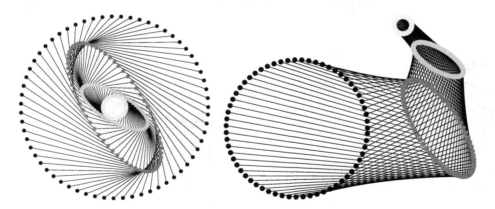

Figure 6.5.
Composing affine maps: The affine maps from (6.9) and (6.10) are applied iteratively, resulting in the left and right images, respectively. The starting object is a set of points on the unit circle centered at the origin. Successive iterations are lighter gray. The same linear map is used for both images; however, a translation has been added to create the right image.

the translation steps away from the origin, thus the rotation action moves the geometry in the $\mathbf{e}_2$-direction even though the translation is strictly in $\mathbf{e}_1$. Same idea as in Example 6.5, but a very different affine map! One interesting artifact of this map: We expect a circle to be mapped to an ellipse, but by iteratively applying the shear and rotation, the ellipse is stretched just right to morph back to a circle!

Rotations can also be made more interesting. In Figure 6.6, you see the letter **S** rotated several times around the origin, which is near the lower left of the letter.

Adding scaling and rotation results in Figure 6.7. The basic affine map for this case is given by

$$\mathbf{x}' = S[R\mathbf{x} + \mathbf{p}]$$

where R rotates by $-20°$, S scales nonuniformly, and $\mathbf{p}$ translates:

$$R = \begin{bmatrix} \cos(-20) & -\sin(-20) \\ \sin(-20) & \cos(-20) \end{bmatrix}, \quad S = \begin{bmatrix} 1.25 & 0 \\ 0 & 1.1 \end{bmatrix}, \quad \mathbf{p} = \begin{bmatrix} 5 \\ 5 \end{bmatrix}.$$

We finish this chapter with Figure 6.8 by the Dutch artist, M.C. Escher [5], who in a very unique way mixed complex geometric issues with a unique style. The figure plays with reflections, which are affine maps.

Figure 6.6.
Rotations: the letter **S** is rotated several times; the origin is at the lower left of the letter.

Figure 6.7.
Rotations: the letter **S** is rotated several times; scalings and translations are also applied.

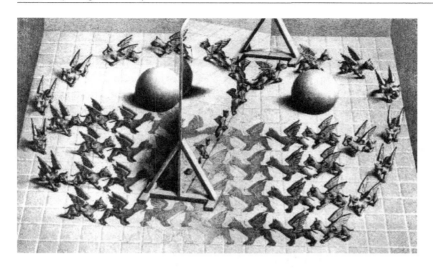

Figure 6.8.
M.C. Escher: Magic Mirror (1949).

Figure 6.8 is itself a 2D object, and so may be subjected to affine maps. Figure 6.9 gives an example. The matrix used here is

$$A = \begin{bmatrix} 0.7 & 0.35 \\ -0.14 & 0.49 \end{bmatrix}. \tag{6.11}$$

No translation is applied.

Figure 6.9.
M.C. Escher: Magic Mirror (1949); affine map applied.

- linear map
- affine map
- translation
- identity matrix
- barycentric combination
- invariant ratios
- rigid body motion

- rotate a point about another point
- reflect a point about a line
- three points mapped to three points
- mapping triangles to triangles

6.7   Exercises

For Exercises 1 and 2 let an affine map be defined by

$$A = \begin{bmatrix} 2 & 1 \\ 1 & 2 \end{bmatrix} \quad \text{and} \quad \mathbf{p} = \begin{bmatrix} 2 \\ 2 \end{bmatrix}.$$

1. Let

$$\mathbf{r} = \begin{bmatrix} 0 \\ 1 \end{bmatrix}, \qquad \mathbf{s} = \begin{bmatrix} 1 \\ 3/2 \end{bmatrix},$$

and $\mathbf{q} = (1/3)\mathbf{r} + (2/3)\mathbf{s}$. Compute $\mathbf{r}', \mathbf{s}', \mathbf{q}'$; e.g., $\mathbf{r}' = A\mathbf{r} + \mathbf{p}$. Show that $\mathbf{q}' = (1/3)\mathbf{r}' + (2/3)\mathbf{s}'$.

2. Let

$$\mathbf{t} = \begin{bmatrix} 0 \\ 1 \end{bmatrix} \quad \text{and} \quad \mathbf{m} = \begin{bmatrix} 2 \\ 1 \end{bmatrix}.$$

Compute $\mathbf{t}'$ and $\mathbf{m}'$. Sketch the lines defined by $\mathbf{t}, \mathbf{m}$ and $\mathbf{t}', \mathbf{m}'$. Do the same for $\mathbf{r}$ and $\mathbf{s}$ from Exercise 1. What does this illustrate?

3. Map the three collinear points

$$\mathbf{x}_1 = \begin{bmatrix} 0 \\ 0 \end{bmatrix}, \quad \mathbf{x}_2 = \begin{bmatrix} 1 \\ 1 \end{bmatrix}, \quad \mathbf{x}_3 = \begin{bmatrix} 2 \\ 2 \end{bmatrix},$$

to points $\mathbf{x}_i'$ by the affine map $A\mathbf{x} + \mathbf{p}$, where

$$A = \begin{bmatrix} 1 & 2 \\ 0 & 1 \end{bmatrix} \quad \text{and} \quad \mathbf{p} = \begin{bmatrix} -4 \\ 0 \end{bmatrix}.$$

What is the ratio of the $\mathbf{x}_i$? What is the ratio of the $\mathbf{x}_i'$?

4. Rotate the point

$$\mathbf{x} = \begin{bmatrix} -2 \\ -2 \end{bmatrix}$$

by $90°$ around the point

$$\mathbf{r} = \begin{bmatrix} -2 \\ 2 \end{bmatrix}.$$

Define the matrix and point for this affine map.

5. Rotate the points

$$\mathbf{p}_0 = \begin{bmatrix} 1 \\ 0 \end{bmatrix}, \quad \mathbf{p}_1 = \begin{bmatrix} 2 \\ 0 \end{bmatrix}, \quad \mathbf{p}_2 = \begin{bmatrix} 3 \\ 0 \end{bmatrix}, \quad \mathbf{p}_3 = \begin{bmatrix} 4 \\ 0 \end{bmatrix}$$

by $45°$ around the point $\mathbf{p}_0$. Define the affine map needed here in terms of a matrix and a point. *Hint: Note that the points are evenly spaced so some economy in calculation is possible.*

6. Reflect the point $\mathbf{x} = \begin{bmatrix} 0 \\ 2 \end{bmatrix}$ about the line $\mathbf{l}(t) = \mathbf{p} + t\mathbf{v}$, $\mathbf{l}(t) = \begin{bmatrix} 0 \\ 0 \end{bmatrix} + t\begin{bmatrix} 1 \\ 2 \end{bmatrix}$.

7. Reflect the points

$$\mathbf{p}_0 = \begin{bmatrix} 2 \\ 1 \end{bmatrix}, \quad \mathbf{p}_1 = \begin{bmatrix} 1 \\ 1 \end{bmatrix}, \quad \mathbf{p}_2 = \begin{bmatrix} 0 \\ 1 \end{bmatrix}, \quad \mathbf{p}_3 = \begin{bmatrix} -1 \\ 1 \end{bmatrix}, \quad \mathbf{p}_4 = \begin{bmatrix} -2 \\ 1 \end{bmatrix}$$

about the line $\mathbf{l}(t) = \mathbf{p}_0 + t\mathbf{v}$, where

$$\mathbf{v} = \begin{bmatrix} -2 \\ 2 \end{bmatrix}.$$

Hint: Note that the points are evenly spaced so some economy in calculation is possible.

8. Given a triangle T with vertices

$$\mathbf{a}_1 = \begin{bmatrix} 0 \\ 0 \end{bmatrix}, \quad \mathbf{a}_2 = \begin{bmatrix} 1 \\ 0 \end{bmatrix}, \quad \mathbf{a}_3 = \begin{bmatrix} 0 \\ 1 \end{bmatrix},$$

and T' with vertices

$$\mathbf{a}_1' = \begin{bmatrix} 1 \\ 0 \end{bmatrix}, \quad \mathbf{a}_2' = \begin{bmatrix} 0 \\ 1 \end{bmatrix}, \quad \mathbf{a}_3' = \begin{bmatrix} 0 \\ 0 \end{bmatrix},$$

what is the affine map that maps T to T'? What are the coordinates of the point $\mathbf{x}'$ corresponding to $\mathbf{x} = [1/2 \ 1/2]^T$?

9. Given a triangle T with vertices

$$\mathbf{a}_1 = \begin{bmatrix} 2 \\ 0 \end{bmatrix}, \quad \mathbf{a}_2 = \begin{bmatrix} 0 \\ 1 \end{bmatrix}, \quad \mathbf{a}_3 = \begin{bmatrix} -2 \\ 0 \end{bmatrix},$$

and T' with vertices

$$\mathbf{a}_1' = \begin{bmatrix} 2 \\ 0 \end{bmatrix}, \quad \mathbf{a}_2' = \begin{bmatrix} 0 \\ -1 \end{bmatrix}, \quad \mathbf{a}_3' = \begin{bmatrix} -2 \\ 0 \end{bmatrix},$$

suppose that the triangle T has been mapped to T' via an affine map. What are the coordinates of the point $\mathbf{x}'$ corresponding to

$$\mathbf{x} = \begin{bmatrix} 0 \\ 0 \end{bmatrix}?$$

10. Construct the affine map that maps any point $\mathbf{x}$ with respect to triangle T to $\mathbf{x}'$ with respect to triangle T' using the vertices in Exercise 9.

11. Let's revisit the coordinate transformation from Exercise 10 in Chapter 1. Construct the affine map which takes a 2D point $\mathbf{x}$ in *NDC coordinates* to the 2D point $\mathbf{x}'$ in a *viewport*. Recall that the extents of the NDC system are defined by the lower-left and upper-right points

$$\mathbf{l}_n = \begin{bmatrix} -1 \\ -1 \end{bmatrix} \quad \text{and} \quad \mathbf{u}_n = \begin{bmatrix} 1 \\ 1 \end{bmatrix},$$

respectively. Suppose we want to map to a viewport with extents

$$\mathbf{l}_v = \begin{bmatrix} 10 \\ 10 \end{bmatrix} \quad \text{and} \quad \mathbf{u}_v = \begin{bmatrix} 30 \\ 20 \end{bmatrix}.$$

After constructing the affine map, find the points in the viewport associated with the NDC points

$$\mathbf{x}_1 = \begin{bmatrix} -1 \\ -1 \end{bmatrix}, \quad \mathbf{x}_2 = \begin{bmatrix} 1 \\ 1 \end{bmatrix}, \quad \mathbf{x}_3 = \begin{bmatrix} -1/2 \\ 1/2 \end{bmatrix}.$$

12. Affine maps transform parallel lines to parallel lines. Do affine maps transform perpendicular lines to perpendicular lines?

13. Which affine maps are rigid body motions?

14. The solution to the problem of reflecting a point across a line is given by (6.6). Why is this a valid combination of points?

7

Eigen Things

Figure 7.1.
The Tacoma Narrows Bridge: a view from the approach shortly before collapsing.

A linear map is described by a matrix, but that does not say much about its geometric properties. When you look at the 2D linear map figures from Chapter 4, you see that they all map a circle, formed from the wings of the Phoenix, to some ellipse—called the action ellipse, thereby stretching and rotating the circle. This stretching

135

Figure 7.2.
The Tacoma Narrows Bridge: a view from shore shortly before collapsing.

and rotating is the geometry of a linear map; it is captured by its eigenvectors and eigenvalues, the subject of this chapter.

Eigenvalues and eigenvectors play an important role in the analysis of mechanical structures. If a bridge starts to sway because of strong winds, then this may be described in terms of certain eigenvalues associated with the bridge's mathematical model. Figures 7.1 and 7.2 show how the Tacoma Narrows Bridge swayed violently during mere 42-mile-per-hour winds on November 7, 1940. It collapsed seconds later. Today, a careful eigenvalue analysis is carried out before any bridge is built. But bridge design is complex and this problem can still occur; the Millennium Bridge in London was swaying enough to cause seasickness in some visitors when it opened in June 2000.

The essentials of all eigentheory are already present in the humble 2D case, the subject of this chapter. A discussion of the higher-dimensional case is given in Section 15.1.

7.1 Fixed Directions

Consider Figure 4.2. You see that the $\mathbf{e}_1$-axis is mapped to itself; so is the $\mathbf{e}_2$-axis. This means that any vector of the form $c\mathbf{e}_1$ or $d\mathbf{e}_2$ is mapped to some multiple of itself. Similarly, in Figure 4.8, you see that all vectors of the form $c\mathbf{e}_1$ are mapped to multiples of each other.

The directions defined by those vectors are called *fixed directions*, for the reason that those directions are not changed by the map. All vectors in the fixed directions change only in length. The fixed directions need not be the coordinate axes.

If a matrix A takes a (nonzero) vector $\mathbf{r}$ to a multiple of itself, then this may be written as

$$A\mathbf{r} = \lambda\mathbf{r} \qquad (7.1)$$

with some real number λ.[1] The value of λ will determine if $\mathbf{r}$ will expand, contract, or reverse direction. From now on, we will disregard the "trivial solution" $\mathbf{r} = \mathbf{0}$ from our considerations.

We next observe that A will treat any multiple of $\mathbf{r}$ in this way as well. Given a matrix A, one might then ask which vectors it treats in this special way. It turns out that there are at most two directions (in 2D), and when we study symmetric matrices (e.g., a scaling) in Section 7.5, we'll see that they are orthogonal to each other. These special vectors are called the *eigenvectors* of A, from the German word *eigen*, meaning special or proper. An eigenvector is mapped to a multiple of itself, and the corresponding factor λ is called its *eigenvalue*. The eigenvalues and eigenvectors of a matrix are the keys to understanding its geometry.

7.2 Eigenvalues

We now develop a way to find the eigenvalues of a 2×2 matrix A. First, we rewrite (7.1) as

$$A\mathbf{r} = \lambda I\mathbf{r},$$

with I being the identity matrix. We may change this to

$$[A - \lambda I]\mathbf{r} = \mathbf{0}. \qquad (7.2)$$

This means that the matrix $[A - \lambda I]$ maps a nonzero vector $\mathbf{r}$ to the zero vector; $[A - \lambda I]$ must be a rank deficient matrix. Then $[A - \lambda I]$'s determinant vanishes:

$$p(\lambda) = \det[A - \lambda I] = 0. \qquad (7.3)$$

As you will see, (7.3) is a quadratic equation in λ, called the *characteristic equation* of A. And $p(\lambda)$ is called the *characteristic polynomial*.

[1] This is the Greek letter *lambda*.

Figure 7.3.
Action of a matrix: behavior of the matrix from Example 7.1.

Example 7.1

Before we proceed further, an example. Let

$$A = \begin{bmatrix} 2 & 1 \\ 1 & 2 \end{bmatrix}.$$

Its action is shown in Figure 7.3.

So let's write out (7.3). It is

$$p(\lambda) = \begin{vmatrix} 2 - \lambda & 1 \\ 1 & 2 - \lambda \end{vmatrix} = 0.$$

We expand the determinant and gather terms of λ to form the characteristic equation

$$p(\lambda) = \lambda^2 - 4\lambda + 3 = 0.$$

This quadratic equation[2] has the roots

$$\lambda_1 = 3, \qquad \lambda_2 = 1.$$

---

[2]Recall that the quadratic equation $a\lambda^2 + b\lambda + c = 0$ has the solutions $\lambda_1 = \frac{-b+\sqrt{b^2-4ac}}{2a}$ and $\lambda_2 = \frac{-b-\sqrt{b^2-4ac}}{2a}$.

Thus, the eigenvalues of a 2×2 matrix are nothing but the zeroes of the quadratic equation

$$p(\lambda) = (\lambda - \lambda_1)(\lambda - \lambda_2) = 0. \qquad (7.4)$$

Commonly, eigenvalues are ordered in decreasing absolute value order: $|\lambda_1| \geq |\lambda_2|$. The eigenvalue λ_1 is called the *dominant eigenvalue*.

From (7.3), we see that the characteristic polynomial evaluated at $\lambda = 0$ results in the determinant of A: $p(0) = \det[A]$. Furthermore, (7.4) shows that the determinant is the product of the eigenvalues:

$$|A| = \lambda_1 \cdot \lambda_2.$$

Check this in Example 7.1. This makes intuitive sense if we consider the determinant as a measure of the change in area of the unit square as it is mapped by A to a parallelogram. The eigenvalues indicate a scaling of certain fixed directions defined by A.

Now let's look at how to compute these fixed directions, or eigenvectors.

7.3   Eigenvectors

Continuing with Example 7.1, we would still like to know the corresponding eigenvectors. We know that one of them will be mapped to three times itself, the other one to itself. Let's call the corresponding eigenvectors $\mathbf{r}_1$ and $\mathbf{r}_2$. The eigenvector $\mathbf{r}_1$ satisfies

$$\begin{bmatrix} 2-3 & 1 \\ 1 & 2-3 \end{bmatrix} \mathbf{r}_1 = \mathbf{0},$$

or

$$\begin{bmatrix} -1 & 1 \\ 1 & -1 \end{bmatrix} \mathbf{r}_1 = \mathbf{0}.$$

This is a *homogeneous* system. (Section 5.8 introduced these systems and a technique to solve them with Gauss elimination.) Such systems have either none or infinitely many solutions. In our case, since the matrix has rank 1, there are infinitely many solutions. Forward elimination results in

$$\begin{bmatrix} -1 & 1 \\ 0 & 0 \end{bmatrix} \mathbf{r}_1 = \mathbf{0}.$$

Assign $r_{2,1} = 1$, then back substitution results in $r_{1,1} = 1$. Any vector of the form

$$\mathbf{r}_1 = c \begin{bmatrix} 1 \\ 1 \end{bmatrix}$$

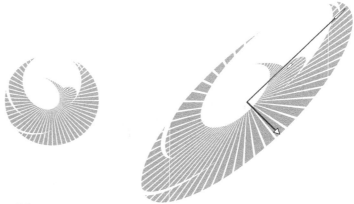

Figure 7.4.
Eigenvectors: the action of the matrix from Example 7.1 and its eigenvectors, scaled
by their corresponding eigenvalues.

will do. And indeed, Figure 7.4 indicates that

$$\begin{bmatrix} 1 \\ 1 \end{bmatrix}$$

is stretched by a factor of three, that is $A\mathbf{r}_1 = 3\mathbf{r}_1$. Of course

$$\begin{bmatrix} -1 \\ -1 \end{bmatrix}$$

is also stretched by a factor of three.

Next, we determine $\mathbf{r}_2$. We get the linear system

$$\begin{bmatrix} 1 & 1 \\ 1 & 1 \end{bmatrix} \mathbf{r}_2 = \mathbf{0}.$$

Again, we have a homogeneous system with infinitely many solutions.
They are all of the form

$$\mathbf{r}_2 = c \begin{bmatrix} -1 \\ 1 \end{bmatrix}.$$

Now recheck Figure 7.4; you see that the vector

$$\begin{bmatrix} 1 \\ -1 \end{bmatrix}$$

is not stretched, and indeed it is mapped to itself!

Typically, eigenvectors are normalized to achieve a degree of uniqueness, and we then have

$$\mathbf{r}_1 = \frac{1}{\sqrt{2}} \begin{bmatrix} 1 \\ 1 \end{bmatrix}, \qquad \mathbf{r}_2 = \frac{1}{\sqrt{2}} \begin{bmatrix} -1 \\ 1 \end{bmatrix}.$$

Let us return to the general 2×2 case and review the ideas thus far. The fixed directions $\mathbf{r}$ of a map A that satisfy $A\mathbf{r} = \lambda\mathbf{r}$ are key to understanding the action of the map. The expression $\det[A - \lambda I] = 0$ is a quadratic polynomial in λ, and its zeroes λ_1 and λ_2 are A's eigenvalues. To find the corresponding eigenvectors, we set up the linear systems $[A - \lambda_1 I]\mathbf{r}_1 = \mathbf{0}$ and $[A - \lambda_2 I]\mathbf{r}_2 = \mathbf{0}$. Both are homogeneous linear systems with infinitely many solutions, corresponding to the eigenvectors $\mathbf{r}_1$ and $\mathbf{r}_2$, which are in the null space of the matrix $[A - \lambda I]$.

Example 7.2

Let's look at another matrix, namely

$$A = \begin{bmatrix} 1 & 2 \\ 0 & 2 \end{bmatrix}.$$

The characteristic equation, $(1 - \lambda)(2 - \lambda) = 0$ results in eigenvalues $\lambda_1 = 1$ and $\lambda_2 = 2$, and the corresponding homogeneous systems

$$\begin{bmatrix} 0 & 2 \\ 0 & 1 \end{bmatrix} \mathbf{r}_1 = \mathbf{0} \quad \text{and} \quad \begin{bmatrix} -1 & 2 \\ 0 & 0 \end{bmatrix} \mathbf{r}_2 = \mathbf{0}.$$

The system for $\mathbf{r}_1$ deserves some special attention. In order to apply back substitution to this system, *column pivoting* is necessary, thus the system becomes

$$\begin{bmatrix} 2 & 0 \\ 1 & 0 \end{bmatrix} \begin{bmatrix} r_{2,1} \\ r_{1,1} \end{bmatrix} = \mathbf{0}.$$

Notice that when column vectors are exchanged, the solution vector components are exchanged as well. One more forward elimination step leads to

$$\begin{bmatrix} 2 & 0 \\ 0 & 0 \end{bmatrix} \begin{bmatrix} r_{2,1} \\ r_{1,1} \end{bmatrix} = \mathbf{0}.$$

Assign $r_{1,1} = 1$, and back substitution results in $r_{2,1} = 0$.

We use the same assignment and back substitution strategy to solve for $\mathbf{r}_2$. In summary, the two homogeneous systems result in (normalized) eigenvectors

$$\mathbf{r}_1 = \begin{bmatrix} 1 \\ 0 \end{bmatrix} \quad \text{and} \quad \mathbf{r}_2 = \frac{1}{\sqrt{5}} \begin{bmatrix} 2 \\ 1 \end{bmatrix},$$

which unlike the eigenvectors of Example 7.1, are not orthogonal. (The matrix in Example 7.1 is symmetric.) We can confirm that $A\mathbf{r}_1 = \mathbf{r}_1$ and $A\mathbf{r}_2 = 2\mathbf{r}_2$.

The eigenvector, $\mathbf{r}_1$ corresponding to the dominant eigenvalue λ_1 is called the *dominant eigenvector*.

7.4　Striving for More Generality

In our examples so far, we only encountered quadratic polynomials with two zeroes. Life is not always that simple: As you might recall from calculus, there are either no, one, or two real zeroes of a quadratic polynomial,[3] as illustrated in Figure 7.5.

Figure 7.5.
Quadratic polynomials: from left to right, no zero, one zero, two real zeroes.

If there are no zeroes, then the corresponding matrix A has no fixed directions. We know one example—rotations. They rotate every vector, leaving no direction unchanged. Let's look at a rotation by $-90°$, given by

$$\begin{bmatrix} 0 & 1 \\ -1 & 0 \end{bmatrix}.$$

Its characteristic equation is

$$\begin{vmatrix} -\lambda & 1 \\ -1 & -\lambda \end{vmatrix} = 0$$

---

[3] Actually, every quadratic polynomial has two zeroes, but they may be complex numbers.

or

$$\lambda^2 + 1 = 0.$$

This has no real solutions, as expected.

A quadratic equation may also have one double root; then there is only one fixed direction. A shear in the $\mathbf{e}_1$-direction provides an example—it maps all vectors in the $\mathbf{e}_1$-direction to themselves. An example is

$$A = \begin{bmatrix} 1 & 1/2 \\ 0 & 1 \end{bmatrix}.$$

The action of this shear is illustrated in Figure 4.8. You clearly see that the $\mathbf{e}_1$-axis is not changed.

The characteristic equation for A is

$$\begin{vmatrix} 1 - \lambda & 1/2 \\ 0 & 1 - \lambda \end{vmatrix} = 0$$

or

$$(1 - \lambda)^2 = 0.$$

It has the double root $\lambda_1 = \lambda_2 = 1$. For the corresponding eigenvector, we have to solve

$$\begin{bmatrix} 0 & 1/2 \\ 0 & 0 \end{bmatrix} \mathbf{r} = \mathbf{0}.$$

In order to apply back substitution to this system, column pivoting is necessary, thus the system becomes

$$\begin{bmatrix} 1/2 & 0 \\ 0 & 0 \end{bmatrix} \begin{bmatrix} r_2 \\ r_1 \end{bmatrix} = \mathbf{0}.$$

Now we set $r_1 = 1$ and proceed with back substitution to find that $r_2 = 0$. Thus

$$\mathbf{r} = \begin{bmatrix} 1 \\ 0 \end{bmatrix}$$

and any multiple of it are solutions. This is quite as expected; those vectors line up along the $\mathbf{e}_1$-direction. The shear A has only one fixed direction.

Is there an easy way to decide if a matrix has real eigenvalues or not? In general, no. But there is one important special case: *all symmetric matrices have real eigenvalues.* In Section 7.5 we'll take a closer look at symmetric matrices.

The last special case to be covered is that of a *zero eigenvalue*.

Example 7.3

Take the projection matrix from Figure 4.11. It was given by

$$A = \begin{bmatrix} 0.5 & 0.5 \\ 0.5 & 0.5 \end{bmatrix}.$$

The characteristic equation is

$$\lambda(\lambda - 1) = 0,$$

resulting in $\lambda_1 = 1$ and $\lambda_2 = 0$. The eigenvector corresponding to λ_2 is found by solving

$$\begin{bmatrix} 0.5 & 0.5 \\ 0.5 & 0.5 \end{bmatrix} \mathbf{r}_1 = \begin{bmatrix} 0 \\ 0 \end{bmatrix},$$

a homogeneous linear system. Forward elimination leads to

$$\begin{bmatrix} 0.5 & 0.5 \\ 0.0 & 0.0 \end{bmatrix} \mathbf{r}_1 = \begin{bmatrix} 0 \\ 0 \end{bmatrix}.$$

Assign $r_{2,1} = 1$, then back substitution results in $r_{1,1} = -1$. The homogeneous system's solutions are of the form

$$\mathbf{r}_1 = c \begin{bmatrix} -1 \\ 1 \end{bmatrix}.$$

Since this matrix maps nonzero vectors (multiples of $\mathbf{r}_1$) to the zero vector, it reduces dimensionality, and thus has rank one. Note that the eigenvector corresponding to the zero eigenvalue defines the *kernel* or *null space* of the matrix.

There is more to learn from the projection matrix in Example 7.3. From Section 4.8 we know that such rank one matrices are idempotent, i.e., $A^2 = A$. One eigenvalue is zero; let λ be the nonzero one, with corresponding eigenvector $\mathbf{r}$. Multiply both sides of (7.1) by A, then

$$A^2\mathbf{r} = \lambda A\mathbf{r}$$
$$\lambda\mathbf{r} = \lambda^2\mathbf{r},$$

and hence $\lambda = 1$. Thus, a 2D projection matrix always has eigenvalues 0 and 1. As a general statement, we may say that a 2×2 matrix with one zero eigenvalue has rank one.

7.5 The Geometry of Symmetric Matrices

Symmetric matrices, those for which $A = A^T$, arise often in practical problems, and two important examples are addressed in this book: conics in Chapter 19 and least squares approximation in Section 12.7. However, many more practical examples exist, coming from fields such as classical mechanics, elasticity theory, quantum mechanics, and thermodynamics.

One nice thing about real symmetric matrices: their eigenvalues are real. And another nice thing: they have an interesting geometric interpretation, as we will see below. As a result, symmetric matrices have become the workhorse for a plethora of numerical algorithms.

We know the two basic equations for eigenvalues and eigenvectors of a symmetric 2×2 matrix A:

$$A\mathbf{r}_1 = \lambda_1 \mathbf{r}_1, \qquad (7.5)$$

$$A\mathbf{r}_2 = \lambda_2 \mathbf{r}_2. \qquad (7.6)$$

Since A is symmetric, we may use (7.5) to write the following:

$$(A\mathbf{r}_1)^T = (\lambda_1 \mathbf{r}_1)^T$$
$$\mathbf{r}_1^T A^T = \mathbf{r}_1^T \lambda_1$$
$$\mathbf{r}_1^T A = \lambda_1 \mathbf{r}_1^T$$

and then after multiplying both sides by $\mathbf{r}_2$,

$$\mathbf{r}_1^T A \mathbf{r}_2 = \lambda_1 \mathbf{r}_1^T \mathbf{r}_2. \qquad (7.7)$$

Multiply both sides of (7.6) by $\mathbf{r}_1^T$

$$\mathbf{r}_1^T A \mathbf{r}_2 = \lambda_2 \mathbf{r}_1^T \mathbf{r}_2 \qquad (7.8)$$

and equating (7.7) and (7.8)

$$\lambda_1 \mathbf{r}_1^T \mathbf{r}_2 = \lambda_2 \mathbf{r}_1^T \mathbf{r}_2$$

or

$$(\lambda_1 - \lambda_2)\mathbf{r}_1^T \mathbf{r}_2 = 0.$$

If $\lambda_1 \neq \lambda_2$ (the standard case), then we conclude that $\mathbf{r}_1^T \mathbf{r}_2 = 0$; in other words, A's two eigenvectors are *orthogonal*. Check this in Example 7.1 and the continuation of this example in Section 7.3.

We may condense (7.5) and (7.6) into one matrix equation,

$$\begin{bmatrix} A\mathbf{r}_1 & A\mathbf{r}_2 \end{bmatrix} = \begin{bmatrix} \lambda_1 \mathbf{r}_1 & \lambda_2 \mathbf{r}_2 \end{bmatrix}. \qquad (7.9)$$

If we define (using the capital Greek *Lambda:* Λ)

$$R = \begin{bmatrix} \mathbf{r}_1 & \mathbf{r}_2 \end{bmatrix} \quad \text{and} \quad \Lambda = \begin{bmatrix} \lambda_1 & 0 \\ 0 & \lambda_2 \end{bmatrix},$$

then (7.9) becomes

$$AR = R\Lambda. \tag{7.10}$$

Example 7.4

For Example 7.1, $AR = R\Lambda$ becomes

$$\begin{bmatrix} 2 & 1 \\ 1 & 2 \end{bmatrix} \begin{bmatrix} 1/\sqrt{2} & -1/\sqrt{2} \\ 1/\sqrt{2} & 1/\sqrt{2} \end{bmatrix} = \begin{bmatrix} 1/\sqrt{2} & -1/\sqrt{2} \\ 1/\sqrt{2} & 1/\sqrt{2} \end{bmatrix} \begin{bmatrix} 3 & 0 \\ 0 & 1 \end{bmatrix}.$$

Verify this identity.

Assume the eigenvectors are normalized so $\mathbf{r}_1^{\mathrm{T}}\mathbf{r}_1 = 1$ and $\mathbf{r}_2^{\mathrm{T}}\mathbf{r}_2 = 1$, and since they are orthogonal, $\mathbf{r}_1^{\mathrm{T}}\mathbf{r}_2 = \mathbf{r}_2^{\mathrm{T}}\mathbf{r}_1 = 0$; thus $\mathbf{r}_1$ and $\mathbf{r}_2$ are orthonormal. These four equations may also be written in matrix form

$$R^{\mathrm{T}}R = I, \tag{7.11}$$

with I the identity matrix. Thus,

$$R^{-1} = R^{\mathrm{T}}$$

and R is an orthogonal matrix. Now (7.10) becomes

$$A = R\Lambda R^{\mathrm{T}}, \tag{7.12}$$

and it is called the *eigendecomposition* of A. Because it is possible to transform A to the diagonal matrix $\Lambda = R^{-1}AR$, A is said to be *diagonalizable*. Matrix decomposition is a fundamental tool in linear algebra for giving insight into the action of a matrix and for building stable and efficient methods to solve linear systems.

What does decomposition (7.12) mean geometrically? Since R is an orthogonal matrix, it is a rotation, a reflection, or a combination of the two. Recall that these linear maps preserve lengths and angles. Its inverse, R^{T}, is the same type of linear map as R, but a reversal of the action of R. The diagonal matrix Λ is a scaling along each of

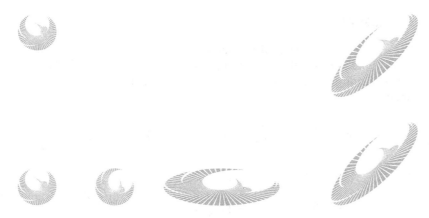

Figure 7.6.
Eigendecomposition: the action of the symmetric matrix A from Example 7.1. Top: I, A. Bottom: I, R^T (rotate $-45°$), ΛR^T (scale), $R\Lambda R^\mathsf{T}$ (rotate $45°$).

the coordinate axes. In Example 7.4, R is a rotation of $45°$ and the eigenvalues are $\lambda_1 = 3$ and $\lambda_2 = 1$. This decomposition of A into rotation, scale, rotation is illustrated in Figure 7.6. (R^T is applied first, so R is actually reversing the action of R^T.)

Recall that there was a degree of freedom in selecting the eigenvectors, so $\mathbf{r}_2$ from above could have been $\mathbf{r}_2 = [1/\sqrt{2} \ \ -1/\sqrt{2}]^\mathsf{T}$, for example. This selection would result in R being a rotation and reflection. Verify this statement. It is always possible to choose the eigenvectors so that R is simply a rotation.

Another way to look at the action of the map A on a vector $\mathbf{x}$ is to write

$$A\mathbf{x} = R\Lambda R^\mathsf{T}\mathbf{x}$$

$$= \begin{bmatrix} \mathbf{r}_1 & \mathbf{r}_2 \end{bmatrix} \Lambda \begin{bmatrix} \mathbf{r}_1^\mathsf{T} \\ \mathbf{r}_2^\mathsf{T} \end{bmatrix} \mathbf{x}$$

$$= \begin{bmatrix} \mathbf{r}_1 & \mathbf{r}_2 \end{bmatrix} \begin{bmatrix} \lambda_1 \mathbf{r}_1^\mathsf{T}\mathbf{x} \\ \lambda_2 \mathbf{r}_2^\mathsf{T}\mathbf{x} \end{bmatrix}$$

$$= \lambda_1 \mathbf{r}_1 \mathbf{r}_1^\mathsf{T}\mathbf{x} + \lambda_2 \mathbf{r}_2 \mathbf{r}_2^\mathsf{T}\mathbf{x}. \tag{7.13}$$

Each matrix $\mathbf{r}_k \mathbf{r}_k^\mathsf{T}$ is a projection onto $\mathbf{r}_k$, as introduced in Section 4.8. As a result, the action of A can be interpreted as a linear combination of projections onto the orthogonal eigenvectors.

Example 7.5

Let's look at the action of the matrix A in Example 7.1 in terms of (7.13), and specifically we will apply the map to $\mathbf{x} = [2 \ 1/2]^{\mathrm{T}}$.

See Example 7.4 for the eigenvalues and eigenvectors for A. The projection matrices are

$$P_1 = \mathbf{r}_1\mathbf{r}_1^{\mathrm{T}} = \begin{bmatrix} 1/2 & 1/2 \\ 1/2 & 1/2 \end{bmatrix} \quad \text{and} \quad P_2 = \mathbf{r}_2\mathbf{r}_2^{\mathrm{T}} = \begin{bmatrix} 1/2 & -1/2 \\ -1/2 & 1/2 \end{bmatrix}.$$

The action of the map is then

$$A\mathbf{x} = 3P_1\mathbf{x} + P_2\mathbf{x} = \begin{bmatrix} 15/4 \\ 15/4 \end{bmatrix} + \begin{bmatrix} 3/4 \\ -3/4 \end{bmatrix} = \begin{bmatrix} 9/2 \\ 3 \end{bmatrix}.$$

This example is illustrated in Figure 7.7.

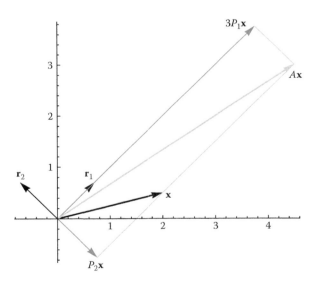

Figure 7.7.
Eigendecomposition: the action of the matrix from Example 7.5 interpreted as a linear combination of projections. The vector $\mathbf{x}$ is projected onto each eigenvector, $\mathbf{r}_1$ and $\mathbf{r}_2$, and scaled by the eigenvalues ($\lambda_1 = 3$, $\lambda_2 = 1$). The action of A is a sum of scaled projection vectors.

We will revisit this projection idea in the section of Principal Components Analysis in Section 16.8; an application will be introduced there.

7.6 Quadratic Forms

Recall from calculus the concept of a *bivariate function:* this is a function f with two arguments x, y; we will use the notation v_1, v_2 for the arguments. Then we write f as

$$f(v_1, v_2) \quad \text{or} \quad f(\mathbf{v}).$$

In this section, we deal with very special bivariate functions that are defined in terms of 2×2 symmetric matrices C:

$$f(\mathbf{v}) = \mathbf{v}^{\mathrm{T}} C \mathbf{v}. \tag{7.14}$$

Such functions are called *quadratic forms* because all terms are quadratic, as we see by expanding (7.14)

$$f(\mathbf{v}) = c_{1,1} v_1^2 + 2 c_{2,1} v_1 v_2 + c_{2,2} v_2^2.$$

The graph of a quadratic form is a 3D point set $[v_1, v_2, f(v_1, v_2)]^{\mathrm{T}}$, forming a quadratic surface. Figure 7.8 illustrates three types of quadratic surfaces: ellipsoid, paraboloid, and hyperboloid. The corresponding matrices and quadratic forms are

$$C_1 = \begin{bmatrix} 2 & 0 \\ 0 & 0.5 \end{bmatrix}, \qquad C_2 = \begin{bmatrix} 2 & 0 \\ 0 & 0 \end{bmatrix}, \quad C_3 = \begin{bmatrix} -2 & 0 \\ 0 & 0.5 \end{bmatrix}, \tag{7.15}$$

$$f_1(\mathbf{v}) = 2v_1^2 + 0.5v_2^2, \quad f_2(\mathbf{v}) = 2v_1^2, \quad f_3(\mathbf{v}) = -2v_1^2 + 0.5v_2^2, \tag{7.16}$$

respectively. Accompanying each quadratic surface is a *contour plot*, which is created from planar slices of the surface at several $f(\mathbf{v}) =$ constant values. Each curve in a planar slice is called a *contour*, and between contours, the area is colored to indicate a range of function values. Each quadratic form type has distinct contours: the (bowl-shaped) ellipsoid contains ellipses, the (taco-shaped) paraboloid contains straight lines, and the (saddle-shaped) hyperboloid contains hyperbolas.[4]

---

[4]These are *conic sections*, to be covered in Chapter 19.

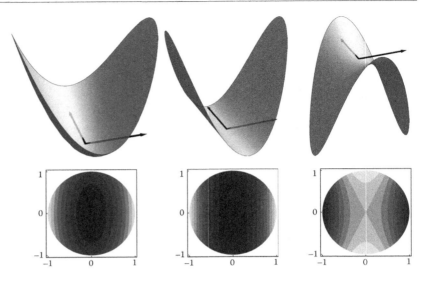

Figure 7.8.
Quadratic forms: ellipsoid, paraboloid, hyperboloid evaluated over the unit circle. The
$[\mathbf{e}_1, \mathbf{e}_2]$-axes are displayed. A contour plot for each quadratic form communicates
additional shape information. Color map extents: min $f(\mathbf{v})$ colored black and max $f(\mathbf{v})$
colored white.

Some more properties of our sample matrices: the determinant and
the eigenvalues, which are easily found by inspection,

$$
\begin{aligned}
|C_1| &= 1 & \lambda_1 &= 2, & \lambda_2 &= 0.5 \\
|C_2| &= 0 & \lambda_1 &= 2, & \lambda_2 &= 0 \\
|C_3| &= -1 & \lambda_1 &= -2, & \lambda_2 &= 0.5.
\end{aligned}
$$

A matrix is called *positive definite* if

$$f(\mathbf{v}) = \mathbf{v}^{\mathrm{T}} A \mathbf{v} > 0 \tag{7.17}$$

for any nonzero vector $\mathbf{v} \in \mathbb{R}^2$. This means that the quadratic form
is positive everywhere except for $\mathbf{v} = \mathbf{0}$. An example is illustrated in
the left part of Figure 7.8. It is an ellipsoid: the matrix is in (7.15)
and the function is in (7.16). (It is hard to see exactly, but the middle
function, the paraboloid, has a line touching the zero plane.)

Positive definite symmetric matrices are a special class of matrices
that arise in a number of applications, and their well-behaved nature
lends them to numerically stable and efficient algorithms.

Geometrically we can get a handle on the positive definite condition (7.17) by first considering only unit vectors. Then, this condition states that the angle between $\mathbf{v}$ and $A\mathbf{v}$ is between $-90°$ and $90°$, indicating that A is somehow constrained in its action on $\mathbf{v}$. It isn't sufficient to only consider unit vectors, though. Therefore, for a general matrix, (7.17) is a difficult condition to verify.

However, for two special matrices, (7.17) is easy to verify. Suppose A is not necessarily symmetric; then $A^{\mathrm{T}}A$ and AA^{T} are symmetric and positive definite. We show this for the former matrix product:

$$\mathbf{v}^{\mathrm{T}}A^{\mathrm{T}}A\mathbf{v} = (A\mathbf{v})^{\mathrm{T}}(A\mathbf{v}) = \mathbf{y}^{\mathrm{T}}\mathbf{y} > 0.$$

These matrices will be at the heart of an important decomposition called the singular value decomposition (SVD), the topic of Chapter 16.

The determinant of a positive definite 2×2 matrix is always positive, and therefore this matrix is always nonsingular. Of course these concepts apply to $n \times n$ matrices, however there are additional requirements on the determinant in order for a matrix to be positive definite. This is discussed in more detail in Chapters 12 and 15.

Let's examine quadratic forms where we restrict C to be positive definite, and in particular $C = A^{\mathrm{T}}A$. If we concentrate on all $\mathbf{v}$ for which (7.14) yields the value 1, we get

$$\mathbf{v}^{\mathrm{T}}C\mathbf{v} = 1.$$

This is a contour of f.

Let's look at this contour for C_1:

$$2v_1^2 + 0.5v_2^2 = 1;$$

it is the equation of an ellipse.

By setting $v_1 = 0$ and solving for v_2, we can identify the $\mathbf{e}_2$-axis extents of the ellipse: $\pm 1/\sqrt{0.5}$. Similarly by setting $v_2 = 0$ we find the $\mathbf{e}_1$-axis extents: $\pm 1/\sqrt{2}$. By definition, the major axis is the longest of the ellipse's two axes, so in this case, it is in the $\mathbf{e}_2$-direction.

As noted above, the eigenvalues for C_1 are $\lambda_1 = 2$ and $\lambda_2 = 0.5$ and the corresponding eigenvectors are $\mathbf{r}_1 = [1 \ \ 0]^{\mathrm{T}}$ and $\mathbf{r}_2 = [0 \ \ 1]^{\mathrm{T}}$. (The eigenvectors of a symmetric matrix are orthogonal.) Thus we have that the minor axis corresponds to the dominant eigenvector. Examining the contour plot in Figure 7.8 (left), we see that the ellipses do indeed have the major axis corresponding to $\mathbf{r}_2$ and minor axis corresponding to $\mathbf{r}_1$. Interpreting the contour plot as a terrain map, we see that the minor axis (dominant eigenvector direction) indicates steeper ascent.

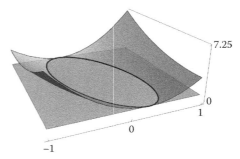

Figure 7.9.

Quadratic form contour: a planar slice of the quadratic form from Example 7.6 defines an ellipse. The quadratic form has been scaled in $\mathbf{e}_3$ to make the contour easier to see.

Example 7.6

Here is an example for which the major and minor axes are not aligned with the coordinate axes:

$$A = \begin{bmatrix} 2 & 0.5 \\ 0 & 1 \end{bmatrix} \quad \text{thus} \quad C_4 = A^{\mathrm{T}}A = \begin{bmatrix} 4 & 1 \\ 1 & 1.25 \end{bmatrix},$$

and this quadratic form is illustrated in Figure 7.9.

The eigendecomposition, $C_4 = R\Lambda R^{\mathrm{T}}$, is defined by

$$R = \begin{bmatrix} -0.95 & -0.30 \\ -0.30 & -0.95 \end{bmatrix} \quad \text{and} \quad \Lambda = \begin{bmatrix} 4.3 & 0 \\ 0 & 0.92 \end{bmatrix},$$

where the eigenvectors are the columns of R.

The ellipse defined by $f_4 = \mathbf{v}^{\mathrm{T}}C_4\mathbf{v} = 1$ is

$$4v_1^2 + 2v_1v_2 + 1.25v_2^2 = 1;$$

it is illustrated in Figure 7.9. The major and minor axis lengths are easiest to determine by using the eigendecomposition to perform a coordinate transformation. The ellipse can be expressed as

$$\mathbf{v}^{\mathrm{T}}R\Lambda R^{\mathrm{T}}\mathbf{v} = 1$$
$$\hat{\mathbf{v}}^{\mathrm{T}}\Lambda\hat{\mathbf{v}} = 1$$
$$\lambda_1\hat{v}_1^2 + \lambda_2\hat{v}_2^2 = 1,$$

aligning the ellipse with the coordinate axes. Thus the minor axis has length $1/\sqrt{\lambda_1} = 1/\sqrt{4.3} = 0.48$ on the $\mathbf{e}_1$ axis and the major axis has length $1/\sqrt{\lambda_2} = 1/\sqrt{0.92} = 1.04$ on the $\mathbf{e}_2$ axis.

Figure 7.10.
Repetitions: a symmetric matrix is applied several times. One eigenvalue is greater than one, causing stretching in one direction. One eigenvalue is less than one, causing compaction in the opposing direction.

Figure 7.11.
Repetitions: a matrix is applied several times. The eigenvalues are not real, therefore the Phoenixes do not line-up along fixed directions.

7.7 Repeating Maps

When we studied matrices, we saw that they always map the unit circle to an ellipse. Nothing keeps us from now mapping the ellipse again using the same map. We can then repeat again, and so on. Figures 7.10 and 7.11 show two such examples.

Figure 7.10 corresponds to the matrix

$$A = \begin{bmatrix} 1 & 0.3 \\ 0.3 & 1 \end{bmatrix}.$$

Being symmetric, it has two real eigenvalues and orthogonal eigenvectors. As the map is repeated several times, the resulting ellipses become more and more stretched: they are elongated in the direction $\mathbf{r}_1$ by $\lambda_1 = 1.3$ and compacted in the direction of $\mathbf{r}_2$ by a factor of $\lambda_2 = 0.7$, with

$$\mathbf{r}_1 = \begin{bmatrix} 1/\sqrt{2} \\ 1/\sqrt{2} \end{bmatrix}, \quad \mathbf{r}_2 = \begin{bmatrix} -1/\sqrt{2} \\ 1/\sqrt{2} \end{bmatrix}.$$

To get some more insight into this phenomenon, consider applying A (now a generic matrix) twice to $\mathbf{r}_1$. We get

$$AA\mathbf{r}_1 = A\lambda_1\mathbf{r}_1 = \lambda_1^2\mathbf{r}_1.$$

In general,

$$A^n\mathbf{r}_1 = \lambda_1^n\mathbf{r}_1. \tag{7.18}$$

The same holds for $\mathbf{r}_2$ and λ_2, of course. So you see that once a matrix has real eigenvectors, they play a more and more prominent role as the matrix is applied repeatedly.

By contrast, the matrix corresponding to Figure 7.11 is given by

$$A = \begin{bmatrix} 0.7 & 0.3 \\ -1 & 1 \end{bmatrix}.$$

As you should verify for yourself, this matrix does not have real eigenvalues. In that sense, it is related to a rotation matrix. If you study Figure 7.11, you will notice a rotational component as we progress—the figures do not line up along any (real) fixed directions.

In Section 15.2 on the power method, we will apply this idea of repeating a map in order to find the eigenvectors.

• fixed direction	• eigentheory of a
• eigenvalue	symmetric matrix
• eigenvector	• matrix with real
• characteristic equation	eigenvalues
• dominant eigenvalue	• diagonalizable matrix
• dominant eigenvector	• eigendecomposition
• homogeneous system	• quadratic form
• kernel or null space	• contour plot
• orthogonal matrix	• positive definite matrix
	• repeated linear map

7.8   Exercises

1. For each of the following matrices, describe the action of the linear map
    and find the eigenvalues and eigenvectors:

    $$A = \begin{bmatrix} 1 & s \\ 0 & 1 \end{bmatrix}, \qquad B = \begin{bmatrix} s & 0 \\ 0 & s \end{bmatrix}, \qquad C = \begin{bmatrix} \cos\theta & -\sin\theta \\ \sin\theta & \cos\theta \end{bmatrix}.$$

    Since we have not been working with complex numbers, if an eigenvalue
    is complex, do not compute the eigenvector.

2. Find the eigenvalues and eigenvectors of the matrix

    $$A = \begin{bmatrix} 1 & -2 \\ -2 & 1 \end{bmatrix}.$$

    Which is the dominant eigenvalue?

3. Find the eigenvalues and eigenvectors of

    $$A = \begin{bmatrix} 1 & -2 \\ -1 & 0 \end{bmatrix}.$$

    Which is the dominant eigenvalue?

4. If a 2×2 matrix has eigenvalues $\lambda_1 = 4$ and $\lambda_2 = 2$, and we apply this
    map to the vertices of the unit square (area 1), what is the area of the
    resulting parallelogram?

5. For 2×2 matrices A and I, what vectors are in the null space of $A - \lambda I$?

6. For the matrix in Example 7.1, identify the four sets of eigenvectors
    that may be constructed (by using the degree of freedom available in
    choosing the direction). Consider each set as column vectors of a matrix
    R, and sketch the action of the matrix. Which sets involve a reflection?

7. Are the eigenvalues of the matrix

    $$A = \begin{bmatrix} a & b \\ b & c \end{bmatrix}$$

    real or complex? Why?

8. Let $\mathbf{u}$ be a unit length $2D$ vector and $P = \mathbf{u}\mathbf{u}^{\mathrm{T}}$. What are the eigen-
    values of P?

9. Show that a 2×2 matrix A and its transpose share the same eigenvalues.

10. What is the eigendecomposition of

    $$A = \begin{bmatrix} 3 & 0 \\ 0 & 4 \end{bmatrix}?$$

    Use the convention that the dominant eigenvalue is λ_1.

11. Express the action of the map

$$A = \begin{bmatrix} 2 & 0 \\ 0 & 4 \end{bmatrix}$$

on the vector $\mathbf{x} = \begin{bmatrix} 1 & 1 \end{bmatrix}^{\mathrm{T}}$ in terms of projections formed from the eigendecomposition.

12. Consider the following quadratic forms $f_i(\mathbf{v}) = \mathbf{v}^{\mathrm{T}} C_i \mathbf{v}$:

$$C_1 = \begin{bmatrix} 3 & 1 \\ 1 & 3 \end{bmatrix}, \qquad C_2 = \begin{bmatrix} 3 & 4 \\ 4 & 1 \end{bmatrix}, \qquad C_3 = \begin{bmatrix} 3 & 3 \\ 3 & 3 \end{bmatrix}.$$

Give the eigenvalues and classification (ellipsoid, paraboloid, hyperboloid) of each.

13. For C_i in Exercise 12, give the equation $f_i(\mathbf{v})$ for each.

14. For C_1 in Exercise 12, what are the semi-major and semi-minor axis lengths for the ellipse $\mathbf{v}^{\mathrm{T}} C_1 \mathbf{v} = 1$? What are the (normalized) axis directions?

15. For the quadratic form, $f(\mathbf{v}) = \mathbf{v}^{\mathrm{T}} C \mathbf{v} = 2v_1^2 + 4v_1 v_2 + 3v_2^2$, what is the matrix C?

16. Is $f(\mathbf{v}) = 2v_1^2 - 4v_1 v_2 + 2v_2 + 3v_2^2$ a quadratic form? Why?

17. Show that

$$\mathbf{x}^{\mathrm{T}} A A^{\mathrm{T}} \mathbf{x} > 0$$

for nonsingular A.

18. Are the eigenvalues of the matrix

$$A = \begin{bmatrix} a & a/2 \\ a/2 & a/2 \end{bmatrix}$$

real or complex? Are any of the eigenvalues negative? Why?

19. If all eigenvalues of a matrix have absolute value less than one, what will happen as you keep repeating the map?

20. For the matrix A in Exercise 2, what are the eigenvalues and eigenvectors for A^2 and A^3?

8

3D Geometry

Figure 8.1.
3D objects: Guggenheim Museum in Bilbao, Spain. Designed by Frank Gehry.

This chapter introduces the essential building blocks of 3D geometry by first extending the 2D tools from Chapters 2 and 3 to 3D. But beyond that, we will also encounter some concepts that are truly 3D, i.e., those that do not have 2D counterparts. With the geometry presented in this chapter, we will be ready to create and analyze

157

simple 3D objects, which then could be utilized to create forms such as in Figure 8.1.

8.1   From 2D to 3D

Moving from 2D to 3D geometry requires a coordinate system with one more dimension. Sketch 8.1 illustrates the $[\mathbf{e}_1, \mathbf{e}_2, \mathbf{e}_3]$-system that consists of the vectors

$$\mathbf{e}_1 = \begin{bmatrix} 1 \\ 0 \\ 0 \end{bmatrix}, \quad \mathbf{e}_2 = \begin{bmatrix} 0 \\ 1 \\ 0 \end{bmatrix}, \quad \text{and} \quad \mathbf{e}_3 = \begin{bmatrix} 0 \\ 0 \\ 1 \end{bmatrix}.$$

Thus, a *vector* in 3D is given as

$$\mathbf{v} = \begin{bmatrix} v_1 \\ v_2 \\ v_3 \end{bmatrix}. \tag{8.1}$$

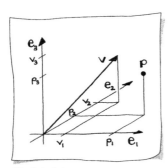

Sketch 8.1.

The $[\mathbf{e}_1, \mathbf{e}_2, \mathbf{e}_3]$-axes, a point, and a vector.

The three *components* of $\mathbf{v}$ indicate the displacement along each axis in the $[\mathbf{e}_1, \mathbf{e}_2, \mathbf{e}_3]$-system. This is illustrated in Sketch 8.1. A 3D vector $\mathbf{v}$ is said to live in 3D space, or $\mathbb{R}^3$, that is $\mathbf{v} \in \mathbb{R}^3$.

A *point* is a reference to a *location*. Points in 3D are given as

$$\mathbf{p} = \begin{bmatrix} p_1 \\ p_2 \\ p_3 \end{bmatrix}. \tag{8.2}$$

The *coordinates* indicate the point's location in the $[\mathbf{e}_1, \mathbf{e}_2, \mathbf{e}_3]$-system, as illustrated in Sketch 8.1. A point $\mathbf{p}$ is said to live in Euclidean 3D-space, or $\mathbb{E}^3$, that is $\mathbf{p} \in \mathbb{E}^3$.

Let's look briefly at some basic 3D vector properties, as we did for 2D vectors. First of all, the 3D *zero vector*:

$$\mathbf{0} = \begin{bmatrix} 0 \\ 0 \\ 0 \end{bmatrix}.$$

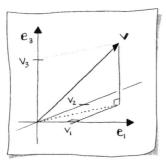

Sketch 8.2.

Length of a 3D vector.

Sketch 8.2 illustrates a 3D vector $\mathbf{v}$ along with its components. Notice the two right triangles. Applying the *Pythagorean theorem* twice, the *length* or *Euclidean norm* of $\mathbf{v}$, denoted as $\|\mathbf{v}\|$, is

$$\|\mathbf{v}\| = \sqrt{v_1^2 + v_2^2 + v_3^2}. \tag{8.3}$$

The length or magnitude of a 3D vector can be interpreted as distance, speed, or force.

Scaling a vector by an amount k yields $\|k\mathbf{v}\| = |k| \|\mathbf{v}\|$. Also, a *normalized vector* has unit length, $\|\mathbf{v}\| = 1$.

Example 8.1

We will get some practice working with 3D vectors. The first task is to normalize the vector

$$\mathbf{v} = \begin{bmatrix} 1 \\ 2 \\ 3 \end{bmatrix}.$$

First calculate the length of $\mathbf{v}$ as

$$\|\mathbf{v}\| = \sqrt{1^2 + 2^2 + 3^2} = \sqrt{14},$$

then the normalized vector $\mathbf{w}$ is

$$\mathbf{w} = \frac{\mathbf{v}}{\|\mathbf{v}\|} = \frac{1}{\sqrt{14}} \begin{bmatrix} 1 \\ 2 \\ 3 \end{bmatrix} \approx \begin{bmatrix} 0.27 \\ 0.53 \\ 0.80 \end{bmatrix}.$$

Check for yourself that $\|\mathbf{w}\| = 1$.

Scale $\mathbf{v}$ by $k = 2$:

$$2\mathbf{v} = \begin{bmatrix} 2 \\ 4 \\ 6 \end{bmatrix}.$$

Now calculate

$$\|2\mathbf{v}\| = \sqrt{2^2 + 4^2 + 6^2} = 2\sqrt{14}.$$

Thus we verified that $\|2\mathbf{v}\| = 2\|\mathbf{v}\|$.

There are infinitely many 3D unit vectors. In Sketch 8.3 a few of these are drawn emanating from the origin. The sketch is a sphere of radius one.

All the rules for combining points and vectors in 2D from Section 2.2 carry over to 3D. The *dot product* of two 3D vectors, $\mathbf{v}$ and $\mathbf{w}$, becomes

$$\mathbf{v} \cdot \mathbf{w} = v_1 w_1 + v_2 w_2 + v_3 w_3.$$

The cosine of the angle θ between the two vectors can be determined as

$$\cos \theta = \frac{\mathbf{v} \cdot \mathbf{w}}{\|\mathbf{v}\| \|\mathbf{w}\|}. \tag{8.4}$$

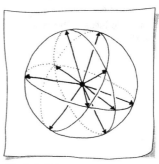

Sketch 8.3.

All 3D unit vectors define a sphere.

8.2 Cross Product

The dot product is a type of multiplication for two vectors that reveals geometric information, namely the angle between them. However, this does not reveal information about their orientation in relation to $\mathbb{R}^3$. Two vectors define a plane. It would be useful to have yet another vector in order to create a 3D coordinate system that is *embedded* in the $[\mathbf{e}_1, \mathbf{e}_2, \mathbf{e}_3]$-system. This is the purpose of another form of vector multiplication called the *cross product*.

In other words, the cross product of $\mathbf{v}$ and $\mathbf{w}$, written as

$$\mathbf{u} = \mathbf{v} \wedge \mathbf{w},$$

produces the vector $\mathbf{u}$, which satisfies the following:

1. The vector $\mathbf{u}$ is perpendicular to $\mathbf{v}$ and $\mathbf{w}$, that is

$$\mathbf{u} \cdot \mathbf{v} = 0 \quad \text{and} \quad \mathbf{u} \cdot \mathbf{w} = 0.$$

2. The orientation of the vector $\mathbf{u}$ follows the *right-hand rule*. This means that if you curl the fingers of your right hand from $\mathbf{v}$ to $\mathbf{w}$, your thumb will point in the direction of $\mathbf{u}$. See Sketch 8.4.

3. The magnitude of $\mathbf{u}$ is the area of the parallelogram defined by $\mathbf{v}$ and $\mathbf{w}$.

Because the cross product produces a vector, it is also called a *vector product*.

Items 1 and 2 determine the direction of $\mathbf{u}$ and item 3 determines the length of $\mathbf{u}$. The cross product is defined as

$$\mathbf{v} \wedge \mathbf{w} = \begin{bmatrix} v_2 w_3 - w_2 v_3 \\ v_3 w_1 - w_3 v_1 \\ v_1 w_2 - w_1 v_2 \end{bmatrix}. \tag{8.5}$$

Item 3 ensures that the cross product of orthogonal and unit length vectors $\mathbf{v}$ and $\mathbf{w}$ results in a vector $\mathbf{u}$ such that $\mathbf{u}, \mathbf{v}, \mathbf{w}$ are *orthonormal*. In other words, $\mathbf{u}$ is unit length and perpendicular to $\mathbf{v}$ and $\mathbf{w}$.

Notice that each component of (8.5) is a 2×2 determinant. For the ith component, omit the ith component of $\mathbf{v}$ and $\mathbf{w}$ and negate the middle determinant:

$$\mathbf{v} \wedge \mathbf{w} = \begin{bmatrix} \begin{vmatrix} v_2 & w_2 \\ v_3 & w_3 \end{vmatrix} \\ -\begin{vmatrix} v_1 & w_1 \\ v_3 & w_3 \end{vmatrix} \\ \begin{vmatrix} v_1 & w_1 \\ v_2 & w_2 \end{vmatrix} \end{bmatrix}.$$

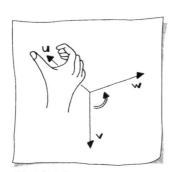

Sketch 8.4.

Characteristics of the cross product.

Example 8.2

Compute the cross product of

$$\mathbf{v} = \begin{bmatrix} 1 \\ 0 \\ 2 \end{bmatrix} \quad \text{and} \quad \mathbf{w} = \begin{bmatrix} 0 \\ 3 \\ 4 \end{bmatrix}.$$

The cross product is

$$\mathbf{u} = \mathbf{v} \wedge \mathbf{w} = \begin{bmatrix} 0 \times 4 - 3 \times 2 \\ 2 \times 0 - 4 \times 1 \\ 1 \times 3 - 0 \times 0 \end{bmatrix} = \begin{bmatrix} -6 \\ -4 \\ 3 \end{bmatrix}. \tag{8.6}$$

Section 4.9 described why the 2×2 determinant, formed from two 2D vectors, is equal to the area P of the parallelogram defined by these two vectors. The analogous result for two vectors in 3D is

$$P = \|\mathbf{v} \wedge \mathbf{w}\|. \tag{8.7}$$

Recall that P is also defined by measuring a height and side length of the parallelogram, as illustrated in Sketch 8.5. The height h is

$$h = \|\mathbf{w}\| \sin \theta,$$

and the side length is $\|\mathbf{v}\|$, which makes

$$P = \|\mathbf{v}\| \|\mathbf{w}\| \sin \theta. \tag{8.8}$$

Equating (8.7) and (8.8) results in

$$\|\mathbf{v} \wedge \mathbf{w}\| = \|\mathbf{v}\| \|\mathbf{w}\| \sin \theta. \tag{8.9}$$

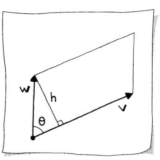

Sketch 8.5.
Area of a parallelogram.

Example 8.3

Compute the area of the parallelogram formed by

$$\mathbf{v} = \begin{bmatrix} 2 \\ 2 \\ 0 \end{bmatrix} \quad \text{and} \quad \mathbf{w} = \begin{bmatrix} 0 \\ 0 \\ 1 \end{bmatrix}.$$

Set up the cross product

$$\mathbf{v} \wedge \mathbf{w} = \begin{bmatrix} 2 \\ -2 \\ 0 \end{bmatrix}.$$

Then the area is

$$P = \|\mathbf{v} \wedge \mathbf{w}\| = 2\sqrt{2}.$$

Since the parallelogram is a rectangle, the area is the product of the edge lengths, so this is the correct result. Verifying (8.9),

$$P = 2\sqrt{2}\sin 90° = 2\sqrt{2}.$$

In order to derive another useful expression in terms of the cross product, square both sides of (8.9). Thus, we have

$$
\begin{aligned}
\|\mathbf{v} \wedge \mathbf{w}\|^2 &= \|\mathbf{v}\|^2\|\mathbf{w}\|^2 \sin^2 \theta \\
&= \|\mathbf{v}\|^2\|\mathbf{w}\|^2(1 - \cos^2 \theta) \\
&= \|\mathbf{v}\|^2\|\mathbf{w}\|^2 - \|\mathbf{v}\|^2\|\mathbf{w}\|^2 \cos^2 \theta \\
&= \|\mathbf{v}\|^2\|\mathbf{w}\|^2 - (\mathbf{v} \cdot \mathbf{w})^2.
\end{aligned}
\tag{8.10}
$$

The last line is referred to as *Lagrange's identity*. We now have an expression for the area of a parallelogram in terms of a dot product.

To get a better feeling for the behavior of the cross product, let's look at some of its properties.

- Parallel vectors result in the zero vector: $\mathbf{v} \wedge c\mathbf{v} = \mathbf{0}$.

- Homogeneous: $c\mathbf{v} \wedge \mathbf{w} = c(\mathbf{v} \wedge \mathbf{w})$.

- Antisymmetric: $\mathbf{v} \wedge \mathbf{w} = -(\mathbf{w} \wedge \mathbf{v})$.

- Nonassociative: $\mathbf{u} \wedge (\mathbf{v} \wedge \mathbf{w}) \neq (\mathbf{u} \wedge \mathbf{v}) \wedge \mathbf{w}$, in general.

- Distributive: $\mathbf{u} \wedge (\mathbf{v} + \mathbf{w}) = \mathbf{u} \wedge \mathbf{v} + \mathbf{u} \wedge \mathbf{w}$.

- Right-hand rule:

$$
\begin{aligned}
\mathbf{e}_1 \wedge \mathbf{e}_2 &= \mathbf{e}_3, \\
\mathbf{e}_2 \wedge \mathbf{e}_3 &= \mathbf{e}_1, \\
\mathbf{e}_3 \wedge \mathbf{e}_1 &= \mathbf{e}_2.
\end{aligned}
$$

- Orthogonality:

$$
\begin{aligned}
\mathbf{v} \cdot (\mathbf{v} \wedge \mathbf{w}) = 0: &\quad \mathbf{v} \wedge \mathbf{w} \quad \text{is orthogonal to} \quad \mathbf{v}. \\
\mathbf{w} \cdot (\mathbf{v} \wedge \mathbf{w}) = 0: &\quad \mathbf{v} \wedge \mathbf{w} \quad \text{is orthogonal to} \quad \mathbf{w}.
\end{aligned}
$$

Example 8.4

Let's test these properties of the cross product with

$$\mathbf{u} = \begin{bmatrix} 1 \\ 1 \\ 1 \end{bmatrix}, \quad \mathbf{v} = \begin{bmatrix} 2 \\ 0 \\ 0 \end{bmatrix}, \quad \mathbf{w} = \begin{bmatrix} 0 \\ 3 \\ 0 \end{bmatrix}.$$

Make your own sketches and don't forget the right-hand rule to guess the resulting vector direction.

- Parallel vectors:

$$\mathbf{v} \wedge 3\mathbf{v} = \begin{bmatrix} 0 \times 0 - 0 \times 0 \\ 0 \times 6 - 0 \times 2 \\ 2 \times 0 - 6 \times 0 \end{bmatrix} = \mathbf{0}.$$

- Homogeneous:

$$4\mathbf{v} \wedge \mathbf{w} = \begin{bmatrix} 0 \times 0 - 3 \times 0 \\ 0 \times 0 - 0 \times 8 \\ 8 \times 3 - 0 \times 0 \end{bmatrix} = \begin{bmatrix} 0 \\ 0 \\ 24 \end{bmatrix},$$

  and

$$4(\mathbf{v} \wedge \mathbf{w}) = 4 \begin{bmatrix} 0 \times 0 - 3 \times 0 \\ 0 \times 0 - 0 \times 2 \\ 2 \times 3 - 0 \times 0 \end{bmatrix} = 4 \begin{bmatrix} 0 \\ 0 \\ 6 \end{bmatrix} = \begin{bmatrix} 0 \\ 0 \\ 24 \end{bmatrix}.$$

- Antisymmetric:

$$\mathbf{v} \wedge \mathbf{w} = \begin{bmatrix} 0 \\ 0 \\ 6 \end{bmatrix} \quad \text{and} \quad -(\mathbf{w} \wedge \mathbf{v}) = - \left(\begin{bmatrix} 0 \\ 0 \\ -6 \end{bmatrix} \right).$$

- Nonassociative:

$$\mathbf{u} \wedge (\mathbf{v} \wedge \mathbf{w}) = \begin{bmatrix} 1 \times 6 - 0 \times 1 \\ 1 \times 0 - 6 \times 1 \\ 1 \times 0 - 0 \times 1 \end{bmatrix} = \begin{bmatrix} 6 \\ -6 \\ 0 \end{bmatrix},$$

  which is not the same as

$$(\mathbf{u} \wedge \mathbf{v}) \wedge \mathbf{w} = \begin{bmatrix} 0 \\ 2 \\ -2 \end{bmatrix} \wedge \begin{bmatrix} 0 \\ 3 \\ 0 \end{bmatrix} = \begin{bmatrix} 6 \\ 0 \\ 0 \end{bmatrix}.$$

- Distributive:

$$\mathbf{u} \wedge (\mathbf{v} + \mathbf{w}) = \begin{bmatrix} 1 \\ 1 \\ 1 \end{bmatrix} \wedge \begin{bmatrix} 2 \\ 3 \\ 0 \end{bmatrix} = \begin{bmatrix} -3 \\ 2 \\ 1 \end{bmatrix},$$

which is equal to

$$(\mathbf{u} \wedge \mathbf{v}) + (\mathbf{u} \wedge \mathbf{w}) = \begin{bmatrix} 0 \\ 2 \\ -2 \end{bmatrix} + \begin{bmatrix} -3 \\ 0 \\ 3 \end{bmatrix} = \begin{bmatrix} -3 \\ 2 \\ 1 \end{bmatrix}.$$

The cross product is an invaluable tool for engineering. One reason: it facilitates the construction of a coordinate independent frame of reference. We present two applications of this concept. In Section 8.6, outward normal vectors to a 3D triangulation, constructed using cross products, are used for rendering. In Section 20.7, a coordinate independent frame is used to move an object in space along a specified curve.

8.3 Lines

Specifying a line with 3D geometry differs a bit from 2D. In terms of points and vectors, two pieces of information define a line; however, we are restricted to specifying

- two points or

- a point and a vector parallel to the line.

The 2D geometry item (from Section 3.1), which specifies only

- a point and a vector perpendicular to the line,

no longer works. It isn't specific enough. (See Sketch 8.6.) In other words, an entire family of lines satisfies this specification; this family lies in a plane. (More on planes in Section 8.4.) As a consequence, the concept of a *normal* to a 3D line does not exist.

Let's look at the mathematical representations of a 3D line. Clearly, from the discussion above, there cannot be an *implicit form*.

The *parametric form* of a 3D line does not differ from the 2D line except for the fact that the given information lives in 3D. A line $\mathbf{l}(t)$ has the form

$$\mathbf{l}(t) = \mathbf{p} + t\mathbf{v}, \tag{8.11}$$

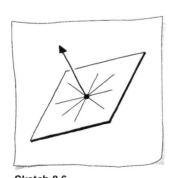

Sketch 8.6.

Point and perpendicular don't define a line.

where $\mathbf{p} \in \mathbb{E}^3$ and $\mathbf{v} \in \mathbb{R}^3$. Points are generated on the line as the parameter t varies.

In 2D, two lines either intersect or they are parallel. In 3D this is not the case; a third possibility is that the lines are *skew*. Sketch 8.7 illustrates skew lines using a cube as a reference frame.

Because lines in 3D can be skew, two lines might not intersect. Revisiting the problem of the intersection of two lines given in parametric form from Section 3.8, we can see the algebraic truth in this statement. Now the two lines are

$$l_1 : \quad \mathbf{l}_1(t) = \mathbf{p} + t\mathbf{v}$$
$$l_2 : \quad \mathbf{l}_2(s) = \mathbf{q} + s\mathbf{w}$$

where $\mathbf{p}, \mathbf{q} \in \mathbb{E}^3$ and $\mathbf{v}, \mathbf{w} \in \mathbb{R}^3$. To find the intersection point, we solve for t or s. Repeating (3.15), we have the linear system

$$\hat{t}\mathbf{v} - \hat{s}\mathbf{w} = \mathbf{q} - \mathbf{p}.$$

However, now there are three equations and still only two unknowns. Thus, the system is *overdetermined*. No solution exists when the lines are skew. But we can find a best approximation, the least squares solution, and that is the topic of Section 12.7. In many applications it is important to know the closest point on a line to another line. This problem is solved in Section 11.2.

We still have the concepts of perpendicular and parallel lines in 3D.

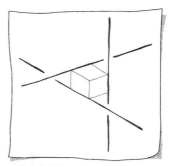

Sketch 8.7.
Skew lines.

8.4 Planes

While exploring the possibility of a 3D implicit line, we encountered a plane. We'll essentially repeat that here, however, with a little change in notation. Suppose we are given a point $\mathbf{p}$ and a vector $\mathbf{n}$ bound to $\mathbf{p}$. The locus of all points $\mathbf{x}$ that satisfy the equation

$$\mathbf{n} \cdot (\mathbf{x} - \mathbf{p}) = 0 \qquad (8.12)$$

defines the *implicit form* of a plane. This is illustrated in Sketch 8.8. The vector $\mathbf{n}$ is called the *normal* to the plane if $\|\mathbf{n}\| = 1$. If this is the case, then (8.12) is called the *point normal plane equation*.

Expanding (8.12), we have

$$n_1 x_1 + n_2 x_2 + n_3 x_3 - (n_1 p_1 + n_2 p_2 + n_3 p_3) = 0.$$

Typically, this is written as

$$A x_1 + B x_2 + C x_3 + D = 0, \qquad (8.13)$$

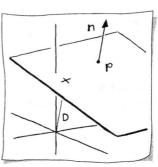

Sketch 8.8.
Point normal plane equation.

where

$$A = n_1$$
$$B = n_2$$
$$C = n_3$$
$$D = -(n_1 p_1 + n_2 p_2 + n_3 p_3).$$

Example 8.5

Compute the implicit form of the plane through the point

$$\mathbf{p} = \begin{bmatrix} 4 \\ 0 \\ 0 \end{bmatrix}$$

that is perpendicular to the vector

$$\mathbf{n} = \begin{bmatrix} 1 \\ 1 \\ 1 \end{bmatrix}.$$

All we need to compute is D:

$$D = -(1 \times 4 + 1 \times 0 + 1 \times 0) = -4.$$

Thus, the plane equation is

$$x_1 + x_2 + x_3 - 4 = 0.$$

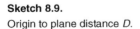

Sketch 8.9.

Origin to plane distance D.

Similar to a 2D implicit line, if the coefficients A, B, C correspond to the normal to the plane, then $|D|$ describes the distance of the plane to the origin. Notice in Sketch 8.8 that this is the *perpendicular distance*. A 2D cross section of the geometry is illustrated in Sketch 8.9. We equate two definitions of the cosine of an angle,[1]

$$\cos(\theta) = \frac{D}{\|\mathbf{p}\|} \quad \text{and} \quad \cos(\theta) = \frac{\mathbf{n} \cdot \mathbf{p}}{\|\mathbf{n}\| \|\mathbf{p}\|},$$

and remember that the normal is unit length, to find that $D = \mathbf{n} \cdot \mathbf{p}$.

---

[1]The equations that follow should refer to $\mathbf{p} - \mathbf{0}$ rather than simply $\mathbf{p}$, but we omit the origin for simplicity.

The point normal form reflects the (perpendicular) distance of a point from a plane. This situation is illustrated in Sketch 8.10. The distance d of an arbitrary point $\hat{\mathbf{x}}$ from the point normal form of the plane is

$$d = A\hat{x}_1 + B\hat{x}_2 + C\hat{x}_3 + D.$$

The reason for this is precisely the same as for the implicit line in Section 3.3. See Section 11.1 for more on this topic.

Suppose we would like to find the distance of many points to a given plane. Then it is computationally more efficient to have the plane in point normal form. The new coefficients will be A', B', C', D'. We normalize $\mathbf{n}$ in (8.12), then we may divide the implicit equation by this factor,

$$\frac{\mathbf{n} \cdot (\mathbf{x} - \mathbf{p})}{\|\mathbf{n}\|} = 0,$$

$$\frac{\mathbf{n} \cdot \mathbf{x}}{\|\mathbf{n}\|} - \frac{\mathbf{n} \cdot \mathbf{p}}{\|\mathbf{n}\|} = 0,$$

resulting in

$$A' = \frac{A}{\|\mathbf{n}\|}, \quad B' = \frac{B}{\|\mathbf{n}\|}, \quad C' = \frac{C}{\|\mathbf{n}\|}, \quad D' = \frac{D}{\|\mathbf{n}\|}.$$

Sketch 8.10.
Point to plane distance.

Example 8.6

Let's continue with the plane from the previous example,

$$x_1 + x_2 + x_3 - 4 = 0.$$

Clearly it is not in point normal form because the length of the vector

$$\mathbf{n} = \begin{bmatrix} 1 \\ 1 \\ 1 \end{bmatrix}$$

is not equal to one. In fact, $\|\mathbf{n}\| = 1/\sqrt{3}$.

The new coefficients of the plane equation are

$$A' = B' = C' = \frac{1}{\sqrt{3}} \qquad D' = \frac{-4}{\sqrt{3}},$$

resulting in the point normal plane equation,

$$\frac{1}{\sqrt{3}}x_1 + \frac{1}{\sqrt{3}}x_2 + \frac{1}{\sqrt{3}}x_3 - \frac{4}{\sqrt{3}} = 0.$$

Determine the distance d of the point

$$\mathbf{q} = \begin{bmatrix} 4 \\ 4 \\ 4 \end{bmatrix}$$

from the plane:

$$d = \frac{1}{\sqrt{3}} \times 4 + \frac{1}{\sqrt{3}} \times 4 + \frac{1}{\sqrt{3}} \times 4 - \frac{4}{\sqrt{3}} = \frac{8}{\sqrt{3}} \approx 4.6.$$

Notice that $d > 0$; this is because the point $\mathbf{q}$ is on the same side of the plane as the normal direction. The distance of the origin to the plane is $d = D' = -4/\sqrt{3}$, which is negative because it is on the opposite side of the plane to which the normal points. This is analogous to the 2D implicit line.

The implicit plane equation is wonderful for determining if a point is in a plane; however, it is not so useful for creating points in a plane. For this we have the *parametric form* of a plane.

The *given* information for defining a parametric representation of a plane usually comes in one of two ways:

- three points, or

- a point and two vectors.

If we start with the first scenario, we choose three points $\mathbf{p}, \mathbf{q}, \mathbf{r}$, then choose one of these points and form two vectors $\mathbf{v}$ and $\mathbf{w}$ bound to that point as shown in Sketch 8.11:

$$\mathbf{v} = \mathbf{q} - \mathbf{p} \quad \text{and} \quad \mathbf{w} = \mathbf{r} - \mathbf{p}. \tag{8.14}$$

Why not just specify one point and a vector in the plane, analogous to the implicit form of a plane? Sketch 8.12 illustrates that this is not enough information to uniquely define a plane. Many planes fit that data.

Two vectors bound to a point are the data we'll use to define a plane $\mathbf{P}$ in parametric form as

$$\mathbf{P}(s,t) = \mathbf{p} + s\mathbf{v} + t\mathbf{w}. \tag{8.15}$$

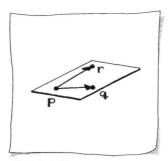

Sketch 8.11.
Parametric plane.

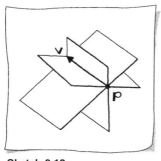

Sketch 8.12.
Family of planes through a point and vector.

The two independent parameters, s and t, determine a point $\mathbf{P}(s,t)$ in the plane.[2] Notice that (8.15) can be rewritten as

$$\begin{aligned}\mathbf{P}(s,t) &= \mathbf{p} + s(\mathbf{q} - \mathbf{p}) + t(\mathbf{r} - \mathbf{p}) \\ &= (1 - s - t)\mathbf{p} + s\mathbf{q} + t\mathbf{r}.\end{aligned} \tag{8.16}$$

As described in Section 17.1, $(1 - s - t, s, t)$ are the *barycentric coordinates* of a point $\mathbf{P}(s,t)$ with respect to the triangle with vertices $\mathbf{p}, \mathbf{q}$, and $\mathbf{r}$, respectively.

Another method for specifying a plane, as illustrated in Sketch 8.13, is as the *bisector of two points*. This is how a plane is defined in Euclidean geometry—the locus of points equidistant from two points. The line between two given points defines the normal to the plane, and the midpoint of this line segment defines a point in the plane. With this information it is most natural to express the plane in implicit form.

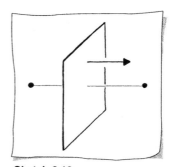

Sketch 8.13.
A plane defined as the bisector of two points.

8.5 Scalar Triple Product

In Section 8.2 we encountered the area P of the parallelogram formed by vectors $\mathbf{v}$ and $\mathbf{w}$ measured as

$$P = \|\mathbf{v} \wedge \mathbf{w}\|.$$

The next natural question is how do we measure the *volume* of the *parallelepiped*, or skew box, formed by three vectors. See Sketch 8.14. The volume is a product of a face area and the corresponding height of the skew box. As illustrated in the sketch, after choosing $\mathbf{v}$ and $\mathbf{w}$ to form the face, the height is $\|\mathbf{u}\| \cos\theta$. Thus, the volume V is

$$V = \|\mathbf{u}\|\|\mathbf{v} \wedge \mathbf{w}\| \cos\theta.$$

Substitute a dot product for $\cos\theta$, then

$$V = \mathbf{u} \cdot (\mathbf{v} \wedge \mathbf{w}). \tag{8.17}$$

This is called the *scalar triple product*, and it is a number representing a signed volume.

The sign reveals something about the orientation of the three vectors. If $\cos\theta > 0$, resulting in a positive volume, then $\mathbf{u}$ is on the

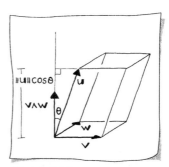

Sketch 8.14.
Scalar triple product for the volume.

---

[2]This is a slight deviation in notation: an uppercase boldface letter rather than a lowercase one denoting a point.

same side of the plane formed by $\mathbf{v}$ and $\mathbf{w}$ as $\mathbf{v} \wedge \mathbf{w}$. If $\cos\theta < 0$, resulting in a negative volume, then $\mathbf{u}$ is on the opposite side of the plane as $\mathbf{v} \wedge \mathbf{w}$. If $\cos\theta = 0$ resulting in zero volume, then $\mathbf{u}$ lies in this plane—the vectors are *coplanar*.

From the discussion above and the antisymmetry of the cross product, we see that the scalar triple product is invariant under *cyclic permutations*. This means that the we get the same volume for the following:

$$
\begin{aligned}
V &= \mathbf{u} \cdot (\mathbf{v} \wedge \mathbf{w}) \\
&= \mathbf{w} \cdot (\mathbf{u} \wedge \mathbf{v}) \\
&= \mathbf{v} \cdot (\mathbf{w} \wedge \mathbf{u}).
\end{aligned}
\tag{8.18}
$$

Example 8.7

Let's compute the volume for a parallelepiped defined by

$$
\mathbf{v} = \begin{bmatrix} 2 \\ 0 \\ 0 \end{bmatrix}, \quad
\mathbf{w} = \begin{bmatrix} 0 \\ 1 \\ 0 \end{bmatrix}, \quad
\mathbf{u} = \begin{bmatrix} 3 \\ 3 \\ 3 \end{bmatrix}.
$$

First we compute the cross product,

$$
\mathbf{y} = \mathbf{v} \wedge \mathbf{w} = \begin{bmatrix} 0 \\ 0 \\ 2 \end{bmatrix},
$$

then the volume $V = \mathbf{u} \cdot \mathbf{y} = 6$. Notice that if $u_3 = -3$, then $V = -6$, demonstrating that the sign of the scalar triple product reveals information about the orientation of the given vectors.

We'll see in Section 9.5 that the parallelepiped given here is simply a rectangular box with dimensions $2 \times 1 \times 3$ that has been sheared. Since shears preserve areas and volumes, this also confirms that the parallelepiped has volume 6.

In Section 4.9 we introduced the 2×2 determinant as a tool to calculate area. The scalar triple product is really just a fancy name for a 3×3 determinant, but we'll get to that in Section 9.8.

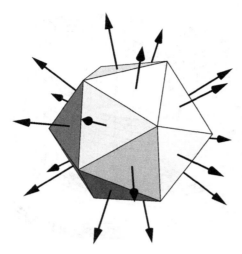

Figure 8.2.
Hedgehog plot: the normal of each facet is drawn at the centroid.

8.6   Application: Lighting and Shading

Let's look at an application of a handful of the tools that we have
developed so far: lighting and shading for computer graphics. One
of the most basic elements needed to calculate the lighting of a 3D
object (model) is the *normal*, which was introduced in Section 8.4.
Although a lighted model might look smooth, it is represented simply
with vertices and planar facets, most often triangles. The normal of
each planar facet is used in conjunction with the light source location
and our eye location to calculate the lighting (color) of each vertex,
and one such method is called the *Phong illumination model*; details of
the method may be found in graphics texts such as [10, 14]. Figure 8.2
illustrates normals drawn emanating from the centroid of the facets.
Determining the color of a facet is called *shading*. A nice example is
illustrated in Figure 8.3.

We calculate the normal by using the *cross product* from Section 8.2.
Suppose we have a triangle defined by points $\mathbf{p}, \mathbf{q}, \mathbf{r}$. Then we form
two vectors, $\mathbf{v}$ and $\mathbf{w}$ as defined in (8.14) from these points. The
normal $\mathbf{n}$ is

$$\mathbf{n} = \frac{\mathbf{v} \wedge \mathbf{w}}{\|\mathbf{v} \wedge \mathbf{w}\|}.$$

The normal is by convention considered to be of unit length. Why

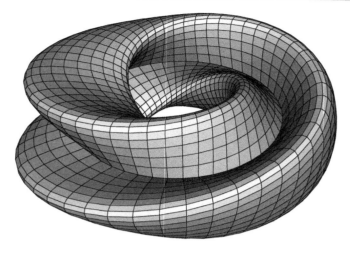

Figure 8.3.

Flat shading: the normal to each planar facet is used to calculate the illumination of each facet.

use $\mathbf{v} \wedge \mathbf{w}$ instead of $\mathbf{w} \wedge \mathbf{v}$? Triangle vertices imply an *orientation*. From this we follow the right-hand rule to determine the normal direction. It is important to have a rule, as just described, so the facets for a model are consistently defined. In turn, the lighting will be consistently calculated at all points on the model.

Figure 8.3 illustrates *flat shading*, which is the fastest and most rough-looking shading method. Each facet is given one color, so a lighting calculation is done at one point, say the centroid, based on the facet's normal. Figure 8.4 illustrates *Gouraud or smooth shading*, which produces smoother-looking models but involves more calculation than flat shading. At each vertex of a triangle, the lighting is calculated. Each of these lighting vectors, $\mathbf{i_p}, \mathbf{i_q}, \mathbf{i_r}$, each vector indicating red, green, and blue components of light, are then interpolated across the triangle. For instance, a point $\mathbf{x}$ in the triangle with barycentric coordinates (u, v, w),

$$\mathbf{x} = u\mathbf{p} + v\mathbf{q} + w\mathbf{r},$$

will be assigned color

$$\mathbf{i_x} = u\mathbf{i_p} + v\mathbf{i_q} + w\mathbf{i_r},$$

which is a handy application of barycentric coordinates!

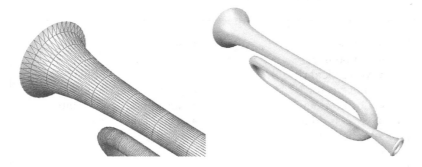

Figure 8.4.
Smooth shading: a normal at each vertex is used to calculate the illumination over each facet. Left: zoomed-in and displayed with triangles. Right: the smooth shaded bugle.

Still unanswered in the smooth shading method: what normals do we assign the vertices in order to achieve a smooth shaded model? If we used the same normal at each vertex, for small enough triangles, we would wind up with (expensive) flat shading. The answer: a *vertex normal* is calculated as the average of the triangle normals for the triangles in the *star* of the vertex. (See Section 17.4 for a review of triangulations.) Better normals can be generated by weighting the contribution of each triangle normal based on the area of the triangle.

The direction of the normal **n** relative to our eye's position can be used to eliminate facets from the rendering pipeline. For a closed surface, such as that in Figure 8.3, the "back-facing" facets will be obscured by the "front-facing" facets. If the centroid of a triangle is **c** and the eye's position is **e**, then form the vector

$$\mathbf{v} = (\mathbf{e} - \mathbf{c})/\|\mathbf{e} - \mathbf{c}\|.$$

If

$$\mathbf{n} \cdot \mathbf{v} < 0$$

then the triangle is back-facing, and we need not render it. This process is called *culling*. A great savings in rendering time can be achieved with culling.

Planar facet normals play an important role in computer graphics, as demonstrated in this section. For more advanced applications, consult a graphics text such as [14].

- 3D vector
- 3D point
- vector length
- unit vector
- dot product
- cross product
- right-hand rule
- orthonormal
- area
- Lagrange's identity
- 3D line
- implicit form of a plane
- parametric form of a plane
- normal
- point normal plane equation
- point-plane distance
- plane-origin distance
- barycentric coordinates
- scalar triple product
- volume
- cyclic permutations of vectors
- triangle normal
- back-facing triangle
- lighting model
- flat and Gouraud shading
- vertex normal
- culling

8.7 Exercises

For the following exercises, use the following points and vectors:

$$\mathbf{p} = \begin{bmatrix} 0 \\ 0 \\ 1 \end{bmatrix}, \mathbf{q} = \begin{bmatrix} 1 \\ 1 \\ 1 \end{bmatrix}, \mathbf{r} = \begin{bmatrix} 4 \\ 2 \\ 4 \end{bmatrix}, \mathbf{v} = \begin{bmatrix} 1 \\ 0 \\ 0 \end{bmatrix}, \mathbf{w} = \begin{bmatrix} 1 \\ 1 \\ 1 \end{bmatrix}, \mathbf{u} = \begin{bmatrix} 0 \\ 0 \\ 1 \end{bmatrix}.$$

1. Normalize the vector $\mathbf{r}$. What is the length of the vector $2\mathbf{r}$?

2. Find the angle between the vectors $\mathbf{v}$ and $\mathbf{w}$.

3. Compute $\mathbf{v} \wedge \mathbf{w}$.

4. Compute the area of the parallelogram formed by vectors $\mathbf{v}$ and $\mathbf{w}$.

5. What is the sine of the angle between $\mathbf{v}$ and $\mathbf{w}$?

6. Find three vectors so that their cross product is associative.

7. Are the lines

$$\mathbf{l}_1(t) = \begin{bmatrix} 0 \\ 0 \\ 1 \end{bmatrix} + t \begin{bmatrix} 0 \\ 0 \\ 1 \end{bmatrix} \quad \text{and} \quad \mathbf{l}_2(s) = \begin{bmatrix} 1 \\ 1 \\ 1 \end{bmatrix} + s \begin{bmatrix} -1 \\ -1 \\ 1 \end{bmatrix}$$

skew?

8. Form the point normal plane equation for a plane through point $\mathbf{p}$ and with normal direction $\mathbf{r}$.

9. Form the point normal plane equation for the plane defined by points $\mathbf{p}$, $\mathbf{q}$, and $\mathbf{r}$.

10. Form a parametric plane equation for the plane defined by points $\mathbf{p}$, $\mathbf{q}$, and $\mathbf{r}$.

11. Form an equation of the plane that bisects the points $\mathbf{p}$ and $\mathbf{q}$.

12. Given the line l defined by point $\mathbf{q}$ and vector $\mathbf{v}$, what is the length of the projection of vector $\mathbf{w}$ bound to $\mathbf{q}$ onto l?

13. Given the line l defined by point $\mathbf{q}$ and vector $\mathbf{v}$, what is the (perpendicular) distance of the point $\mathbf{q} + \mathbf{w}$ (where $\mathbf{w}$ is a vector) to the line l?

14. What is $\mathbf{w} \wedge 6\mathbf{w}$?

15. For the plane in Exercise 8, what is the distance of this plane to the origin?

16. For the plane in Exercise 8, what is the distance of the point $\mathbf{q}$ to this plane?

17. Find the volume of the parallelepiped defined by vectors $\mathbf{v}$, $\mathbf{w}$, and $\mathbf{u}$.

18. Decompose $\mathbf{w}$ into
$$\mathbf{w} = \mathbf{u}_1 + \mathbf{u}_2,$$
where $\mathbf{u}_1$ and $\mathbf{u}_2$ are perpendicular. Additionally, find $\mathbf{u}_3$ to complete an orthogonal frame. *Hint: Orthogonal projections are the topic of Section 2.8.*

19. Given the triangle formed by points $\mathbf{p}, \mathbf{q}, \mathbf{r}$, and colors

$$\mathbf{i_p} = \begin{bmatrix} 1 \\ 0 \\ 0 \end{bmatrix}, \quad \mathbf{i_q} = \begin{bmatrix} 0 \\ 1 \\ 0 \end{bmatrix}, \quad \mathbf{i_r} = \begin{bmatrix} 1 \\ 0 \\ 0 \end{bmatrix},$$

what color $\mathbf{i_c}$ would be assigned to the centroid of this triangle using Gouraud shading? Note: the color vectors are scaled between $[0, 1]$. The colors black, white, red, green, blue are represented by

$$\begin{bmatrix} 0 \\ 0 \\ 0 \end{bmatrix}, \quad \begin{bmatrix} 1 \\ 1 \\ 1 \end{bmatrix}, \quad \begin{bmatrix} 1 \\ 0 \\ 0 \end{bmatrix}, \quad \begin{bmatrix} 0 \\ 1 \\ 0 \end{bmatrix}, \quad \begin{bmatrix} 0 \\ 0 \\ 1 \end{bmatrix},$$

respectively.

9

Linear Maps in 3D

Figure 9.1.
Flight simulator: 3D linear maps are necessary to create the twists and turns in a flight simulator. (Image is from the NASA website http://www.nasa.gov.)

The flight simulator is an important part in the training of airplane pilots. It has a real cockpit, but what you see outside the windows is

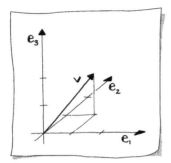

Sketch 9.1.
A vector in the [$\mathbf{e}_1$, $\mathbf{e}_2$, $\mathbf{e}_3$]-coordinate system.

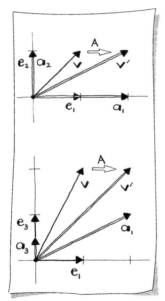

Sketch 9.2.
The matrix A maps $\mathbf{v}$ in the [$\mathbf{e}_1$, $\mathbf{e}_2$, $\mathbf{e}_3$]-coordinate system to the vector $\mathbf{v}'$ in the [$\mathbf{a}_1$, $\mathbf{a}_2$, $\mathbf{a}_3$]-coordinate system.

computer imagery. As you take a right turn, the terrain below changes accordingly; as you dive downwards, it comes closer to you. When you change the (simulated) position of your plane, the simulation software must recompute a new view of the terrain, clouds, or other aircraft. This is done through the application of 3D affine and linear maps.[1] Figure 9.1 shows an image that was generated by an actual flight simulator. For each frame of the simulated scene, complex 3D computations are necessary, most of them consisting of the types of maps discussed in this section.

9.1 Matrices and Linear Maps

The general concept of a linear map in 3D is the same as that for a 2D map. Let $\mathbf{v}$ be a vector in the standard [$\mathbf{e}_1$, $\mathbf{e}_2$, $\mathbf{e}_3$]-coordinate system, i.e.,

$$\mathbf{v} = v_1\mathbf{e}_1 + v_2\mathbf{e}_2 + v_3\mathbf{e}_3.$$

(See Sketch 9.1 for an illustration.)

Let another coordinate system, the [$\mathbf{a}_1$, $\mathbf{a}_2$, $\mathbf{a}_3$]-coordinate system, be given by the origin $\mathbf{o}$ and three vectors $\mathbf{a}_1, \mathbf{a}_2, \mathbf{a}_3$. What vector $\mathbf{v}'$ in the [$\mathbf{a}_1$, $\mathbf{a}_2$, $\mathbf{a}_3$]-system corresponds to $\mathbf{v}$ in the [$\mathbf{e}_1$, $\mathbf{e}_2$, $\mathbf{e}_3$]-system? Simply the vector with the same coordinates relative to the [$\mathbf{a}_1$, $\mathbf{a}_2$, $\mathbf{a}_3$]-system. Thus,

$$\mathbf{v}' = v_1\mathbf{a}_1 + v_2\mathbf{a}_2 + v_3\mathbf{a}_3. \tag{9.1}$$

This is illustrated by Sketch 9.2 and the following example.

Example 9.1

Let

$$\mathbf{v} = \begin{bmatrix} 1 \\ 1 \\ 2 \end{bmatrix}, \quad \mathbf{a}_1 = \begin{bmatrix} 2 \\ 0 \\ 1 \end{bmatrix}, \quad \mathbf{a}_2 = \begin{bmatrix} 0 \\ 1 \\ 0 \end{bmatrix}, \quad \mathbf{a}_3 = \begin{bmatrix} 0 \\ 0 \\ 1/2 \end{bmatrix}.$$

Then

$$\mathbf{v}' = 1 \cdot \begin{bmatrix} 2 \\ 0 \\ 1 \end{bmatrix} + 1 \cdot \begin{bmatrix} 0 \\ 1 \\ 0 \end{bmatrix} + 2 \cdot \begin{bmatrix} 0 \\ 0 \\ 1/2 \end{bmatrix} = \begin{bmatrix} 2 \\ 1 \\ 2 \end{bmatrix}.$$

---
[1] Actually, perspective maps are also needed here. They will be discussed in Section 10.5.

You should recall that we had the same configuration earlier for the 2D case—(9.1) corresponds directly to (4.2) of Section 4.1. In Section 4.2, we then introduced the matrix form. That is now an easy project for this chapter—nothing changes except the matrices will be 3×3 instead of 2×2. In 3D, a matrix equation looks like this:

$$\mathbf{v}' = A\mathbf{v}, \tag{9.2}$$

i.e., just the same as for the 2D case. Written out in detail, there is a difference:

$$\begin{bmatrix} v_1' \\ v_2' \\ v_3' \end{bmatrix} = \begin{bmatrix} a_{1,1} & a_{1,2} & a_{1,3} \\ a_{2,1} & a_{2,2} & a_{2,3} \\ a_{3,1} & a_{3,2} & a_{3,3} \end{bmatrix} \begin{bmatrix} v_1 \\ v_2 \\ v_3 \end{bmatrix}. \tag{9.3}$$

All matrix properties from Chapter 4 carry over almost verbatim.

Example 9.2

Returning to our example, it is quite easy to condense it into a matrix equation:

$$\begin{bmatrix} 2 & 0 & 0 \\ 0 & 1 & 0 \\ 1 & 0 & 1/2 \end{bmatrix} \begin{bmatrix} 1 \\ 1 \\ 2 \end{bmatrix} = \begin{bmatrix} 2 \\ 1 \\ 2 \end{bmatrix}.$$

Again, if we multiply a matrix A by a vector $\mathbf{v}$, the ith component of the result vector is obtained as the dot product of the ith row of A and $\mathbf{v}$.

The matrix A represents a *linear map*. Given the vector $\mathbf{v}$ in the $[\mathbf{e}_1, \mathbf{e}_2, \mathbf{e}_3]$-system, there is a vector $\mathbf{v}'$ in the $[\mathbf{a}_1, \mathbf{a}_2, \mathbf{a}_3]$-system such that $\mathbf{v}'$ has the same components in the $[\mathbf{a}_1, \mathbf{a}_2, \mathbf{a}_3]$-system as did $\mathbf{v}$ in the $[\mathbf{e}_1, \mathbf{e}_2, \mathbf{e}_3]$-system. The matrix A finds the components of $\mathbf{v}'$ relative to the $[\mathbf{e}_1, \mathbf{e}_2, \mathbf{e}_3]$-system.

With the 2×2 matrices of Section 4.2, we introduced the *transpose* A^{T} of a matrix A. We will need this for 3×3 matrices, and it is obtained by interchanging rows and columns, i.e.,

$$\begin{bmatrix} \mathbf{2} & \mathbf{3} & \mathbf{-4} \\ 3 & 9 & -4 \\ -1 & -9 & 4 \end{bmatrix}^T = \begin{bmatrix} \mathbf{2} & 3 & -1 \\ \mathbf{3} & 9 & -9 \\ \mathbf{-4} & -4 & 4 \end{bmatrix}.$$

The boldface row of A has become the boldface column of A^{T}. As a concise formula,

$$a_{i,j}^{\mathrm{T}} = a_{j,i}.$$

9.2   Linear Spaces

The set of all 3D vectors is referred to as a 3D *linear space* or *vector space*, and it is denoted as $\mathbb{R}^3$. We associate with $\mathbb{R}^3$ the operation of forming linear combinations. This means that if $\mathbf{v}$ and $\mathbf{w}$ are two vectors in this linear space, then any vector

$$\mathbf{u} = s\mathbf{v} + t\mathbf{w} \tag{9.4}$$

is also in this space. The vector $\mathbf{u}$ is then said to be a *linear combination* of $\mathbf{v}$ and $\mathbf{w}$. This is also called the *linearity property*. Notice that the linear combination (9.4) combines scalar multiplication and vector addition. These are the key operations necessary for a linear space.

Select two vectors $\mathbf{v}$ and $\mathbf{w}$ and consider all vectors $\mathbf{u}$ that may be expressed as (9.4) with arbitrary scalars s, t. Clearly, all vectors $\mathbf{u}$ form a subset of all 3D vectors. But beyond that, they form a linear space themselves—a 2D space. For if two vectors $\mathbf{u}_1$ and $\mathbf{u}_2$ are in this space, then they can be written as

$$\mathbf{u}_1 = s_1\mathbf{v} + t_1\mathbf{w} \quad \text{and} \quad \mathbf{u}_2 = s_2\mathbf{v} + t_2\mathbf{w},$$

and thus any linear combination of them can be written as

$$\alpha\mathbf{u}_1 + \beta\mathbf{u}_2 = (\alpha s_1 + \beta s_2)\mathbf{v} + (\alpha t_1 + \beta t_2)\mathbf{w},$$

which is again in the same space. We call the set of all vectors of the form (9.4) a *subspace* of the linear space of all 3D vectors. The term subspace is justified since not all 3D vectors are in it. Take for instance the vector $\mathbf{n} = \mathbf{v} \wedge \mathbf{w}$, which is perpendicular to both $\mathbf{v}$ and $\mathbf{w}$. There is no way to write this vector as a linear combination of $\mathbf{v}$ and $\mathbf{w}$!

We say our subspace has dimension 2 since it is generated, or *spanned*, by two vectors. These vectors have to be noncollinear; otherwise, they just define a line, or a 1D (1-dimensional) subspace. (In Section 2.8, we needed the concept of a subspace in order to find the orthogonal projection of $\mathbf{w}$ onto $\mathbf{v}$. Thus the projection lived in the one-dimensional subspace formed by $\mathbf{v}$.)

If two vectors are collinear, then they are also called *linearly dependent*. If $\mathbf{v}$ and $\mathbf{w}$ are linearly dependent, then $\mathbf{v} = s\mathbf{w}$. Conversely, if they are not collinear, they are called *linearly independent*. If $\mathbf{v}_1, \mathbf{v}_2, \mathbf{v}_3$ are linearly independent, then we will not have a solution set s_1, s_2 for

$$\mathbf{v}_3 = s_1\mathbf{v}_1 + s_2\mathbf{v}_2,$$

and the only way to express the zero vector,

$$0 = s_1\mathbf{v}_1 + s_2\mathbf{v}_2 + s_3\mathbf{v}_3$$

is if $s_1 = s_2 = s_3 = 0$. Three linearly independent vectors in $\mathbb{R}^3$ span the entire space and the vectors are said to form a basis for $\mathbb{R}^3$.

Given two linearly independent vectors $\mathbf{v}$ and $\mathbf{w}$, how do we decide if another vector $\mathbf{u}$ is in the subspace spanned by $\mathbf{v}$ and $\mathbf{w}$? Simple: check the volume of the parallelepiped formed by the three vectors, which is equivalent to calculating the scalar triple product (8.17) and checking if it is zero (within a round-off tolerance). In Section 9.8, we'll introduce the 3×3 determinant, which is a matrix-oriented calculation of volume.

We'll revisit this topic in a more abstract setting for n-dimensional vectors in Chapter 14.

9.3   Scalings

A scaling is a linear map that enlarges or reduces vectors:

$$\mathbf{v}' = \begin{bmatrix} s_{1,1} & 0 & 0 \\ 0 & s_{2,2} & 0 \\ 0 & 0 & s_{3,3} \end{bmatrix} \mathbf{v}. \tag{9.5}$$

If all scale factors $s_{i,i}$ are larger than one, then all vectors are enlarged, as is done in Figure 9.2. If all $s_{i,i}$ are positive yet less than one, all vectors are shrunk.

Example 9.3

In this example,

$$\begin{bmatrix} s_{1,1} & 0 & 0 \\ 0 & s_{2,2} & 0 \\ 0 & 0 & s_{3,3} \end{bmatrix} = \begin{bmatrix} 1/3 & 0 & 0 \\ 0 & 1 & 0 \\ 0 & 0 & 3 \end{bmatrix},$$

we shrink in the $\mathbf{e}_1$-direction, leave the $\mathbf{e}_2$-direction unchanged, and stretch the $\mathbf{e}_3$-direction. See Figure 9.3.

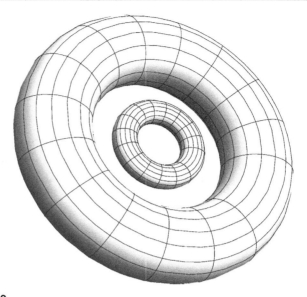

Figure 9.2.
Scalings in 3D: the large torus is scaled by 1/3 in each coordinate to form the small torus.

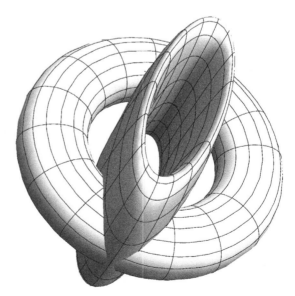

Figure 9.3.
Nonuniform scalings in 3D: the "standard" torus is scaled by 1/3, 1, 3 in the e_1-, e_2-, e_3-directions, respectively.

Negative numbers for the $s_{i,i}$ will cause a flip in addition to a scale. So, for instance

$$\begin{bmatrix} -2 & 0 & 0 \\ 0 & 1 & 0 \\ 0 & 0 & -1 \end{bmatrix}$$

will stretch and reverse in the $\mathbf{e}_1$-direction, leave the $\mathbf{e}_2$-direction unchanged, and will reverse in the $\mathbf{e}_3$-direction.

How do scalings affect volumes? If we map the *unit cube*, given by the three vectors $\mathbf{e}_1, \mathbf{e}_2, \mathbf{e}_3$ with a scaling, we get a rectangular box. Its side lengths are $s_{1,1}$ in the $\mathbf{e}_1$-direction, $s_{2,2}$ in the $\mathbf{e}_2$-direction, and $s_{3,3}$ in the $\mathbf{e}_3$-direction. Hence, its volume is given by $s_{1,1}s_{2,2}s_{3,3}$. A scaling thus changes the volume of an object by a factor that equals the product of the diagonal elements of the scaling matrix.[2] For 2×2 matrices in Chapter 4, we developed a geometric understanding of the map through illustrations of the action ellipse. For example, a nonuniform scaling is illustrated in Figure 4.3. In 3D, the same idea works as well. Now we can examine what happens to 3D unit vectors forming a sphere. They are mapped to an ellipsoid—the *action ellipsoid!* The action ellipsoid corresponding to Figure 9.2 is simply a sphere that is smaller than the unit sphere. The action ellipsoid corresponding to Figure 9.3 has its major axis in the $\mathbf{e}_3$-direction and its minor axis in the $\mathbf{e}_1$-direction. In Chapter 16, we will relate the shape of the ellipsoid to the linear map.

9.4 Reflections

If we reflect a vector about the $\mathbf{e}_2, \mathbf{e}_3$-plane, then its first component should change in sign:

$$\begin{bmatrix} v_1 \\ v_2 \\ v_3 \end{bmatrix} \longrightarrow \begin{bmatrix} -v_1 \\ v_2 \\ v_3 \end{bmatrix},$$

as shown in Sketch 9.3.

This reflection is achieved by a scaling matrix:

$$\begin{bmatrix} -v_1 \\ v_2 \\ v_3 \end{bmatrix} = \begin{bmatrix} -1 & 0 & 0 \\ 0 & 1 & 0 \\ 0 & 0 & 1 \end{bmatrix} \begin{bmatrix} v_1 \\ v_2 \\ v_3 \end{bmatrix}.$$

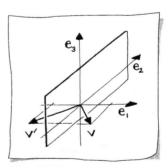

Sketch 9.3.
Reflection of a vector about the $\mathbf{e}_2, \mathbf{e}_3$-plane.

---

[2]We have shown this only for the unit cube, but it is true for any other object as well.

The following is also a reflection, as Sketch 9.4 shows:

$$\begin{bmatrix} v_1 \\ v_2 \\ v_3 \end{bmatrix} \longrightarrow \begin{bmatrix} v_3 \\ v_2 \\ v_1 \end{bmatrix}.$$

It interchanges the first and third component of a vector, and is thus a reflection about the plane $x_1 = x_3$. (This is an implicit plane equation, as discussed in Section 8.4.)

This map is achieved by the following matrix equation:

$$\begin{bmatrix} v_3 \\ v_2 \\ v_1 \end{bmatrix} = \begin{bmatrix} 0 & 0 & 1 \\ 0 & 1 & 0 \\ 1 & 0 & 0 \end{bmatrix} \begin{bmatrix} v_1 \\ v_2 \\ v_3 \end{bmatrix}.$$

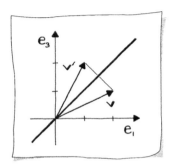

Sketch 9.4.

Reflection of a vector about the $x_1 = x_3$ plane.

In Section 11.5, we develop a more general reflection matrix, called the Householder matrix. Instead of reflecting about a coordinate plane, with this matrix, we can reflect about a given (unit) normal. This matrix is central to the Householder method for solving a linear system in Section 13.1.

By their very nature, reflections do not change volumes—but they do change their signs. See Section 9.8 for more details.

9.5 Shears

What map takes a cube to the parallelepiped (skew box) of Sketch 9.5? The answer: a shear. Shears in 3D are more complicated than the 2D shears from Section 4.7 because there are so many more directions to shear. Let's look at some of the shears more commonly used.

Consider the shear that maps $\mathbf{e}_1$ and $\mathbf{e}_2$ to themselves, and that also maps $\mathbf{e}_3$ to

$$\mathbf{a}_3 = \begin{bmatrix} a \\ b \\ 1 \end{bmatrix}.$$

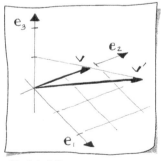

Sketch 9.5.

A 3D shear parallel to the $\mathbf{e}_1$, $\mathbf{e}_2$-plane.

The shear matrix S_1 that accomplishes the desired task is easily found:

$$S_1 = \begin{bmatrix} 1 & 0 & a \\ 0 & 1 & b \\ 0 & 0 & 1 \end{bmatrix}.$$

It is illustrated in Sketch 9.5 with $a = 1$ and $b = 1$, and in Figure 9.4. Thus this map shears parallel to the $[\mathbf{e}_1, \mathbf{e}_2]$-plane. Suppose we apply

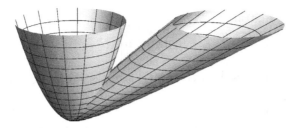

Figure 9.4.
Shears in 3D: a paraboloid is sheared in the $\mathbf{e}_1$- and $\mathbf{e}_2$-directions. The $\mathbf{e}_3$-direction
runs through the center of the left paraboloid.

this shear to a vector $\mathbf{v}$ resulting in

$$\mathbf{v}' = S\mathbf{v} = \begin{bmatrix} v_1 + av_3 \\ v_2 + bv_3 \\ v_3 \end{bmatrix}.$$

An $a_{i,j}$ element is a factor by which the jth component of $\mathbf{v}$ affects
the ith component of $\mathbf{v}'$.

What shear maps $\mathbf{e}_2$ and $\mathbf{e}_3$ to themselves, and also maps

$$\begin{bmatrix} a \\ b \\ c \end{bmatrix} \quad \text{to} \quad \begin{bmatrix} a \\ 0 \\ 0 \end{bmatrix}?$$

This shear is given by the matrix

$$S_2 = \begin{bmatrix} 1 & 0 & 0 \\ \frac{-b}{a} & 1 & 0 \\ \frac{-c}{a} & 0 & 1 \end{bmatrix}. \tag{9.6}$$

One quick check gives:

$$\begin{bmatrix} 1 & 0 & 0 \\ \frac{-b}{a} & 1 & 0 \\ \frac{-c}{a} & 0 & 1 \end{bmatrix} \begin{bmatrix} a \\ b \\ c \end{bmatrix} = \begin{bmatrix} a \\ 0 \\ 0 \end{bmatrix};$$

thus our map does what it was meant to do: it shears parallel to the
$[\mathbf{e}_2, \mathbf{e}_3]$-plane. (This is the shear of the Gauss elimination step that
we will encounter in Section 12.2.)

Although it is possible to shear in any direction, it is more common to shear parallel to a coordinate axis or coordinate plane. Try
constructing a matrix for a shear parallel to the $[\mathbf{e}_1, \mathbf{e}_3]$-plane.

The shear matrix

$$\begin{bmatrix} 1 & a & b \\ 0 & 1 & 0 \\ 0 & 0 & 1 \end{bmatrix}$$

shears parallel to the $\mathbf{e}_1$-axis. Matrices for the other axes follow similarly.

How does a shear affect volume? For a geometric feeling, notice the simple shear S_1 from above. It maps the unit cube to a skew box with the same base and the same height—thus it does not change volume! All shears are volume preserving. After reading Section 9.8, revisit these shear matrices and check the volumes for yourself.

9.6 Rotations

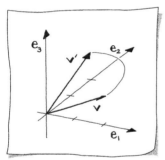

Sketch 9.6.
Rotation example.

Suppose you want to rotate a vector $\mathbf{v}$ around the $\mathbf{e}_3$-axis by $90°$ to a vector $\mathbf{v}'$. Sketch 9.6 illustrates such a rotation:

$$\mathbf{v} = \begin{bmatrix} 2 \\ 0 \\ 1 \end{bmatrix} \quad \rightarrow \quad \mathbf{v}' = \begin{bmatrix} 0 \\ 2 \\ 1 \end{bmatrix}.$$

A rotation around $\mathbf{e}_3$ by different angles would result in different vectors, but they all will have one thing in common: their third components will not be changed by the rotation. Thus, if we rotate a vector around $\mathbf{e}_3$, the rotation action will change only its first and second components. This suggests another look at the 2D rotation matrices from Section 4.6. Our desired rotation matrix R_3 looks much like the one from (4.16):

$$R_3 = \begin{bmatrix} \cos\alpha & -\sin\alpha & 0 \\ \sin\alpha & \cos\alpha & 0 \\ 0 & 0 & 1 \end{bmatrix}. \tag{9.7}$$

Figure 9.5 illustrates the letter **L** rotated through several angles about the $\mathbf{e}_3$-axis.

Example 9.4

Let us verify that R_3 performs as promised with $\alpha = 90°$:

$$\begin{bmatrix} 0 & -1 & 0 \\ 1 & 0 & 0 \\ 0 & 0 & 1 \end{bmatrix} \begin{bmatrix} 2 \\ 0 \\ 1 \end{bmatrix} = \begin{bmatrix} 0 \\ 2 \\ 1 \end{bmatrix},$$

so it works!

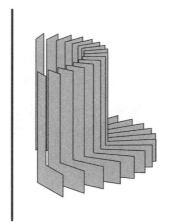

Figure 9.5.
Rotations in 3D: the letter **L** rotated about the e_3-axis.

Similarly, we may rotate around the $\mathbf{e}_2$-axis; the corresponding matrix is

$$R_2 = \begin{bmatrix} \cos\alpha & 0 & \sin\alpha \\ 0 & 1 & 0 \\ -\sin\alpha & 0 & \cos\alpha \end{bmatrix}. \tag{9.8}$$

Notice the pattern here. The rotation matrix for a rotation about the $\mathbf{e}_i$-axis is characterized by the ith row being $\mathbf{e}_i^{\mathrm{T}}$ and the ith column being $\mathbf{e}_i$. For completeness, the last rotation matrix about the $\mathbf{e}_1$-axis:

$$R_1 = \begin{bmatrix} 1 & 0 & 0 \\ 0 & \cos\alpha & -\sin\alpha \\ 0 & \sin\alpha & \cos\alpha \end{bmatrix}. \tag{9.9}$$

The direction of rotation by a positive angle follows the right-hand rule: curl your fingers with the rotation, and your thumb points in the direction of the rotation axis.

If you examine the column vectors of a rotation matrix, you will see that each one is a unit length vector and they are orthogonal to each other. Thus, the column vectors form an orthonormal set of vectors, and a rotation matrix is an orthogonal matrix. (These properties hold for the row vectors of the matrix too.) As a result, we have that

$$R^{\mathrm{T}}R = I$$
$$R^{\mathrm{T}} = R^{-1}$$

Additionally, if R rotates by θ, then R^{-1} rotates by $-\theta$.

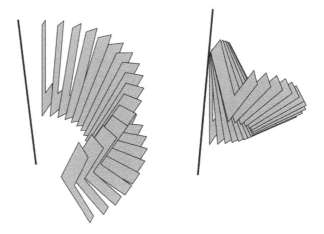

Figure 9.6.
Rotations in 3D: the letter **L** is rotated about axes that are not the coordinate axes. On the right the point on the **L** that touches the rotation axes does not move.

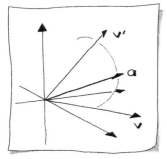

Sketch 9.7.
Rotation about an arbitrary vector.

How about a rotation by α degrees around an arbitrary vector **a**? The principle is illustrated in Sketch 9.7. The derivation of the following matrix is more tedious than called for here, so we just give the result:

$$R = \begin{bmatrix} a_1^2 + C(1 - a_1^2) & a_1a_2(1 - C) - a_3S & a_1a_3(1 - C) + a_2S \\ a_1a_2(1 - C) + a_3S & a_2^2 + C(1 - a_2^2) & a_2a_3(1 - C) - a_1S \\ a_1a_3(1 - C) - a_2S & a_2a_3(1 - C) + a_1S & a_3^2 + C(1 - a_3^2) \end{bmatrix},$$

$$(9.10)$$

where we have set $C = \cos\alpha$ and $S = \sin\alpha$. It is necessary that $\|\mathbf{a}\| = 1$ in order for the rotation to take place without scaling. Figure 9.6 illustrates two examples of rotations about an arbitrary axis.

Example 9.5

With a complicated result such as (9.10), a sanity check is not a bad idea. So let $\alpha = 90°$,

$$\mathbf{a} = \begin{bmatrix} 0 \\ 0 \\ 1 \end{bmatrix} \quad \text{and} \quad \mathbf{v} = \begin{bmatrix} 1 \\ 0 \\ 0 \end{bmatrix}.$$

This means that we want to rotate **v** around **a**, or the $\mathbf{e}_3$-axis, by 90°

as shown in Sketch 9.8. In advance, we know what R should be. In (9.10), $C = 0$ and $S = 1$, and we calculate

$$R = \begin{bmatrix} 0 & -1 & 0 \\ 1 & 0 & 0 \\ 0 & 0 & 1 \end{bmatrix},$$

which is the expected matrix. We obtain

$$\mathbf{v}' = \begin{bmatrix} 0 \\ 1 \\ 0 \end{bmatrix}.$$

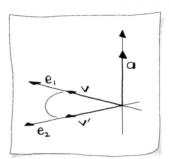

Sketch 9.8.
A simple example of a rotation about a vector.

With some confidence that (9.10) works, let's try a more complicated example.

Example 9.6

Let $\alpha = 90°$,

$$\mathbf{a} = \begin{bmatrix} \frac{1}{\sqrt{3}} \\ \frac{1}{\sqrt{3}} \\ \frac{1}{\sqrt{3}} \end{bmatrix} \quad \text{and} \quad \mathbf{v} = \begin{bmatrix} 1 \\ 0 \\ 0 \end{bmatrix}.$$

With $C = 0$ and $S = 1$ in (9.10), we calculate

$$R = \begin{bmatrix} \frac{1}{3} & \frac{1}{3} - \frac{1}{\sqrt{3}} & \frac{1}{3} + \frac{1}{\sqrt{3}} \\ \frac{1}{3} + \frac{1}{\sqrt{3}} & \frac{1}{3} & \frac{1}{3} - \frac{1}{\sqrt{3}} \\ \frac{1}{3} - \frac{1}{\sqrt{3}} & \frac{1}{3} + \frac{1}{\sqrt{3}} & \frac{1}{3} \end{bmatrix},$$

We obtain

$$\mathbf{v}' = \begin{bmatrix} \frac{1}{3} \\ \frac{1}{3} + \frac{1}{\sqrt{3}} \\ \frac{1}{3} - \frac{1}{\sqrt{3}} \end{bmatrix}.$$

Convince yourself that $\|\mathbf{v}'\| = \|\mathbf{v}\|$.

Continue this example with the vector

$$\mathbf{v} = \begin{bmatrix} 1 \\ 1 \\ 1 \end{bmatrix}.$$

Surprised by the result?

It should be intuitively clear that rotations do not change volumes. Recall from 2D that rotations are *rigid body motions.*

9.7   Projections

Projections that are linear maps are *parallel projections.* There are two categories. If the projection direction is perpendicular to the projection plane then it is an *orthogonal projection*, otherwise it is an *oblique projection.*Two examples are illustrated in Figure 9.7, in which one of the key properties of projections is apparent: flattening. The orthogonal and oblique projection matrices that produced this figure are

$$\begin{bmatrix} 1 & 0 & 0 \\ 0 & 1 & 0 \\ 0 & 0 & 0 \end{bmatrix} \quad \text{and} \quad \begin{bmatrix} 1 & 0 & 1/\sqrt{2} \\ 0 & 1 & 1/\sqrt{2} \\ 0 & 0 & 0 \end{bmatrix},$$

respectively.

Projections are essential in computer graphics to view 3D geometry on a 2D screen. A parallel projection is a linear map, as opposed to a perspective projection, which is not. A parallel projection preserves relative dimensions of an object, thus it is used in drafting to produce accurate views of a design.

Recall from 2D, Section 4.8, that a projection reduces dimensionality and it is an idempotent map. It flattens geometry because a projection matrix P is rank deficient; in 3D this means that a vector

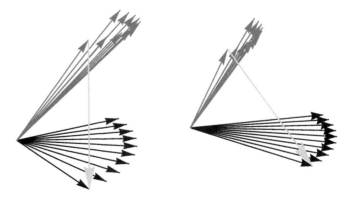

Figure 9.7.

Projections in 3D: on the left is an orthogonal projection, and on the right is an oblique projection of $45°$.

is projected into a subspace, which can be a (2D) plane or (1D) line. The idempotent property leaves a vector in the subspace of the map unchanged by the map, $P\mathbf{v} = P^2\mathbf{v}$. Let's see how to construct an orthogonal projection in 3D.

First we choose the subspace U into which we would like to project. If we want to project onto a line (1D subspace), specify a unit vector $\mathbf{u}_1$. If we want to project into a plane (2D subspace), specify two orthonormal vectors $\mathbf{u}_1, \mathbf{u}_2$. Now form a matrix A_k from the vectors defining the k-dimensional subspace U:

$$A_1 = \mathbf{u}_1 \quad \text{or} \quad A_2 = \begin{bmatrix} \mathbf{u}_1 & \mathbf{u}_2 \end{bmatrix}.$$

The projection matrix P_k is then defined as

$$P_k = A_k A_k^{\mathrm{T}}.$$

It follows that P_1 is very similar to the projection matrix from Section 4.8 except the projection line is in 3D,

$$P_1 = A_1 A_1^{\mathrm{T}} = \begin{bmatrix} u_{1,1}\mathbf{u}_1 & u_{2,1}\mathbf{u}_1 & u_{3,1}\mathbf{u}_1 \end{bmatrix}.$$

Projection into a plane takes the form,

$$P_2 = A_2 A_2^{\mathrm{T}} = \begin{bmatrix} \mathbf{u}_1 & \mathbf{u}_2 \end{bmatrix} \begin{bmatrix} \mathbf{u}_1^{\mathrm{T}} \\ \mathbf{u}_2^{\mathrm{T}} \end{bmatrix}. \tag{9.11}$$

Expanding this, we see the columns of P_2 are linear combinations of $\mathbf{u}_1$ and $\mathbf{u}_2$,

$$P_2 = \begin{bmatrix} u_{1,1}\mathbf{u}_1 + u_{1,2}\mathbf{u}_2 & u_{2,1}\mathbf{u}_1 + u_{2,2}\mathbf{u}_2 & u_{3,1}\mathbf{u}_1 + u_{3,2}\mathbf{u}_2 \end{bmatrix}.$$

The action of P_1 and P_2 is thus

$$P_1\mathbf{v} = (\mathbf{u} \cdot \mathbf{v})\mathbf{u}, \tag{9.12}$$
$$P_2\mathbf{v} = (\mathbf{u}_1 \cdot \mathbf{v})\mathbf{u}_1 + (\mathbf{u}_2 \cdot \mathbf{v})\mathbf{u}_2. \tag{9.13}$$

An application of these projections is demonstrated in the Gram-Schmidt orthonormal coordinate frame construction in Section 11.8.

Example 9.7

Let's construct an orthogonal projection P_2 into the $[\mathbf{e}_1, \mathbf{e}_2]$-plane. Although this example is easy enough to write down the matrix

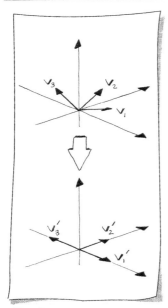

directly, let's construct it with (9.11),

$$P_2 = \begin{bmatrix} \mathbf{e}_1 & \mathbf{e}_2 \end{bmatrix} \begin{bmatrix} \mathbf{e}_1^{\mathrm{T}} \\ \mathbf{e}_2^{\mathrm{T}} \end{bmatrix}$$

$$= \begin{bmatrix} 1 & 0 & 0 \\ 0 & 1 & 0 \\ 0 & 0 & 0 \end{bmatrix}$$

The action achieved by this linear map is

$$\begin{bmatrix} v_1 \\ v_2 \\ 0 \end{bmatrix} = \begin{bmatrix} 1 & 0 & 0 \\ 0 & 1 & 0 \\ 0 & 0 & 0 \end{bmatrix} \begin{bmatrix} v_1 \\ v_2 \\ v_3 \end{bmatrix}.$$

See Sketch 9.9 and Figure 9.7 (left).

Sketch 9.9.
Projection example.

The example above is very simple, and we can immediately see that the projection direction is $\mathbf{d} = [0 \ 0 \ \pm 1]^{\mathrm{T}}$. This vector satisfies the equation

$$P_2 \mathbf{d} = \mathbf{0},$$

and we see that the projection direction is in the kernel of the map.

The idempotent property for P_2 is easily understood by noticing that $A_2^{\mathrm{T}} A_2$ is simply the 2×2 identity matrix,

$$P_2^2 = A_2 A_2^{\mathrm{T}} A_2 A_2^{\mathrm{T}}$$

$$= \begin{bmatrix} \mathbf{u}_1 & \mathbf{u}_2 \end{bmatrix} \begin{bmatrix} \mathbf{u}_1^{\mathrm{T}} \\ \mathbf{u}_2^{\mathrm{T}} \end{bmatrix} \begin{bmatrix} \mathbf{u}_1 & \mathbf{u}_2 \end{bmatrix} \begin{bmatrix} \mathbf{u}_1^{\mathrm{T}} \\ \mathbf{u}_2^{\mathrm{T}} \end{bmatrix}$$

$$= \begin{bmatrix} \mathbf{u}_1 & \mathbf{u}_2 \end{bmatrix} I \begin{bmatrix} \mathbf{u}_1^{\mathrm{T}} \\ \mathbf{u}_2^{\mathrm{T}} \end{bmatrix}$$

$$= P_2.$$

In addition to being idempotent, orthogonal projection matrices are symmetric. The action of the map is $P\mathbf{v}$ and this vector is orthogonal to $\mathbf{v} - P\mathbf{v}$, thus

$$0 = (P\mathbf{v})^{\mathrm{T}}(\mathbf{v} - P\mathbf{v})$$

$$= \mathbf{v}^{\mathrm{T}}(P^{\mathrm{T}} - P^{\mathrm{T}} P)\mathbf{v},$$

from which we conclude that $P = P^{\mathrm{T}}$.

We will examine oblique projections in the context of affine maps in Section 10.4. Finally, we note that a projection has a significant effect on the volume of an object. Since everything is flat after a projection, it has zero 3D volume.

9.8 Volumes and Linear Maps: Determinants

Most linear maps change volumes; some don't. Since this is an important aspect of the action of a map, this section will discuss the effect of a linear map on volume. The unit cube in the $[\mathbf{e}_1, \mathbf{e}_2, \mathbf{e}_3]$-system has volume one. A linear map A will change that volume to that of the skew box spanned by the images of $\mathbf{e}_1, \mathbf{e}_2, \mathbf{e}_3$, i.e., by the volume spanned by the vectors $\mathbf{a}_1, \mathbf{a}_2, \mathbf{a}_3$—the column vectors of A. What is the volume spanned by $\mathbf{a}_1, \mathbf{a}_2, \mathbf{a}_3$?

First, let's look at what we have done so far with areas and volumes. Recall the 2×2 determinant from Section 4.9. Through Sketch 4.8, the area of a 2D parallelogram was shown to be equivalent to a determinant. In fact, in Section 8.2 it was shown that the cross product can be used to calculate this area for a parallelogram embedded in 3D. With a very geometric approach, the *scalar triple product* of Section 8.5 gives us the means to calculate the volume of a parallelepiped by simply using a "base area times height" calculation. Let's revisit that formula and look at it from the perspective of linear maps.

So, using linear maps, we want to illustrate that the volume of the parallelepiped, or skew box, simply reduces to a 3D determinant calculation. Proceeding directly with a sketch in the 3D case would be difficult to follow. For 3D, let's augment the determinant idea with the tools from Section 5.4. There we demonstrated how shears— area-preserving linear maps—can be used to transform a matrix to upper triangular. These are the forward elimination steps of Gauss elimination.

First, let's introduce a 3×3 determinant of a matrix A. It is easily remembered as an alternating sum of 2×2 determinants:

$$|A| = a_{1,1} \begin{vmatrix} a_{2,2} & a_{2,3} \\ a_{3,2} & a_{3,3} \end{vmatrix} - a_{2,1} \begin{vmatrix} a_{1,2} & a_{1,3} \\ a_{3,2} & a_{3,3} \end{vmatrix} + a_{3,1} \begin{vmatrix} a_{1,2} & a_{1,3} \\ a_{2,2} & a_{2,3} \end{vmatrix}. \quad (9.14)$$

The representation in (9.14) is called the *cofactor expansion*. Each (signed) 2×2 determinant is the *cofactor* of the $a_{i,j}$ it is paired with in the sum. The sign comes from the factor $(-1)^{i+j}$. For example, the cofactor of $a_{2,1}$ is

$$(-1)^{2+1} \begin{vmatrix} a_{1,2} & a_{1,3} \\ a_{3,2} & a_{3,3} \end{vmatrix}.$$

The cofactor is also written as $(-1)^{i+j} M_{i,j}$ where $M_{i,j}$ is called the *minor* of $a_{i,j}$. As a result, (9.14) is also known as *expansion by minors*. We'll look into this method more in Section 12.6.

If (9.14) is expanded, then an interesting form for writing the determinant arises. The formula is nearly impossible to remember, but the following trick is not. Copy the first two columns after the last column. Next, form the product of the three "diagonals" and add them. Then, form the product of the three "antidiagonals" and subtract them. The three "plus" products may be written as:

$$
\begin{matrix}
a_{1,1} & a_{1,2} & a_{1,3} & \square & \square \\
\square & a_{2,2} & a_{2,3} & a_{2,1} & \square \\
\square & \square & a_{3,3} & a_{3,1} & a_{3,2}
\end{matrix}
$$

and the three "minus" products as:

$$
\begin{matrix}
\square & \square & a_{1,3} & a_{1,1} & a_{1,2} \\
\square & a_{2,2} & a_{2,3} & a_{2,1} & \square \\
a_{3,1} & a_{3,2} & a_{3,3} & \square & \square
\end{matrix}.
$$

The complete formula for the 3×3 determinant is

$$
\begin{aligned}
|A| = {} & a_{1,1}a_{2,2}a_{3,3} + a_{1,2}a_{2,3}a_{3,1} + a_{1,3}a_{2,1}a_{3,2} \\
& - a_{3,1}a_{2,2}a_{1,3} - a_{3,2}a_{2,3}a_{1,1} - a_{3,3}a_{2,1}a_{1,2}.
\end{aligned}
$$

Example 9.8

What is the volume spanned by the three vectors

$$
\mathbf{a}_1 = \begin{bmatrix} 4 \\ 0 \\ 0 \end{bmatrix}, \quad \mathbf{a}_2 = \begin{bmatrix} -1 \\ 4 \\ 4 \end{bmatrix}, \quad \mathbf{a}_3 = \begin{bmatrix} 0.1 \\ -0.1 \\ 0.1 \end{bmatrix}?
$$

All we have to do is to compute

$$
\det[\mathbf{a}_1, \mathbf{a}_2, \mathbf{a}_3] = 4 \begin{vmatrix} 4 & -0.1 \\ 4 & 0.1 \end{vmatrix}
$$

$$
= 4(4 \times 0.1 - (-0.1) \times 4) = 3.2.
$$

(Here we used an alternative notation, det, for the determinant.) In this computation, we did not write down zero terms.

As we have seen in Section 9.5, a 3D shear preserves volume. Therefore, we can apply a series of shears to the matrix A, resulting in a new matrix

$$\tilde{A} = \begin{bmatrix} \tilde{a}_{1,1} & \tilde{a}_{1,2} & \tilde{a}_{1,3} \\ 0 & \tilde{a}_{2,2} & \tilde{a}_{2,3} \\ 0 & 0 & \tilde{a}_{3,3} \end{bmatrix}.$$

The determinant of $\tilde{A}$ is

$$|\tilde{A}| = \tilde{a}_{1,1}\tilde{a}_{2,2}\tilde{a}_{3,3}, \tag{9.15}$$

with of course, $|A| = |\tilde{A}|$.

Let's continue with Example 9.8. One simple row operation, $\text{row}_3 = \text{row}_3 - \text{row}_2$, will achieve the upper triangular matrix

$$\tilde{A} = \begin{bmatrix} 4 & -1 & 0.1 \\ 0 & 4 & -0.1 \\ 0 & 0 & 0.2 \end{bmatrix},$$

and we can determine that $|\tilde{A}| = |A|$.

For 3×3 matrices, we don't actually calculate the volume of three vectors by proceeding with the forward elimination steps, or shears.[3] We would just directly calculate the 3×3 determinant from (9.14). What is interesting about this development is now we can illustrate, as in Sketch 9.10, how the determinant defines the volume of the skew box. The first two column vectors of $\tilde{A}$ lie in the $[\mathbf{e}_1, \mathbf{e}_2]$-plane. Their determinant defines the area of the parallelogram that they span; this determinant is $\tilde{a}_{1,1}\tilde{a}_{2,2}$. The height of the skew box is simply the $\mathbf{e}_3$ component of $\tilde{\mathbf{a}}_3$. Thus, we have an easy to visualize interpretation of the 3×3 determinant. And, from a slightly different perspective, we have revisited the geometric development of the determinant as the scalar triple product (Section 8.5).

Let's conclude this section with some *rules for determinants*. Suppose we have two 3×3 matrices, A and B. The column vectors of A are $\begin{bmatrix} \mathbf{a}_1 & \mathbf{a}_2 & \mathbf{a}_3 \end{bmatrix}$.

- The determinant of the transpose matrix equals that of the matrix: $|A| = |A^{\text{T}}|$. This property allows us to interchange the terms "row" or "column" when working with determinants.

- Exchanging two columns, creating a noncyclic permutation, changes the sign of the determinant: $\begin{vmatrix} \mathbf{a}_2 & \mathbf{a}_1 & \mathbf{a}_3 \end{vmatrix} = -|A|$.

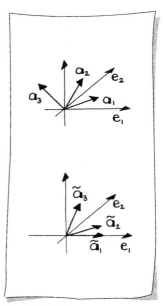

Sketch 9.10.

Determinant and volume in 3D.

---

[3]However, we will use forward elimination for $n \times n$ systems. The sign of the determinant in (9.15) needs to be adjusted if pivoting is included in forward elimination. See Section 12.6 for details.

- Multiplying one column by a scalar c results in the determinant being multiplied by c: $\begin{vmatrix} c\mathbf{a_1} & \mathbf{a_2} & \mathbf{a_3} \end{vmatrix} = c|A|$.

- As an extension of the previous item: $|cA| = c^3|A|$.

- If A has a row of zeroes then $|A| = 0$.

- If A has two identical rows then $|A| = 0$.

- The sum of determinants is not the determinant of the sum, $|A| + |B| \neq |A + B|$, in general.

- The product of determinants is the determinant of the product: $|AB| = |A||B|$.

- Multiples of rows can be added together without changing the determinant. For example, the shears of Gauss elimination do not change the determinant, as we observed in the simple example above.

  The determinant of the shear matrix S_2 in (9.6) is one, thus $|S_2A| = |S_2||A| = |A|$.

- A being invertible is equivalent to $|A| \neq 0$ (see Section 9.10).

- If A is invertible then
$$|A^{-1}| = \frac{1}{|A|}.$$

9.9 Combining Linear Maps

If we apply a linear map A to a vector $\mathbf{v}$ and then apply a map B to the result, we may write this as

$$\mathbf{v'} = BA\mathbf{v}.$$

Matrix multiplication is defined just as in the 2D case; the element $c_{i,j}$ of the product matrix $C = BA$ is obtained as the dot product of the ith row of B with the jth column of A. A handy way to write the matrices so as to keep the dot products in order is

$$\begin{array}{c|c} & A \\ \hline B & C \end{array}.$$

Instead of a complicated formula, an example should suffice.

$$
\begin{array}{ccc|ccc}
 & & & 1 & 5 & -4 \\
 & & & -1 & -2 & 0 \\
 & & & 2 & 3 & -4 \\
\hline
\mathbf{0} & \mathbf{0} & -1 & -2 & \mathbf{-3} & 4 \\
1 & -2 & 0 & 3 & 9 & -4 \\
-2 & 1 & 1 & -1 & -9 & 4
\end{array}
$$

We have computed the dot product of the boldface row of B and column of A to produce the boldface entry of C. In this example, B and A are 3×3 matrices, and thus the result is another 3×3 matrix. In the example in Section 9.1, a 3×3 matrix A is multiplied by a 3×1 matrix (vector) $\mathbf{v}$ resulting in a 3×1 matrix or vector. Thus two matrices need not be the same size in order to multiply them. There is a rule, however! Suppose we are to multiply two matrices A and B together as AB. The sizes of A and B are

$$ m \times n \quad \text{and} \quad n \times p, \tag{9.16} $$

respectively. The resulting matrix will be of size $m \times p$—the "outside" dimensions in (9.16). In order to form AB, it is necessary that the "inside" dimensions, both n here, be equal. The matrix multiplication scheme from Section 4.2 simplifies hand-calculations by illustrating the resulting dimensions.

As in the 2D case, matrix multiplication does not commute! That is, $AB \neq BA$ in most cases. An interesting difference between 2D and 3D is the fact that in 2D, rotations *did* commute; however, in 3D they do not. For example, in 2D, rotating first by α and then by β is no different from doing it the other way around. In 3D, that is not the case. Let's look at an example to illustrate this point.

Example 9.9

Let's look at a rotation by $-90°$ around the $\mathbf{e}_1$-axis with matrix R_1 and a rotation by $-90°$ around the $\mathbf{e}_3$-axis with matrix R_3:

$$
R_1 = \begin{bmatrix} 1 & 0 & 0 \\ 0 & 0 & 1 \\ 0 & -1 & 0 \end{bmatrix} \quad \text{and} \quad R_3 = \begin{bmatrix} 0 & 1 & 0 \\ -1 & 0 & 0 \\ 0 & 0 & 1 \end{bmatrix}.
$$

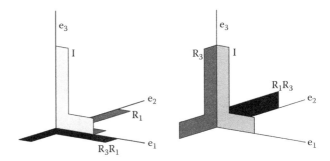

Figure 9.8.
Combining 3D rotations: left and right, the original **L** is labeled *I* for identity matrix. On the left, R_1 is applied and then R_3, the result is labeled R_3R_1. On the right, R_3 is applied and then R_1, the result is labeled R_1R_3. This shows that 3D rotations do not commute.

Figure 9.8 illustrates what the algebra tells us:

$$R_3R_1 = \begin{bmatrix} 0 & 0 & 1 \\ -1 & 0 & 0 \\ 0 & -1 & 0 \end{bmatrix} \quad \text{is not equal to} \quad R_1R_3 = \begin{bmatrix} 0 & 1 & 0 \\ 0 & 0 & 1 \\ 1 & 0 & 0 \end{bmatrix}.$$

Also helpful for understanding what is happening, is to track the transformation of a point **p** on **L**. Form the vector $\mathbf{v} = \mathbf{p} - \mathbf{o}$, and let's track

$$\mathbf{v} = \begin{bmatrix} 0 \\ 0 \\ 1 \end{bmatrix}.$$

In Figure 9.8 on the left, observe the transformation of **v**:

$$R_1\mathbf{v} = \begin{bmatrix} 0 \\ 1 \\ 0 \end{bmatrix} \quad \text{and} \quad R_3R_1\mathbf{v} = \begin{bmatrix} 1 \\ 0 \\ 0 \end{bmatrix}.$$

Now, on the right, observe the transformation of **v**:

$$R_3\mathbf{v} = \begin{bmatrix} 0 \\ 0 \\ 1 \end{bmatrix} \quad \text{and} \quad R_1R_3\mathbf{v} = \begin{bmatrix} 0 \\ 1 \\ 0 \end{bmatrix}.$$

So it does matter which rotation we perform first!

9.10 Inverse Matrices

In Section 5.9, we saw how inverse matrices undo linear maps. A linear map A takes a vector $\mathbf{v}$ to its image $\mathbf{v}'$. The inverse map, A^{-1}, will take $\mathbf{v}'$ back to $\mathbf{v}$, i.e., $A^{-1}\mathbf{v}' = \mathbf{v}$ or $A^{-1}A\mathbf{v} = \mathbf{v}$. Thus, the combined action of $A^{-1}A$ has no effect on any vector $\mathbf{v}$, which we can write as

$$A^{-1}A = I, \tag{9.17}$$

where I is the 3×3 identity matrix. If we applied A^{-1} to $\mathbf{v}$ first, and then applied A, there would not be any action either; in other words,

$$AA^{-1} = I, \tag{9.18}$$

too.

A matrix is not always invertible. For example, the projections from Section 9.7 are rank deficient, and therefore not invertible. This is apparent from Sketch 9.9: once we flatten the vectors $\mathbf{a}_i$ to $\mathbf{a}_i'$ in 2D, there isn't enough information available in the $\mathbf{a}_i'$ to return them to 3D.

As we discovered in Section 9.6 on rotations, orthogonal matrices, which are constructed from a set of orthonormal vectors, possess the nice property $R^{\mathrm{T}} = R^{-1}$. Forming the reverse rotation is simple and requires no computation; this provides for a huge savings in computer graphics where rotating objects is a common operation.

Scaling also has an inverse, which is simple to compute. If

$$S = \begin{bmatrix} s_{1,1} & 0 & 0 \\ 0 & s_{2,2} & 0 \\ 0 & 0 & s_{3,3} \end{bmatrix}, \quad \text{then} \quad S^{-1} = \begin{bmatrix} 1/s_{1,1} & 0 & 0 \\ 0 & 1/s_{2,2} & 0 \\ 0 & 0 & 1/s_{3,3} \end{bmatrix}.$$

Here are more rules for matrices. These involve calculating with inverse matrices.

$$A^{-n} = (A^{-1})^n = \underbrace{A^{-1} \cdot \ldots \cdots A^{-1}}_{n \text{ times}}$$

$$(A^{-1})^{-1} = A$$

$$(kA)^{-1} = \frac{1}{k}A^{-1}$$

$$(AB)^{-1} = B^{-1}A^{-1}$$

See Section 12.4 for details on calculating A^{-1}.

9.11   More on Matrices

A handful of matrix properties are explained and illustrated in Chapter 4. Here we restate them so they are conveniently together. These properties hold for $n \times n$ matrices (the topic of Chapter 12):

- preserve scalings: $A(c\mathbf{v}) = cA\mathbf{v}$

- preserve summations: $A(\mathbf{u} + \mathbf{v}) = A\mathbf{u} + A\mathbf{v}$

- preserve linear combinations: $A(a\mathbf{u} + b\mathbf{v}) = aA\mathbf{u} + bA\mathbf{v}$

- distributive law: $A\mathbf{v} + B\mathbf{v} = (A + B)\mathbf{v}$

- commutative law for addition: $A + B = B + A$

- no commutative law for multiplication: $AB \neq BA$.

- associative law for addition: $A + (B + C) = (A + B) + C$

- associative law for multiplication: $A(BC) = (AB)C$

- distributive law: $A(B + C) = AB + AC$
$$(B + C)A = BA + CA$$

Scalar laws:

- $a(B + C) = aB + aC$

- $(a + b)C = aC + bC$

- $(ab)C = a(bC)$

- $a(BC) = (aB)C = B(aC)$

Laws involving determinants:

- $|A| = |A^{\mathrm{T}}|$
- $|AB| = |A| \cdot |B|$
- $|A| + |B| \neq |A + B|$
- $|cA| = c^n |A|$

Laws involving exponents:

- $A^r = \underbrace{A \cdot \ldots \cdot A}_{r \text{ times}}$
- $A^{r+s} = A^r A^s$
- $A^{rs} = (A^r)^s$
- $A^0 = I$

Laws involving the transpose:

- $[A + B]^{\mathrm{T}} = A^{\mathrm{T}} + B^{\mathrm{T}}$
- $A^{\mathrm{T}\,\mathrm{T}} = A$
- $[cA]^{\mathrm{T}} = cA^{\mathrm{T}}$
- $[AB]^{\mathrm{T}} = B^{\mathrm{T}} A^{\mathrm{T}}$

- 3D linear map
- transpose matrix
- linear space
- vector space
- subspace
- linearity property
- linear combination
- linearly independent
- linearly dependent
- scale
- action ellipsoid
- rotation
- rigid body motions
- shear
- reflection
- projection
- idempotent
- orthographic projection
- oblique projection
- determinant
- volume
- scalar triple product
- cofactor expansion
- expansion by minors
- inverse matrix
- multiply matrices
- noncommutative property of matrix multiplication
- rules of matrix arithmetic

9.12   Exercises

1. Let $\mathbf{v}' = 3\mathbf{a}_1 + 2\mathbf{a}_2 + \mathbf{a}_3$, where

$$\mathbf{a}_1 = \begin{bmatrix} 1 \\ 1 \\ 1 \end{bmatrix}, \quad \mathbf{a}_2 = \begin{bmatrix} 2 \\ 0 \\ 0 \end{bmatrix}, \quad \mathbf{a}_3 = \begin{bmatrix} 0 \\ 3 \\ 0 \end{bmatrix}.$$

   What is $\mathbf{v}'$? Write this equation in matrix form.

2. What is the transpose of the matrix

$$A = \begin{bmatrix} 1 & 5 & -4 \\ -1 & -2 & 0 \\ 2 & 3 & -4 \end{bmatrix}?$$

3. Given a 2D linear subspace formed by vectors $\mathbf{w}$ and $\mathbf{v}$, is $\mathbf{u}$ an element of that subspace?

$$\mathbf{v} = \begin{bmatrix} 1 \\ 0 \\ 0 \end{bmatrix}, \quad \mathbf{w} = \begin{bmatrix} 1 \\ 1 \\ 1 \end{bmatrix}, \quad \mathbf{u} = \begin{bmatrix} 0 \\ 0 \\ 1 \end{bmatrix}.$$

4. Given

$$\mathbf{v} = \begin{bmatrix} 3 \\ 2 \\ -1 \end{bmatrix}, \quad \mathbf{w} = \begin{bmatrix} 1 \\ -1 \\ 2 \end{bmatrix}, \quad \mathbf{u} = \begin{bmatrix} 7 \\ 3 \\ 0 \end{bmatrix},$$

   is $\mathbf{u}$ in the subspace defined by $\mathbf{v}$ and $\mathbf{w}$?

5. Is the vector $\mathbf{u} = \mathbf{v} \wedge \mathbf{w}$ in the subspace defined by $\mathbf{v}$ and $\mathbf{w}$?

6. Let $\mathcal{V}_1$ be the one-dimensional subspace defined by

$$\mathbf{v} = \begin{bmatrix} 1 \\ 1 \\ 0 \end{bmatrix}.$$

   What vector $\mathbf{w}'$ in $\mathcal{V}_1$ is closest to

$$\mathbf{w} = \begin{bmatrix} 1 \\ 0 \\ 1 \end{bmatrix}?$$

7. The vectors $\mathbf{v}$ and $\mathbf{w}$ form a 2D subspace of $\mathbb{R}^3$. Are they linearly dependent?

8. What is the kernel of the matrix formed from linearly independent vectors $\mathbf{v}_1, \mathbf{v}_2, \mathbf{v}_3$?

9. Describe the linear map given by the matrix

$$\begin{bmatrix} 1 & 0 & 0 \\ 0 & 0 & -1 \\ 0 & 1 & 0 \end{bmatrix}$$

by stating if it is volume preserving and stating the action of the map. *Hint: Examine where the e_i-axes are mapped.*

10. What matrix scales by 2 in the e_1-direction, scales by $1/4$ in the e_2-direction, and reverses direction and scales by 4 in the e_3-direction? Map the unit cube with this matrix. What is the volume of the resulting parallelepiped?

11. What is the matrix that reflects a vector about the plane $x_1 = x_2$? Map the unit cube with this matrix. What is the volume of the resulting parallelepiped?

12. What is the shear matrix that maps

$$\begin{bmatrix} a \\ b \\ c \end{bmatrix} \text{ to } \begin{bmatrix} 0 \\ 0 \\ c \end{bmatrix}?$$

Map the unit cube with this matrix. What is the volume of the resulting parallelepiped?

13. What matrix rotates around the e_1-axis by α degrees?

14. What matrix rotates by $45°$ around the vector $\begin{bmatrix} -1 \\ 0 \\ -1 \end{bmatrix}$?

15. Construct the orthogonal projection matrix P that projects onto the line spanned by

$$\mathbf{u} = \begin{bmatrix} 1/\sqrt{3} \\ 1/\sqrt{3} \\ 1/\sqrt{3} \end{bmatrix}$$

and what is the action of the map, $\mathbf{v}' = P\mathbf{v}$? What is the action of the map on the following two vectors:

$$\mathbf{v}_1 = \begin{bmatrix} 1 \\ 1 \\ 1 \end{bmatrix} \quad \text{and} \quad \mathbf{v}_2 = \begin{bmatrix} 0 \\ 0 \\ 1 \end{bmatrix}?$$

What is the rank of this matrix? What is the determinant?

16. Construct the projection matrix P that projects into the plane spanned by

$$\mathbf{u}_1 = \begin{bmatrix} 1/\sqrt{2} \\ 1/\sqrt{2} \\ 0 \end{bmatrix} \quad \text{and} \quad \mathbf{u}_2 = \begin{bmatrix} 0 \\ 0 \\ 1 \end{bmatrix}.$$

What is the action of the map, $\mathbf{v}' = P\mathbf{v}$? What is the action of the map on the following vectors:

$$\mathbf{v}_1 = \begin{bmatrix} 1 \\ 1 \\ 1 \end{bmatrix}, \quad \mathbf{v}_2 = \begin{bmatrix} 1 \\ 0 \\ 0 \end{bmatrix}, \quad \mathbf{v}_3 = \begin{bmatrix} 1 \\ -1 \\ 0 \end{bmatrix}?$$

What is the rank of this matrix? What is the determinant?

17. Given the projection matrix in Exercise 16, what is the projection direction?

18. Given the projection matrix

$$A = \begin{bmatrix} 1 & 0 & -1 \\ 0 & 1 & 0 \\ 0 & 0 & 0 \end{bmatrix},$$

what is the projection direction? What type of projection is it?

19. What is the cofactor expansion of

$$A = \begin{bmatrix} 1 & 2 & 3 \\ 2 & 0 & 0 \\ 1 & 0 & 1 \end{bmatrix}?$$

What is $|A|$?

20. For scalar values c_1, c_2, c_3 and a matrix $A = [\mathbf{a}_1 \ \mathbf{a}_2 \ \mathbf{a}_3]$, what is the determinant of $A_1 = [c_1\mathbf{a}_1 \ \mathbf{a}_2 \ \mathbf{a}_3]$, $A_2 = [c_1\mathbf{a}_1 \ c_2\mathbf{a}_2 \ \mathbf{a}_3]$, and $A_3 = [c_1\mathbf{a}_1 \ c_2\mathbf{a}_2 \ c_3\mathbf{a}_3]$?

21. The matrix

$$A = \begin{bmatrix} 1 & 2 & 3 \\ 2 & 0 & 0 \\ 1 & 1 & 1 \end{bmatrix}$$

is invertible and $|A| = 2$. What is $|A^{-1}|$?

22. Compute

$$\begin{bmatrix} 0 & 0 & 1 \\ 1 & -2 & 0 \\ -2 & 1 & 1 \end{bmatrix} \begin{bmatrix} 1 & 5 & -4 \\ -1 & -2 & 0 \\ 2 & 3 & -4 \end{bmatrix}.$$

23. What is AB and BA given

$$A = \begin{bmatrix} 1 & 2 & 3 \\ 2 & 0 & 0 \\ 1 & 1 & 1 \end{bmatrix} \quad \text{and} \quad B = \begin{bmatrix} 0 & 1 & 0 \\ 1 & 1 & 1 \\ 1 & 0 & 0 \end{bmatrix}?$$

24. What is AB and BA given

$$A = \begin{bmatrix} 1 & 2 & 3 \\ 2 & 0 & 0 \\ 1 & 1 & 1 \end{bmatrix} \quad \text{and} \quad B = \begin{bmatrix} 1 & 1 \\ 2 & 2 \\ 0 & 1 \end{bmatrix}?$$

25. Find the inverse for each of the following matrices:

rotation: scale: projection:

$$
\begin{bmatrix} 1/\sqrt{2} & 0 & 1/\sqrt{2} \\ 0 & 1 & 0 \\ -1/\sqrt{2} & 0 & 1/\sqrt{2} \end{bmatrix}, \quad
\begin{bmatrix} 1/2 & 0 & 0 \\ 0 & 1/4 & 0 \\ 0 & 0 & 2 \end{bmatrix}, \quad
\begin{bmatrix} 1 & 0 & -1 \\ 0 & 1 & 0 \\ 0 & 0 & 0 \end{bmatrix}.
$$

26. For what type of matrix is $A^{-1} = A^{\mathrm{T}}$?

27. What is the inverse of the matrix

$$
A = \begin{bmatrix} 5 & 8 \\ 2 & 2 \\ 3 & 4 \end{bmatrix}?
$$

28. If

$$
B = \begin{bmatrix} 0 & 1 & 0 \\ 1 & 1 & 1 \\ 1 & 0 & 0 \end{bmatrix} \quad \text{and} \quad B^{-1} = \begin{bmatrix} 0 & 0 & 1 \\ 1 & 0 & 0 \\ -1 & 1 & -1 \end{bmatrix},
$$

what is $(3B)^{-1}$?

29. For matrix A in Exercise 2, what is $(A^{\mathrm{T}})^{\mathrm{T}}$?

10

Affine Maps in 3D

Figure 10.1.
Affine maps in 3D: fighter jets twisting and turning through 3D space.

Affine maps in 3D are a primary tool for modeling and computer graphics. Figure 10.1 illustrates the use of various affine maps. This chapter goes a little further than just affine maps by introducing projective maps—the maps used to create realistic 3D images.

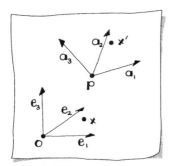

Sketch 10.1.
An affine map in 3D.

10.1 Affine Maps

Linear maps relate vectors to vectors. Affine maps relate points to points. A 3D affine map is written just as a 2D one, namely as

$$\mathbf{x}' = \mathbf{p} + A(\mathbf{x} - \mathbf{o}), \tag{10.1}$$

where $\mathbf{x}, \mathbf{o}, \mathbf{p}, \mathbf{x}'$ are 3D points and A is a 3×3 matrix. In general, we will assume that the origin of $\mathbf{x}$'s coordinate system has three zero coordinates, and drop the $\mathbf{o}$ term:

$$\mathbf{x}' = \mathbf{p} + A\mathbf{x}. \tag{10.2}$$

Sketch 10.1 gives an example. Recall, the column vectors of A are the vectors $\mathbf{a}_1, \mathbf{a}_2, \mathbf{a}_3$. The point $\mathbf{p}$ tells us where to move the origin of the $[\mathbf{e}_1, \mathbf{e}_2, \mathbf{e}_3]$-system; again, the real action of an affine map is captured by the matrix. Thus, by studying matrix actions, or linear maps, we will learn more about affine maps.

We now list some of the important properties of 3D affine maps. They are straightforward generalizations of the 2D cases, and so we just give a brief listing.

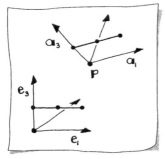

Sketch 10.2.
Affine maps leave ratios invariant. This map is a rigid body motion.

1. Affine maps leave *ratios* invariant (see Sketch 10.2).

2. Affine maps take *parallel planes* to parallel planes (see Figure 10.2).

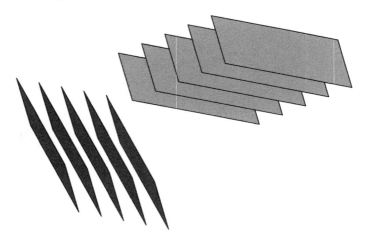

Figure 10.2.
Affine map property: parallel planes get mapped to parallel planes via an affine map.

3. Affine maps take *intersecting planes* to intersecting planes. In particular, the intersection line of the mapped planes is the map of the original intersection line.

4. Affine maps leave *barycentric combinations* invariant. If

$$\mathbf{x} = c_1\mathbf{p}_1 + c_2\mathbf{p}_2 + c_3\mathbf{p}_3 + c_4\mathbf{p}_4,$$

where $c_1 + c_2 + c_3 + c_4 = 1$, then after an affine map we have

$$\mathbf{x}' = c_1\mathbf{p}_1' + c_2\mathbf{p}_2' + c_3\mathbf{p}_3' + c_4\mathbf{p}_4'.$$

For example, the *centroid* of a tetrahedron will be mapped to the centroid of the mapped tetrahedron (see Sketch 10.3).

Most 3D maps do not offer much over their 2D counterparts—but some do. We will go through all of them in detail now.

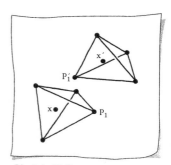

Sketch 10.3.
The centroid is mapped to the centroid.

10.2 Translations

A translation is simply (10.2) with $A = I$, the 3×3 identity matrix:

$$I = \begin{bmatrix} 1 & 0 & 0 \\ 0 & 1 & 0 \\ 0 & 0 & 1 \end{bmatrix},$$

that is

$$\mathbf{x}' = \mathbf{p} + I\mathbf{x}.$$

Thus, the new $[\mathbf{a}_1, \mathbf{a}_2, \mathbf{a}_3]$-system has its coordinate axes parallel to the $[\mathbf{e}_1, \mathbf{e}_2, \mathbf{e}_3]$-system. The term $I\mathbf{x} = \mathbf{x}$ needs to be interpreted as a *vector* in the $[\mathbf{e}_1, \mathbf{e}_2, \mathbf{e}_3]$-system for this to make sense. Figure 10.3 shows an example of repeated 3D translations.

Just as in 2D, a translation is a rigid body motion. The volume of an object is not changed.

10.3 Mapping Tetrahedra

A 3D affine map is determined by four point pairs $\mathbf{p}_i \rightarrow \mathbf{p}_i'$ for $i = 1, 2, 3, 4$. In other words, an affine map is determined by a tetrahedron and its image. What is the image of an arbitrary point $\mathbf{x}$ under this affine map?

Affine maps leave *barycentric combinations* unchanged. This will be the key to finding $\mathbf{x}'$, the image of $\mathbf{x}$. If we can write $\mathbf{x}$ in the form

$$\mathbf{x} = u_1\mathbf{p}_1 + u_2\mathbf{p}_2 + u_3\mathbf{p}_3 + u_4\mathbf{p}_4, \qquad (10.3)$$

Figure 10.3.
Translations in 3D: three translated teapots.

then we know that the image has the same relationship with the $\mathbf{p}'_i$:

$$\mathbf{x}' = u_1\mathbf{p}'_1 + u_2\mathbf{p}'_2 + u_3\mathbf{p}'_3 + u_4\mathbf{p}'_4. \tag{10.4}$$

So all we need to do is find the u_i! These are called the *barycentric coordinates* of $\mathbf{x}$ with respect to the $\mathbf{p}_i$, quite in analogy to the triangle case (see Section 6.5).

We observe that (10.3) is short for three individual coordinate equations. Together with the barycentric combination condition

$$u_1 + u_2 + u_3 + u_4 = 1,$$

we have four equations for the four unknowns $u_1, \ldots, u_4$, which we can solve by consulting Chapter 12.

Example 10.1

Let the original tetrahedron be given by the four points $\mathbf{p}_i$

$$\begin{bmatrix} 0 \\ 0 \\ 0 \end{bmatrix}, \quad \begin{bmatrix} 1 \\ 0 \\ 0 \end{bmatrix}, \quad \begin{bmatrix} 0 \\ 1 \\ 0 \end{bmatrix}, \quad \begin{bmatrix} 0 \\ 0 \\ 1 \end{bmatrix}.$$

Let's assume we want to map this tetrahedron to the four points $\mathbf{p}_i'$

$$\begin{bmatrix} 0 \\ 0 \\ 0 \end{bmatrix}, \quad \begin{bmatrix} -1 \\ 0 \\ 0 \end{bmatrix}, \quad \begin{bmatrix} 0 \\ -1 \\ 0 \end{bmatrix}, \quad \begin{bmatrix} 0 \\ 0 \\ -1 \end{bmatrix}.$$

This is a pretty straightforward map if you consult Sketch 10.4.

Let's see where the point

$$\mathbf{x} = \begin{bmatrix} 1 \\ 1 \\ 1 \end{bmatrix}$$

ends up. First, we find that

$$\begin{bmatrix} 1 \\ 1 \\ 1 \end{bmatrix} = -2 \begin{bmatrix} 0 \\ 0 \\ 0 \end{bmatrix} + \begin{bmatrix} 1 \\ 0 \\ 0 \end{bmatrix} + \begin{bmatrix} 0 \\ 1 \\ 0 \end{bmatrix} + \begin{bmatrix} 0 \\ 0 \\ 1 \end{bmatrix},$$

i.e., the barycentric coordinates of $\mathbf{x}$ with respect to the original $\mathbf{p}_i$ are $(-2, 1, 1, 1)$. Note how they sum to one. Now it is simple to compute the image of $\mathbf{x}$; compute $\mathbf{x}'$ using the same barycentric coordinates with respect to the $\mathbf{p}_i'$:

$$\mathbf{x}' = -2 \begin{bmatrix} 0 \\ 0 \\ 0 \end{bmatrix} + \begin{bmatrix} -1 \\ 0 \\ 0 \end{bmatrix} + \begin{bmatrix} 0 \\ -1 \\ 0 \end{bmatrix} + \begin{bmatrix} 0 \\ 0 \\ -1 \end{bmatrix} = \begin{bmatrix} -1 \\ -1 \\ -1 \end{bmatrix}.$$

A different approach would be to find the 3×3 matrix A and point $\mathbf{p}$ that describe the affine map. Construct a coordinate system from the $\mathbf{p}_i$ tetrahedron. One way to do this is to choose $\mathbf{p}_1$ as the origin[1] and the three axes are defined as $\mathbf{p}_i - \mathbf{p}_1$ for $i = 2, 3, 4$. The coordinate system of the $\mathbf{p}_i'$ tetrahedron must be based on the same indices. Once we have defined A and $\mathbf{p}$ then we will be able to map $\mathbf{x}$ by this map:

$$\mathbf{x}' = A[\mathbf{x} - \mathbf{p}_1] + \mathbf{p}_1'.$$

Thus, the point $\mathbf{p} = \mathbf{p}_1'$. In order to determine A, let's write down some known relationships. Referring to Sketch 10.5, we know

$$A[\mathbf{p}_2 - \mathbf{p}_1] = \mathbf{p}_2' - \mathbf{p}_1',$$
$$A[\mathbf{p}_3 - \mathbf{p}_1] = \mathbf{p}_3' - \mathbf{p}_1',$$
$$A[\mathbf{p}_4 - \mathbf{p}_1] = \mathbf{p}_4' - \mathbf{p}_1',$$

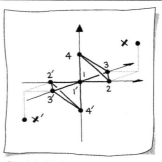

Sketch 10.4.
An example tetrahedron map.

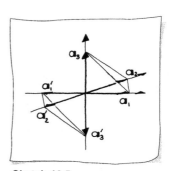

Sketch 10.5.
The relationship between tetrahedra.

---

[1] Any of the four $\mathbf{p}_i$ would do, so for the sake of concreteness, we choose the first one.

which may be written in matrix form as

$$A \begin{bmatrix} \mathbf{p}_2 - \mathbf{p}_1 & \mathbf{p}_3 - \mathbf{p}_1 & \mathbf{p}_4 - \mathbf{p}_1 \end{bmatrix} = \begin{bmatrix} \mathbf{p}'_2 - \mathbf{p}'_1 & \mathbf{p}'_3 - \mathbf{p}'_1 & \mathbf{p}'_4 - \mathbf{p}'_1 \end{bmatrix}.$$
(10.5)

Thus,

$$A = \begin{bmatrix} \mathbf{p}'_2 - \mathbf{p}'_1 & \mathbf{p}'_3 - \mathbf{p}'_1 & \mathbf{p}'_4 - \mathbf{p}'_1 \end{bmatrix} \begin{bmatrix} \mathbf{p}_2 - \mathbf{p}_1 & \mathbf{p}_3 - \mathbf{p}_1 & \mathbf{p}_4 - \mathbf{p}_1 \end{bmatrix}^{-1},$$
(10.6)

and A is defined.

Example 10.2

Revisiting Example 10.1, we now want to construct the matrix A. By selecting $\mathbf{p}_1$ as the origin for the $\mathbf{p}_i$ tetrahedron coordinate system there is no translation; $\mathbf{p}_1$ is the origin in the $[\mathbf{e}_1, \mathbf{e}_2, \mathbf{e}_3]$-system and $\mathbf{p}'_1 = \mathbf{p}_1$. We now compute A. (A is the product matrix in the bottom right position):

$$
\begin{array}{ccc|ccc}
 & & & 1 & 0 & 0 \\
 & & & 0 & 1 & 0 \\
 & & & 0 & 0 & 1 \\
\hline
-1 & 0 & 0 & -1 & 0 & 0 \\
0 & -1 & 0 & 0 & -1 & 0 \\
0 & 0 & -1 & 0 & 0 & -1
\end{array}
$$

In order to compute $\mathbf{x}'$, we have

$$\mathbf{x}' = \begin{bmatrix} -1 & 0 & 0 \\ 0 & -1 & 0 \\ 0 & 0 & -1 \end{bmatrix} \begin{bmatrix} 1 \\ 1 \\ 1 \end{bmatrix} = \begin{bmatrix} -1 \\ -1 \\ -1 \end{bmatrix}.$$

This is the same result as in Example 10.1.

10.4 Parallel Projections

We looked at orthogonal parallel projections as basic linear maps in Sections 4.8 and 9.7. Everything we draw is a projection of necessity—paper is 2D, after all, whereas most interesting objects are 3D. Figure 10.4 gives an example. Here we will look at projections in the context of 3D affine maps that map 3D points onto a plane.

Figure 10.4.
Projections in 3D: a 3D helix is projected into two different 2D planes.

As illustrated in Sketch 10.6, a parallel projection is defined by a *direction* of projection $\mathbf{d}$ and a *projection plane* P. A point $\mathbf{x}$ is projected into P, and is represented as $\mathbf{x}_p$ in the sketch. This information in turn defines a *projection angle* θ between $\mathbf{d}$ and the line joining the perpendicular projection point $\mathbf{x}_o$ in P. This angle is used to categorize parallel projections as *orthogonal* or *oblique*.

Orthogonal (also called orthographic) projections are special; their projection direction is perpendicular to the plane. There are special names for many particular projection angles; see a computer graphics text such as [10] for more details.

Let $\mathbf{x}$ be the 3D point to be projected, let $\mathbf{v}$ indicate the projection direction, and the projection plane is defined by point $\mathbf{q}$ and normal $\mathbf{n}$, as illustrated in Sketch 10.7. For some point $\mathbf{x}'$ in the plane, the plane equation is $[\mathbf{x}' - \mathbf{q}] \cdot \mathbf{n} = 0$. The intersection point is on the line defined by $\mathbf{x}$ and $\mathbf{v}$ and it is given by $\mathbf{x}' = \mathbf{p} + t\mathbf{v}$. We need to find t, and this is achieved by inserting the line equation into the plane

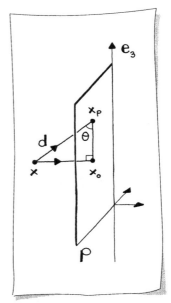

Sketch 10.6.
Oblique and orthographic parallel projections.

equation and solving for t,

$$[\mathbf{x} + t\mathbf{v} - \mathbf{q}] \cdot \mathbf{n} = 0,$$
$$[\mathbf{x} - \mathbf{q}] \cdot \mathbf{n} + t\mathbf{v} \cdot \mathbf{n} = 0,$$
$$t = \frac{[\mathbf{q} - \mathbf{x}] \cdot \mathbf{n}}{\mathbf{v} \cdot \mathbf{n}}.$$

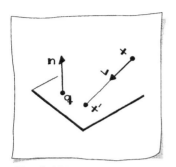

Sketch 10.7.
Projecting a point on a plane.

The intersection point $\mathbf{x}$ is now computed as

$$\mathbf{x}' = \mathbf{x} + \frac{[\mathbf{q} - \mathbf{x}] \cdot \mathbf{n}}{\mathbf{v} \cdot \mathbf{n}}\mathbf{v}. \tag{10.7}$$

How do we write (10.7) as an affine map in the form $A\mathbf{x} + \mathbf{p}$? Without much effort, we find

$$\mathbf{x}' = \mathbf{x} - \frac{\mathbf{n} \cdot \mathbf{x}}{\mathbf{v} \cdot \mathbf{n}}\mathbf{v} + \frac{\mathbf{q} \cdot \mathbf{n}}{\mathbf{v} \cdot \mathbf{n}}\mathbf{v}.$$

We know that we may write dot products in matrix form (see Section 4.11):

$$\mathbf{x}' = \mathbf{x} - \frac{\mathbf{n}^{\mathrm{T}}\mathbf{x}}{\mathbf{v} \cdot \mathbf{n}}\mathbf{v} + \frac{\mathbf{q} \cdot \mathbf{n}}{\mathbf{v} \cdot \mathbf{n}}\mathbf{v}.$$

Next, we observe that

$$\left[\mathbf{n}^{\mathrm{T}}\mathbf{x}\right]\mathbf{v} = \mathbf{v}\left[\mathbf{n}^{\mathrm{T}}\mathbf{x}\right].$$

Since matrix multiplication is associative (see Section 4.12), we also have

$$\mathbf{v}\left[\mathbf{n}^{\mathrm{T}}\mathbf{x}\right] = \left[\mathbf{v}\mathbf{n}^{\mathrm{T}}\right]\mathbf{x}.$$

Notice that $\mathbf{v}\mathbf{n}^{\mathrm{T}}$ is a 3×3 matrix. Now we can write

$$\mathbf{x}' = \left[I - \frac{\mathbf{v}\mathbf{n}^{\mathrm{T}}}{\mathbf{v} \cdot \mathbf{n}}\right]\mathbf{x} + \frac{\mathbf{q} \cdot \mathbf{n}}{\mathbf{v} \cdot \mathbf{n}}\mathbf{v}, \tag{10.8}$$

where I is the 3×3 identity matrix. This is of the form $\mathbf{x}' = A\mathbf{x} + \mathbf{p}$ and hence is an affine map.[2]

Let's check the properties of (10.8). The projection matrix A, formed from $\mathbf{v}$ and $\mathbf{n}$ has rank two and thus reduces dimensionality, as designed. From the derivation of projection, it is intuitively clear

---

[2]Technically, we should replace $\mathbf{x}$ with $\mathbf{x} - \mathbf{o}$ to have a vector and replace $\alpha\mathbf{v}$ with $\alpha\mathbf{v} + \mathbf{o}$ to have a point, where $\alpha = (\mathbf{q} \cdot \mathbf{n})/(\mathbf{v} \cdot \mathbf{n})$.

that once $\mathbf{x}$ has been mapped into the projection plane, to $\mathbf{x}'$, it will remain there. We can also show the map is idempotent algebraically,

$$
\begin{aligned}
A^2 &= \left(I - \frac{\mathbf{vn}^{\mathrm{T}}}{\mathbf{vn}}\right)\left(I - \frac{\mathbf{vn}^{\mathrm{T}}}{\mathbf{vn}}\right) \\
&= I^2 - 2\frac{\mathbf{vn}^{\mathrm{T}}}{\mathbf{vn}} + \left(\frac{\mathbf{vn}^{\mathrm{T}}}{\mathbf{vn}}\right)^2 \\
&= A - \frac{\mathbf{vn}^{\mathrm{T}}}{\mathbf{vn}} + \left(\frac{\mathbf{vn}^{\mathrm{T}}}{\mathbf{vn}}\right)^2
\end{aligned}
$$

Expanding the squared term, we find that

$$
\frac{\mathbf{vn}^{\mathrm{T}}\mathbf{vn}^{\mathrm{T}}}{(\mathbf{v}\cdot\mathbf{n})^2} = \frac{\mathbf{vn}^{\mathrm{T}}}{\mathbf{v}\cdot\mathbf{n}}
$$

and thus $A^2 = A$. We can also show that repeating the affine map is idempotent as well:

$$
\begin{aligned}
A(A\mathbf{x} + \mathbf{p}) + \mathbf{p} &= A^2\mathbf{x} + A\mathbf{p} + \mathbf{p} \\
&= A\mathbf{x} + A\mathbf{p} + \mathbf{p}.
\end{aligned}
$$

Let $\alpha = (\mathbf{q}\cdot\mathbf{n})/(\mathbf{v}\cdot\mathbf{n})$, and examining the middle term,

$$
\begin{aligned}
A\mathbf{p} &= \left(I - \frac{\mathbf{vn}^{\mathrm{T}}}{\mathbf{vn}}\right)\alpha\mathbf{v} \\
&= \alpha\mathbf{v} - \alpha\mathbf{v}\left(\frac{\mathbf{n}^{\mathrm{T}}\mathbf{v}}{\mathbf{vn}}\right) \\
&= 0
\end{aligned}
$$

Therefore, $A(A\mathbf{x} + \mathbf{p}) + \mathbf{p} = A\mathbf{x} + \mathbf{p}$, and we have shown that indeed, the affine map is idempotent.

Example 10.3

Suppose we are given the projection plane $x_1 + x_2 + x_3 - 1 = 0$, a point $\mathbf{x}$ (not in the plane), and a direction $\mathbf{v}$ given by

$$
\mathbf{x} = \begin{bmatrix} 3 \\ 2 \\ 4 \end{bmatrix} \quad \text{and} \quad \mathbf{v} = \begin{bmatrix} 0 \\ 0 \\ -1 \end{bmatrix}.
$$

If we project $\mathbf{x}$ along $\mathbf{v}$ onto the plane, what is $\mathbf{x}'$? Sketch 10.8 illustrates this geometry. First, we need the plane's normal direction. Calling it $\mathbf{n}$, we have

$$\mathbf{n} = \begin{bmatrix} 1 \\ 1 \\ 1 \end{bmatrix}.$$

Now, choose a point $\mathbf{q}$ in the plane. Let's choose

$$\mathbf{q} = \begin{bmatrix} 1 \\ 0 \\ 0 \end{bmatrix}$$

for simplicity. Now we are ready to calculate the quantities in (10.8):

$$\mathbf{v} \cdot \mathbf{n} = -1,$$

$$\mathbf{v}\mathbf{n}^{\mathrm{T}} = \begin{array}{c|ccc} & 1 & 1 & 1 \\ \hline 0 & 0 & 0 & 0 \\ 0 & 0 & 0 & 0 \\ -1 & -1 & -1 & -1 \end{array},$$

$$\frac{\mathbf{q} \cdot \mathbf{n}}{\mathbf{v} \cdot \mathbf{n}}\mathbf{v} = \begin{bmatrix} 0 \\ 0 \\ 1 \end{bmatrix}.$$

Putting all the pieces together:

$$\mathbf{x}' = \left[I - \begin{bmatrix} 0 & 0 & 0 \\ 0 & 0 & 0 \\ 1 & 1 & 1 \end{bmatrix} \right] \begin{bmatrix} 3 \\ 2 \\ 4 \end{bmatrix} + \begin{bmatrix} 0 \\ 0 \\ 1 \end{bmatrix} = \begin{bmatrix} 3 \\ 2 \\ -4 \end{bmatrix}.$$

Just to double-check, enter $\mathbf{x}'$ into the plane equation

$$3 + 2 - 4 - 1 = 0,$$

and we see that

$$\begin{bmatrix} 3 \\ 2 \\ -4 \end{bmatrix} = \begin{bmatrix} 3 \\ 2 \\ 4 \end{bmatrix} + 8 \begin{bmatrix} 0 \\ 0 \\ -1 \end{bmatrix},$$

which together verify that this is the correct point. Sketch 10.8 should convince you that this is indeed the correct answer.

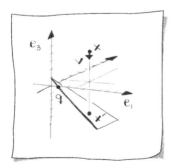

Sketch 10.8.
A projection example.

Which of the two possibilities, (10.7) or the affine map (10.8) should you use? Clearly, (10.7) is more straightforward and less involved. Yet in some computer graphics or CAD system environments, it may be desirable to have all maps in a unified format, i.e., $A\mathbf{x} + \mathbf{p}$. We'll revisit this unified format idea in Section 10.5.

10.5   Homogeneous Coordinates and Perspective Maps

There is a way to condense the form $\mathbf{x}' = A\mathbf{x} + \mathbf{p}$ of an affine map into just one matrix multiplication

$$\underline{\mathbf{x}}' = M\underline{\mathbf{x}}. \tag{10.9}$$

This is achieved by setting

$$M = \begin{bmatrix} a_{1,1} & a_{1,2} & a_{1,3} & p_1 \\ a_{2,1} & a_{2,2} & a_{2,3} & p_2 \\ a_{3,1} & a_{3,2} & a_{3,3} & p_3 \\ 0 & 0 & 0 & 1 \end{bmatrix}$$

and

$$\underline{\mathbf{x}} = \begin{bmatrix} x_1 \\ x_2 \\ x_3 \\ 1 \end{bmatrix}, \quad \underline{\mathbf{x}}' = \begin{bmatrix} x_1' \\ x_2' \\ x_3' \\ 1 \end{bmatrix}.$$

The 4D point $\underline{\mathbf{x}}$ is called the *homogeneous form* of the affine point $\mathbf{x}$. You should verify for yourself that (10.9) is indeed the same affine map as before.

The homogeneous representation of a vector $\underline{\mathbf{v}}$ must have the form,

$$\underline{\mathbf{v}} = \begin{bmatrix} v_1 \\ v_2 \\ v_3 \\ 0 \end{bmatrix}.$$

This form allows us to apply the linear map to the vector,

$$\underline{\mathbf{v}}' = M\underline{\mathbf{v}}.$$

By having a zero fourth component, we disregard the translation, which we know has no effect on vectors. Recall that a vector is defined as the difference of two points.

This method of condensing transfomation information into one matrix is implemented in the popular computer graphics *Application*

Programmer's Interface (*API*), OpenGL [15]. It is very convenient and efficient to have all this information (plus more, as we will see), in one data structure.

The homogeneous form is more general than just adding a fourth coordinate $x_4 = 1$ to a point. If, perhaps as the result of some computation, the fourth coordinate does not equal one, one gets from the homogeneous point $\underline{\mathbf{x}}$ to its affine counterpart $\mathbf{x}$ by dividing through by x_4. Thus, one affine point has infinitely many homogeneous representations!

Example 10.4

This example shows two homogeneous representations of one affine point. (The symbol $\approx$ should be read "corresponds to.")

$$\begin{bmatrix} 1 \\ -1 \\ 3 \end{bmatrix} \approx \begin{bmatrix} 10 \\ -10 \\ 30 \\ 10 \end{bmatrix} \approx \begin{bmatrix} -2 \\ 2 \\ -6 \\ -2 \end{bmatrix}.$$

Using the homogeneous matrix form of (10.9), the matrix M for the point into a plane projection from (10.8) becomes

$\begin{bmatrix} \mathbf{v} \cdot \mathbf{n} & 0 & 0 \\ 0 & \mathbf{v} \cdot \mathbf{n} & 0 \\ 0 & 0 & \mathbf{v} \cdot \mathbf{n} \end{bmatrix} - \mathbf{v}\mathbf{n}^{\mathrm{T}}$	$(\mathbf{q} \cdot \mathbf{n})\mathbf{v}$
$\quad\quad 0 \quad\quad 0 \quad\quad 0$	$\mathbf{v} \cdot \mathbf{n}$

Here, the element $m_{4,4} = \mathbf{v} \cdot \mathbf{n}$. Thus, $\underline{x}_4 = \mathbf{v} \cdot \mathbf{n}$, and we will have to divide $\underline{\mathbf{x}}$'s coordinates by $\underline{x}_4$ in order to obtain the corresponding affine point.

A simple change in our equations will lead us from parallel projections onto a plane to *perspective projections*. Instead of using a constant direction $\mathbf{v}$ for all projections, now the direction depends on the point $\mathbf{x}$. More precisely, let it be the line from $\mathbf{x}$ to the origin of our coordinate system. Then, as shown in Sketch 10.9, $\mathbf{v} = -\mathbf{x}$, and (10.7) becomes

$$\mathbf{x}' = \mathbf{x} + \frac{[\mathbf{q} - \mathbf{x}] \cdot \mathbf{n}}{\mathbf{x} \cdot \mathbf{n}}\mathbf{x},$$

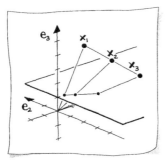

Sketch 10.9.

Perspective projection.

which quickly simplifies to

$$\mathbf{x}' = \frac{\mathbf{q} \cdot \mathbf{n}}{\mathbf{x} \cdot \mathbf{n}} \mathbf{x}. \tag{10.10}$$

In homogeneous form, this is described by the following matrix

$$M : \begin{array}{|c|c|} \hline I[\mathbf{q} \cdot \mathbf{n}] & \mathbf{o} \\ \hline 0 \quad\quad 0 \quad\quad 0 & \mathbf{x} \cdot \mathbf{n} \\ \hline \end{array}.$$

Perspective projections are not affine maps anymore! To see this, a simple example will suffice.

Example 10.5

Take the plane $x_3 = 1$; let

$$\mathbf{q} = \begin{bmatrix} 0 \\ 0 \\ 1 \end{bmatrix}$$

be a point on the plane. Now $\mathbf{q} \cdot \mathbf{n} = 1$ and $\mathbf{x} \cdot \mathbf{n} = x_3$, resulting in the map

$$\mathbf{x}' = \frac{1}{x_3} \mathbf{x}.$$

Take the three points

$$\mathbf{x}_1 = \begin{bmatrix} 2 \\ 0 \\ 4 \end{bmatrix}, \quad \mathbf{x}_2 = \begin{bmatrix} 3 \\ -1 \\ 3 \end{bmatrix}, \quad \mathbf{x}_3 = \begin{bmatrix} 4 \\ -2 \\ 2 \end{bmatrix}.$$

This example is illustrated in Sketch 10.9. Note that $\mathbf{x}_2 = \frac{1}{2}\mathbf{x}_1 + \frac{1}{2}\mathbf{x}_3$, i.e., $\mathbf{x}_2$ is the midpoint of $\mathbf{x}_1$ and $\mathbf{x}_3$.

Their images are

$$\mathbf{x}_1' = \begin{bmatrix} 1/2 \\ 0 \\ 1 \end{bmatrix}, \quad \mathbf{x}_2' = \begin{bmatrix} 1 \\ -1/3 \\ 1 \end{bmatrix}, \quad \mathbf{x}_3' = \begin{bmatrix} 2 \\ -1 \\ 1 \end{bmatrix}.$$

The perspective map destroyed the midpoint relation! Now,

$$\mathbf{x}_2' = \frac{2}{3}\mathbf{x}_1' + \frac{1}{3}\mathbf{x}_3'.$$

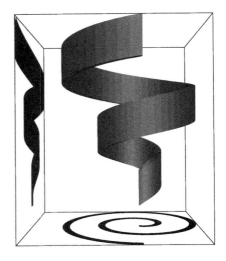

Figure 10.5.
Parallel projection: a 3D helix and two orthographic projections on the left and bottom walls of the bounding cube—not visible due to the orthographic projection used for the whole scene.

Figure 10.6.
Perspective projection: a 3D helix and two orthographic projections on the left and bottom walls of the bounding cube—visible due to the perspective projection used for the whole scene.

Thus, the ratio of three points is changed by perspective maps. As a consequence, two parallel lines will not be mapped to parallel lines. Because of this effect, perspective maps are a good model for how we perceive 3D space around us. Parallel lines do seemingly intersect in a distance, and are thus not *perceived* as being parallel! Figure 10.5 is a parallel projection and Figure 10.6 illustrates the same geometry with a perspective projection. Notice in the perspective image, the sides of the bounding cube that move into the page are no longer parallel.

As we saw above, $m_{4,4}$ allows us to specify perspective projections. The other elements of the bottom row of M are used for *projective maps*, a more general mapping than a perspective projection. The topic of this chapter is affine maps, so we'll leave a detailed discussion of these elements to another source: A mathematical treatment of this map is supplied by [6] and a computer graphics treatment is supplied by [14]. In short, these entries are used in computer graphics for mapping a *viewing volume*[3] in the shape of a frustum to one in the shape of a cube, while preserving the perspective projection effect.

---

[3] The dimension and shape of the viewing volume defines *what* will be displayed and *how* it will be displayed (orthographic or perspective).

Figure 10.7.
Perspective maps: an experiment by A. Dürer.

Algorithms in the graphics pipeline are very much simplified by only dealing with geometry known to be in a cube.

The study of perspective goes back to the fourteenth century— before that, artists simply could not draw realistic 3D images. One of the foremost researchers in the area of perspective maps was A. Dürer.[4] See Figure 10.7 for one of his experiments.

---

[4]From *The Complete Woodcuts of Albrecht Dürer*, edited by W. Durth, Dover Publications Inc., New York, 1963.

- affine map
- translation
- affine map properties
- barycentric combination
- invariant ratios
- barycentric coordinates
- centroid
- mapping four points to four points

- parallel projection
- orthogonal projection
- oblique projection
- line and plane intersection
- idempotent
- dyadic matrix
- homogeneous coordinates
- perspective projection
- rank

10.6   Exercises

We'll use four points

$$\mathbf{x}_1 = \begin{bmatrix} 1 \\ 0 \\ 0 \end{bmatrix}, \quad \mathbf{x}_2 = \begin{bmatrix} 0 \\ 1 \\ 0 \end{bmatrix}, \quad \mathbf{x}_3 = \begin{bmatrix} 0 \\ 0 \\ -1 \end{bmatrix}, \quad \mathbf{x}_4 = \begin{bmatrix} 0 \\ 0 \\ 1 \end{bmatrix},$$

four points

$$\mathbf{y}_1 = \begin{bmatrix} -1 \\ 0 \\ 0 \end{bmatrix}, \quad \mathbf{y}_2 = \begin{bmatrix} 0 \\ -1 \\ 0 \end{bmatrix}, \quad \mathbf{y}_3 = \begin{bmatrix} 0 \\ 0 \\ -1 \end{bmatrix}, \quad \mathbf{y}_4 = \begin{bmatrix} 0 \\ 0 \\ 1 \end{bmatrix},$$

and also the plane through $\mathbf{q}$ with normal $\mathbf{n}$:

$$\mathbf{q} = \begin{bmatrix} 1 \\ 0 \\ 0 \end{bmatrix}, \quad \mathbf{n} = \frac{1}{5} \begin{bmatrix} 3 \\ 0 \\ 4 \end{bmatrix}.$$

1. What are the two parts of an affine map?

2. An affine map $\mathbf{x}_i \rightarrow \mathbf{y}_i; i = 1, 2, 3, 4$ is uniquely defined. What is it?

3. What is the image of

$$\mathbf{p} = \begin{bmatrix} 1 \\ 1 \\ 1 \end{bmatrix}$$

   under the map from Exercise 2? Use two ways to compute it.

4. What are the geometric properties of the affine map from Exercises 2 and 3?

5. An affine map $\mathbf{y}_i \rightarrow \mathbf{x}_i; i = 1, 2, 3, 4$ is uniquely defined. What is it?

6. Using a direction

$$\mathbf{v} = \frac{1}{4} \begin{bmatrix} 2 \\ 0 \\ 2 \end{bmatrix},$$

what are the images of the $\mathbf{x}_i$ when projected in this direction onto the plane defined at the beginning of the exercises?

7. Using the same $\mathbf{v}$ as in Exercise 6, what are the images of the $\mathbf{y}_i$?

8. What are the images of the $\mathbf{x}_i$ when projected onto the plane by a perspective projection through the origin?

9. What are the images of the $\mathbf{y}_i$ when projected onto the plane by a perspective projection through the origin?

10. Compute the centroid $\mathbf{c}$ of the $\mathbf{x}_i$ and then the centroid $\mathbf{c}'$ of their perspective images (from Exercise 8). Is $\mathbf{c}'$ the image of $\mathbf{c}$ under the perspective map?

11. We claimed that (10.8) reduces to (10.10). This necessitates that

$$\left[I - \frac{\mathbf{v}\mathbf{n}^{\mathrm{T}}}{\mathbf{n} \cdot \mathbf{v}} \right] \mathbf{x} = \mathbf{0}.$$

Show that this is indeed true.

12. What is the affine map that rotates the point $\mathbf{q}$ (defined above) $90°$ about the line defined as

$$\mathbf{l}(t) = \begin{bmatrix} 1 \\ 1 \\ 0 \end{bmatrix} + t \begin{bmatrix} 1 \\ 0 \\ 0 \end{bmatrix}?$$

Hint: This is a simple construction, and does not require (9.10).

13. Suppose we have the unit cube with "lower-left" and "upper-right" points

$$\mathbf{l} = \begin{bmatrix} 0 \\ 0 \\ 0 \end{bmatrix} \quad \text{and} \quad \mathbf{u} = \begin{bmatrix} 1 \\ 1 \\ 1 \end{bmatrix},$$

respectively. What is the affine map that scales this cube uniformly by two, rotates it $-45°$ around the $\mathbf{e}_2$-axis, and then positions it so that $\mathbf{l}$ is mapped to

$$\mathbf{l}' = \begin{bmatrix} -2 \\ -2 \\ -2 \end{bmatrix}?$$

What is $\mathbf{u}'$?

14. Suppose we have a diamond-shaped geometric figure defined by the following vertices,

$$\mathbf{v}_1 = \begin{bmatrix} 1 \\ 2 \\ 2 \end{bmatrix}, \quad \mathbf{v}_2 = \begin{bmatrix} 2 \\ 1 \\ 2 \end{bmatrix}, \quad \mathbf{v}_3 = \begin{bmatrix} 3 \\ 2 \\ 2 \end{bmatrix}, \quad \mathbf{v}_4 = \begin{bmatrix} 2 \\ 3 \\ 2 \end{bmatrix},$$

that is positioned in the $x_3 = 2$ plane. We want to rotate the diamond about its centroid $\mathbf{c}$ (with a positive angle) so it is positioned in the $x_1 = c_1$ plane. What is the affine map that achieves this and what are the mapped vertices $\mathbf{v}_i'$?

11

Interactions in 3D

Figure 11.1.
Ray tracing: 3D intersections are a key element of rendering a ray-traced image.
(Courtesy of Ben Steinberg, Arizona State University.)

The tools of points, lines, and planes are our most basic 3D geometry
building blocks. But in order to build real objects, we must be able
to compute with these building blocks. For example, if we are given

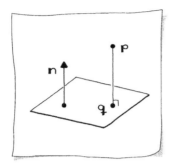

Sketch 11.1.

A point and a plane.

a plane and a straight line, we would be interested in the intersection of those two objects. This chapter outlines the basic algorithms for these types of problems. The ray traced image of Figure 11.1 was generated by using the tools developed in this chapter.[1]

11.1 Distance Between a Point and a Plane

Let a plane be given by its implicit form $\mathbf{n} \cdot \mathbf{x} + c = 0$. If we also have a point $\mathbf{p}$, what is its distance d to the plane, and what is its closest point $\mathbf{q}$ on the plane? See Sketch 11.1 for the geometry. Notice how close this problem is to the foot of a point from Section 3.7.

Clearly, the vector $\mathbf{p} - \mathbf{q}$ must be perpendicular to the plane, i.e., parallel to the plane's normal $\mathbf{n}$. Thus, $\mathbf{p}$ can be written as

$$\mathbf{p} = \mathbf{q} + t\mathbf{n};$$

if we find t, our problem is solved. This is easy, since $\mathbf{q}$ must also satisfy the plane equation:

$$\mathbf{n} \cdot [\mathbf{p} - t\mathbf{n}] + c = 0.$$

Thus,

$$t = \frac{c + \mathbf{n} \cdot \mathbf{p}}{\mathbf{n} \cdot \mathbf{n}}. \tag{11.1}$$

It is good practice to assure that $\mathbf{n}$ is normalized, i.e., $\mathbf{n} \cdot \mathbf{n} = 1$, and then

$$t = c + \mathbf{n} \cdot \mathbf{p}. \tag{11.2}$$

Note that $t = 0$ is equivalent to $\mathbf{n} \cdot \mathbf{p} + c = 0$; in that case, $\mathbf{p}$ is on the plane to begin with!

Example 11.1

Consider the plane

$$x_1 + x_2 + x_3 - 1 = 0$$

and the point

$$\mathbf{p} = \begin{bmatrix} 2 \\ 2 \\ 3 \end{bmatrix},$$

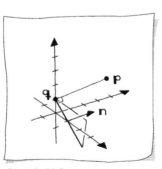

Sketch 11.2.

Example point and a plane.

as shown in Sketch 11.2.

---

[1]See Section 11.3 for a description of the technique used to generate this image.

According to (11.1), we find $t = 2$. Thus,

$$\mathbf{q} = \begin{bmatrix} 2 \\ 2 \\ 3 \end{bmatrix} - 2 \times \begin{bmatrix} 1 \\ 1 \\ 1 \end{bmatrix} = \begin{bmatrix} 0 \\ 0 \\ 1 \end{bmatrix}.$$

The vector $\mathbf{p} - \mathbf{q}$ is given by

$$\mathbf{p} - \mathbf{q} = t\mathbf{n}.$$

Thus, the length of $\mathbf{p} - \mathbf{q}$, or the *distance* of $\mathbf{p}$ to the plane, is given by $t\|\mathbf{n}\|$. If $\mathbf{n}$ is normalized, then $\|\mathbf{p} - \mathbf{q}\| = t$; this means that we simply insert $\mathbf{p}$ into the plane equation and obtain for the distance d:

$$d = c + \mathbf{n} \cdot \mathbf{p}.$$

It is also clear that if $t > 0$, then $\mathbf{n}$ points toward $\mathbf{p}$, and away from it if $t < 0$ (see Sketch 11.3) where the plane is drawn "edge on." Compare with the almost identical Sketch 3.3.

Again, a numerical caveat: If a point is very close to a plane, it becomes very hard numerically to decide which side it is on.

The material in this section should look familiar. We have developed similar equations for lines in Sections 3.3 and 3.6, and the distance of a point to a plane was introduced in Section 8.4.

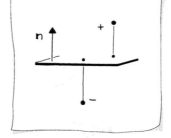

Sketch 11.3.
Points around a plane.

11.2 Distance Between Two Lines

Two 3D lines typically do not meet—such lines are called *skew*. It might be of interest to know how close they are to meeting; in other words, what is the *distance* between the lines? See Sketch 11.4 for an illustration.

Let the two lines $\mathbf{l}_1$ and $\mathbf{l}_2$ be given by

$$\mathbf{x}_1(s_1) = \mathbf{p}_1 + s_1\mathbf{v}_1, \quad \text{and}$$
$$\mathbf{x}_2(s_2) = \mathbf{p}_2 + s_2\mathbf{v}_2,$$

respectively. Let $\mathbf{x}_1$ be the point on $\mathbf{l}_1$ closest to $\mathbf{l}_2$, also let $\mathbf{x}_2$ be the point on $\mathbf{l}_2$ closest to $\mathbf{l}_1$. It should be clear that the vector $\mathbf{x}_2 - \mathbf{x}_1$ is perpendicular to both $\mathbf{l}_1$ and $\mathbf{l}_2$. Thus

$$[\mathbf{x}_2 - \mathbf{x}_1]\mathbf{v}_1 = 0,$$
$$[\mathbf{x}_2 - \mathbf{x}_1]\mathbf{v}_2 = 0,$$

or

$$[\mathbf{p}_2 - \mathbf{p}_1]\mathbf{v}_1 = s_1\mathbf{v}_1 \cdot \mathbf{v}_1 - s_2\mathbf{v}_1 \cdot \mathbf{v}_2,$$
$$[\mathbf{p}_2 - \mathbf{p}_1]\mathbf{v}_2 = s_1\mathbf{v}_1 \cdot \mathbf{v}_2 - s_2\mathbf{v}_2 \cdot \mathbf{v}_2.$$

These are two equations in the two unknowns s_1 and s_2, and are thus readily solved using the methods from Chapter 5.

Example 11.2

Let $\mathbf{l}_1$ be given by

$$\mathbf{x}_1(s_1) = \begin{bmatrix} 0 \\ 0 \\ 0 \end{bmatrix} + s_1 \begin{bmatrix} 1 \\ 0 \\ 0 \end{bmatrix}.$$

This means, of course, that $\mathbf{l}_1$ is the $\mathbf{e}_1$-axis. For $\mathbf{l}_2$, we assume

$$\mathbf{x}_2(s_2) = \begin{bmatrix} 0 \\ 1 \\ 1 \end{bmatrix} + s_2 \begin{bmatrix} 0 \\ 1 \\ 0 \end{bmatrix}.$$

This line is parallel to the $\mathbf{e}_2$-axis; both lines are shown in Sketch 11.4. Our linear system becomes

$$\begin{bmatrix} 1 & 0 \\ 0 & -1 \end{bmatrix} \begin{bmatrix} s_1 \\ s_2 \end{bmatrix} = \begin{bmatrix} 0 \\ 1 \end{bmatrix}$$

with solution $s_1 = 0$ and $s_2 = -1$. Inserting these values, we have

$$\mathbf{x}_1(0) = \begin{bmatrix} 0 \\ 0 \\ 0 \end{bmatrix} \quad \text{and} \quad \mathbf{x}_2(-1) = \begin{bmatrix} 0 \\ 0 \\ 1 \end{bmatrix}.$$

These are the two points of closest proximity; the distance between the lines is $\|\mathbf{x}_1(0) - \mathbf{x}_2(-1)\| = 1$.

Two 3D lines intersect if the two points $\mathbf{x}_1$ and $\mathbf{x}_2$ are identical. However, floating point calculations can introduce round-off error, so it is necessary to accept closeness within a tolerance: $\|\mathbf{x}_1 - \mathbf{x}_2\|^2 <$ tolerance.

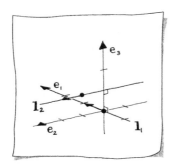

Sketch 11.4.

Distance between two lines.

A condition for two 3D lines to intersect is found from the observation that the three vectors $\mathbf{v}_1, \mathbf{v}_2, \mathbf{p}_2 - \mathbf{p}_1$ must be coplanar, or linearly dependent. This would lead to the determinant condition (see Section 9.8)

$$\det[\mathbf{v}_1, \mathbf{v}_2, \mathbf{p}_2 - \mathbf{p}_1] = 0.$$

From a numerical viewpoint, it is safer to compare the distance between the points $\mathbf{x}_1$ and $\mathbf{x}_2$; in the field of computer-aided design, one usually has known tolerances (e.g., 0.001 mm) for distances. It is much harder to come up with a meaningful tolerance for a determinant since it describes a volume.

11.3 Lines and Planes: Intersections

One of the basic techniques in computer graphics is called *ray tracing.* Figure 11.1 illustrates this technique. A scene is given as an assembly of planes (usually restricted to triangles). A computer-generated image needs to compute proper lighting; this is done by tracing light rays through the scene. The ray intersects a plane, it is reflected, then it intersects the next plane, etc. Sketch 11.5 gives an example.

The basic problem to be solved is this: Given a plane $\mathbf{P}$ and a line $\mathbf{l}$, what is their *intersection point* $\mathbf{x}$? It is most convenient to represent the plane by assuming that we know a point $\mathbf{q}$ on it as well as its normal vector $\mathbf{n}$ (see Sketch 11.6). Then the unknown point $\mathbf{x}$, being on $\mathbf{P}$, must satisfy

$$[\mathbf{x} - \mathbf{q}] \cdot \mathbf{n} = 0. \tag{11.3}$$

By definition, the intersection point is also on the line (the ray, in computer graphics jargon), given by a point $\mathbf{p}$ and a vector $\mathbf{v}$:

$$\mathbf{x} = \mathbf{p} + t\mathbf{v}. \tag{11.4}$$

At this point, we do not know the correct value for t; once we have it, our problem is solved.

The solution is obtained by substituting the expression for $\mathbf{x}$ from (11.4) into (11.3):

$$[\mathbf{p} + t\mathbf{v} - \mathbf{q}] \cdot \mathbf{n} = 0.$$

Thus,

$$[\mathbf{p} - \mathbf{q}] \cdot \mathbf{n} + t\mathbf{v} \cdot \mathbf{n} = 0$$

and

$$t = \frac{[\mathbf{q} - \mathbf{p}] \cdot \mathbf{n}}{\mathbf{v} \cdot \mathbf{n}}. \tag{11.5}$$

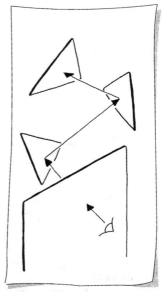

Sketch 11.5.
A ray is traced through a scene.

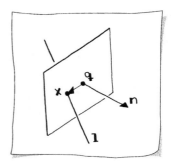

Sketch 11.6.
A line and a plane.

The intersection point $\mathbf{x}$ is now computed as

$$\mathbf{x} = \mathbf{p} + \frac{[\mathbf{q} - \mathbf{p}] \cdot \mathbf{n}}{\mathbf{v} \cdot \mathbf{n}} \mathbf{v}. \tag{11.6}$$

(This equation should look familiar as we encountered this problem when examining projections in Section 10.4, but we include it here to keep this chapter self-contained.)

Example 11.3

Take the plane

$$x_1 + x_2 + x_3 - 1 = 0$$

and the line

$$\mathbf{p}(t) = \begin{bmatrix} 1 \\ 1 \\ 2 \end{bmatrix} + t \begin{bmatrix} 0 \\ 0 \\ 1 \end{bmatrix}$$

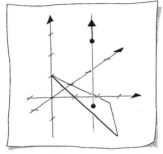

Sketch 11.7.

Intersecting a line and a plane.

as shown in Sketch 11.7.

We need a point $\mathbf{q}$ on the plane; set $x_1 = x_2 = 0$ and solve for x_3, resulting in $x_3 = 1$. This amounts to intersecting the plane with the $\mathbf{e}_3$-axis. From (11.5), we find $t = -3$ and then (11.6) gives the intersection point as

$$\mathbf{x} = \begin{bmatrix} 1 \\ 1 \\ 2 \end{bmatrix} - 3 \begin{bmatrix} 0 \\ 0 \\ 1 \end{bmatrix} = \begin{bmatrix} 1 \\ 1 \\ -1 \end{bmatrix}.$$

It never hurts to carry out a sanity check: Verify that this $\mathbf{x}$ does indeed satisfy the plane equation.

A word of caution: In (11.5), we happily divide by the dot product $\mathbf{v} \cdot \mathbf{n}$—but that better not be zero![2] If it is, then the ray "grazes" the plane, i.e., it is parallel to it, and no intersection exists.

The same problem—intersecting a plane with a line—may be solved if the plane is given in *parametric form* (see (8.15)). Then the unknown intersection point $\mathbf{x}$ must satisfy

$$\mathbf{x} = \mathbf{q} + u_1 \mathbf{r}_1 + u_2 \mathbf{r}_2.$$

---

[2]Keep in mind that real numbers are rarely *equal* to zero. A tolerance needs to be used; 0.001 should work if both $\mathbf{n}$ and $\mathbf{v}$ are normalized. However, such tolerances are driven by applications.

Since we know that $\mathbf{x}$ is also on the line $\mathbf{l}$, we may set

$$\mathbf{p} + t\mathbf{v} = \mathbf{q} + u_1\mathbf{r}_1 + u_2\mathbf{r}_2.$$

This equation is short for three individual equations, one for each coordinate. We thus have three equations in three unknowns u_1, u_2, t,

$$\begin{bmatrix} \mathbf{r}_1 & \mathbf{r}_2 & -\mathbf{v} \end{bmatrix} \begin{bmatrix} u_1 \\ u_2 \\ t \end{bmatrix} = \begin{bmatrix} \mathbf{p} - \mathbf{q} \end{bmatrix}, \tag{11.7}$$

and solve them with Gauss elimination. (The basic idea of this method was presented in Chapter 5, and 3×3 and larger systems are covered in Chapter 12.)

11.4 Intersecting a Triangle and a Line

A plane is, by definition, an unbounded object. In many applications, planes are parts of objects; one is interested in only a small part of a plane. For example, the six faces of a cube are bounded planes, so are the four faces of a tetrahedron.

We will now examine the case of a 3D *triangle* as an example of a bounded plane. If we intersect a 3D triangle with a line (a ray), then we are not interested in an intersection point *outside* the triangle— only an interior one will count.

Let the triangle be given by three points $\mathbf{p}_1, \mathbf{p}_2, \mathbf{p}_3$ and the line by a point $\mathbf{p}$ and a direction $\mathbf{v}$ (see Sketch 11.8).

The plane may be written in parametric form as

$$\mathbf{x}(u_1, u_2) = \mathbf{p}_1 + u_1(\mathbf{p}_2 - \mathbf{p}_1) + u_2(\mathbf{p}_3 - \mathbf{p}_1).$$

We thus arrive at

$$\mathbf{p} + t\mathbf{v} = \mathbf{p}_1 + u_1(\mathbf{p}_2 - \mathbf{p}_1) + u_2(\mathbf{p}_3 - \mathbf{p}_1),$$

a linear system in the unknowns t, u_1, u_2. The solution is inside the triangle if both u_1 and u_2 are between zero and one and their sum is less than or equal to one. This is so since we may view $(u_1, u_2, 1 - u_1 - u_2)$ as *barycentric coordinates* of the triangle defined by $\mathbf{p}_2, \mathbf{p}_3, \mathbf{p}_1$, respectively. These are positive exactly for points inside the triangle. See Section 17.1 for an in-depth look at barycentric coordinates in a triangle.

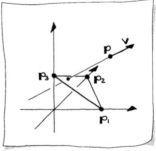

Sketch 11.8.

Intersecting a triangle and a line.

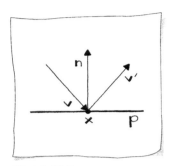

Sketch 11.9.
A reflection.

11.5   Reflections

The next problem is that of line or plane *reflection*. Given a point $\mathbf{x}$ on a plane $\mathbf{P}$ and an "incoming" direction $\mathbf{v}$, what is the reflected or "outgoing" direction $\mathbf{v}'$? See Sketch 11.9, where we look at the plane $\mathbf{P}$ "edge on." We assume that $\mathbf{v}$, $\mathbf{v}'$, and $\mathbf{n}$ are of unit length.

From physics, you might recall that the angle between $\mathbf{v}$ and the plane normal $\mathbf{n}$ must equal that of $\mathbf{v}'$ and $\mathbf{n}$, except for a sign change. We conveniently record this fact using a dot product:

$$-\mathbf{v} \cdot \mathbf{n} = \mathbf{v}' \cdot \mathbf{n}. \tag{11.8}$$

The normal vector $\mathbf{n}$ is thus the *angle bisector* of $\mathbf{v}$ and $\mathbf{v}'$. From inspection of Sketch 11.9, we also infer the symmetry property

$$c\mathbf{n} = \mathbf{v}' - \mathbf{v} \tag{11.9}$$

for some real number c. This means that some multiple of the normal vector may be written as the sum $\mathbf{v}' + (-\mathbf{v})$.

We now solve (11.9) for $\mathbf{v}'$ and insert into (11.8):

$$-\mathbf{v} \cdot \mathbf{n} = [c\mathbf{n} + \mathbf{v}] \cdot \mathbf{n},$$

and solve for c:

$$c = -2\mathbf{v} \cdot \mathbf{n}. \tag{11.10}$$

Here, we made use of the fact that $\mathbf{n}$ is a unit vector and thus $\mathbf{n} \cdot \mathbf{n} = 1$.

The reflected vector $\mathbf{v}'$ is now given by using our value for c in (11.9):

$$\mathbf{v}' = \mathbf{v} - [2\mathbf{v} \cdot \mathbf{n}]\mathbf{n}. \tag{11.11}$$

In the special case of $\mathbf{v}$ being perpendicular to the plane, i.e., $\mathbf{v} = -\mathbf{n}$, we obtain $\mathbf{v}' = -\mathbf{v}$ as expected. Also note that the point of reflection does not enter the equations at all.

We may rewrite (11.11) as

$$\mathbf{v}' = \mathbf{v} - 2[\mathbf{v}^{\mathrm{T}}\mathbf{n}]\mathbf{n},$$

which in turn may be reformulated to

$$\mathbf{v}' = \mathbf{v} - 2[\mathbf{n}\mathbf{n}^{\mathrm{T}}]\mathbf{v}. \tag{11.12}$$

You see this after multiplying out all products involved. Note that $\mathbf{n}\mathbf{n}^{\mathrm{T}}$ is an orthogonal projection matrix, just as we encountered in

Section 4.8. It is a symmetric matrix and an example of a *dyadic matrix*; this type of matrix has rank one. Sketch 11.10 illustrates the action of (11.12).

Now we are in a position to formulate a reflection as a linear map. It is of the form $\mathbf{v}' = H\mathbf{v}$ with

$$H = I - 2\mathbf{n}\mathbf{n}^{\mathrm{T}}. \tag{11.13}$$

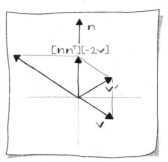

Sketch 11.10.
Reflection.

The matrix H is known as a *Householder matrix*. We'll look at this matrix again in Section 13.1 in the context of solving a linear system.

In computer graphics, this reflection problem is an integral part of calculating the lighting (color) at each vertex. A brief introduction to the lighting model may be found in Section 8.6. The light is positioned somewhere on $\mathbf{v}$, and after the light hits the point $\mathbf{x}$, the reflected light travels in the direction of $\mathbf{v}'$. We use a dot product to measure the cosine of the angle between the direction of our eye $(\mathbf{e} - \mathbf{x})$ and the reflection vector. The smaller the angle, the more reflected light hits our eye. This type of lighting is called *specular reflection*; it is the element of light that is dependent on our position and it produces a highlight—a shiny spot.

11.6   Intersecting Three Planes

Suppose we are given three planes with implicit equations

$$\mathbf{n}_1 \cdot \mathbf{x} + c_1 = 0,$$
$$\mathbf{n}_2 \cdot \mathbf{x} + c_2 = 0,$$
$$\mathbf{n}_3 \cdot \mathbf{x} + c_3 = 0.$$

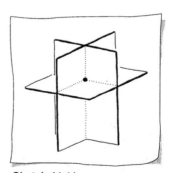

Sketch 11.11.
Intersecting three planes.

Where do they intersect? The answer is some point $\mathbf{x}$ that lies on each of the planes. See Sketch 11.11 for an illustration.

The solution is surprisingly simple; just condense the three plane equations into matrix form:

$$\begin{bmatrix} \mathbf{n}_1^{\mathrm{T}} \\ \mathbf{n}_2^{\mathrm{T}} \\ \mathbf{n}_3^{\mathrm{T}} \end{bmatrix} \begin{bmatrix} x_1 \\ x_2 \\ x_3 \end{bmatrix} = \begin{bmatrix} -c_1 \\ -c_2 \\ -c_3 \end{bmatrix}. \tag{11.14}$$

We have three equations in the three unknowns x_1, x_2, x_3.

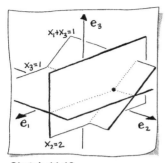

Sketch 11.12.

Intersecting three planes example.

Example 11.4

This example is shown in Sketch 11.12. The equations of the planes in that sketch are

$$x_1 + x_3 = 1, \quad x_3 = 1, \quad x_2 = 2.$$

The linear system is

$$\begin{bmatrix} 1 & 0 & 1 \\ 0 & 0 & 1 \\ 0 & 1 & 0 \end{bmatrix} \begin{bmatrix} x_1 \\ x_2 \\ x_3 \end{bmatrix} = \begin{bmatrix} 1 \\ 1 \\ 2 \end{bmatrix}.$$

Solving it by Gauss elimination (see Chapter 12), we obtain

$$\begin{bmatrix} x_1 \\ x_2 \\ x_3 \end{bmatrix} = \begin{bmatrix} 0 \\ 2 \\ 1 \end{bmatrix}.$$

While simple to solve, the three-planes problem does not always have a solution. Two lines in 2D do not intersect if they are parallel, or in other words, their normal vectors are parallel or linearly dependent. The situation is analogous in 3D. If the normal vectors $\mathbf{n}_1, \mathbf{n}_2, \mathbf{n}_3$ are linearly dependent, then there is no solution to the intersection problem.

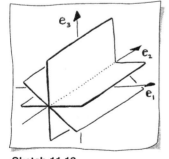

Sketch 11.13.

Three planes intersecting in one line.

Example 11.5

The normal vectors are

$$\mathbf{n}_1 = \begin{bmatrix} 1 \\ 0 \\ 0 \end{bmatrix}, \quad \mathbf{n}_2 = \begin{bmatrix} 1 \\ 0 \\ 1 \end{bmatrix}, \quad \mathbf{n}_3 = \begin{bmatrix} 0 \\ 0 \\ 1 \end{bmatrix}.$$

Since $\mathbf{n}_2 = \mathbf{n}_1 + \mathbf{n}_3$, they are indeed linearly dependent, and thus the planes defined by them do not intersect in one point (see Sketch 11.13).

11.7 Intersecting Two Planes

Odd as it may seem, intersecting two planes is harder than intersecting three of them. The problem is this: Two planes are given in their implicit form

$$\mathbf{n} \cdot \mathbf{x} + c = 0,$$
$$\mathbf{m} \cdot \mathbf{x} + d = 0.$$

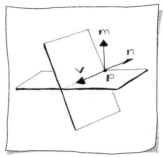

Find their intersection, which is a line. We would like the solution to be of the form

$$\mathbf{x}(t) = \mathbf{p} + t\mathbf{v}.$$

This situation is depicted in Sketch 11.14.

The direction vector $\mathbf{v}$ of this line is easily found; since it lies in each of the planes, it must be perpendicular to both their normal vectors:

$$\mathbf{v} = \mathbf{n} \wedge \mathbf{m}.$$

Sketch 11.14.
Intersecting two planes.

We still need a point $\mathbf{p}$ on the line.

To this end, we come up with an auxiliary plane that intersects both given planes. The intersection point is clearly on the desired line. Define the third plane by

$$\mathbf{v} \cdot \mathbf{x} = 0.$$

This plane passes through the origin and has normal vector $\mathbf{v}$; that is, it is perpendicular to the desired line (see Sketch 11.15).

We now solve the three-plane intersection problem for the two given planes and the auxiliary plane for the missing point $\mathbf{p}$, and our line is determined.

In the case $c = d = 0$, both given planes pass through the origin, and it can serve as the point $\mathbf{p}$.

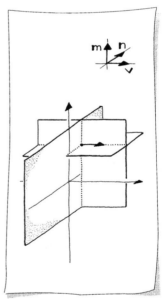

Sketch 11.15.
The auxiliary plane (shaded).

11.8 Creating Orthonormal Coordinate Systems

Often times when working in 3D, life is made easier by creating a local coordinate system. We have seen one example of this already: digitizing. In Section 1.5 we alluded to a coordinate frame as a means to "capture" a cat. Let's look at that example as a motivation for creating an *orthonormal coordinate system* with the Gram-Schmidt method.

The digitizer needs an orthonormal coordinate frame in order to store coordinates for the cat. The digitizer comes with its own frame—the $[\mathbf{e}_1, \mathbf{e}_2, \mathbf{e}_3]$-coordinate frame, stored internally with respect to its base and arm, but as a user, you may choose your own frame of reference that we will call the $[\mathbf{b}_1, \mathbf{b}_2, \mathbf{b}_3]$-coordinate frame. For example, when using this digitized model, it might be convenient to have the $\mathbf{b}_1$ axis generally aligned with the head to tail direction. Placing the origin of your coordinate frame close to the cat is also advantageous for numerical stability with respect to round-off error and accuracy. (The digitizer has its highest accuracy within a small radius from its base.) An orthonormal frame facilitates data capture and the transformation between coordinate frames. It is highly unlikely that you can manually create an orthonormal frame, so the digitizer will do it for you, but it needs some help.

With the digitizing arm, let's choose the point $\mathbf{p}$ to be the origin of our input coordinate frame, the point $\mathbf{q}$ will be along the $\mathbf{b}_1$ axis, and the point $\mathbf{r}$ will be close to the $\mathbf{b}_2$ axis. From these three points, we form two vectors,

$$\mathbf{v}_1 = \mathbf{q} - \mathbf{p} \quad \text{and} \quad \mathbf{v}_2 = \mathbf{r} - \mathbf{p}.$$

A simple cross product will supply us with a vector normal to the table: $\mathbf{v}_3 = \mathbf{v}_1 \wedge \mathbf{v}_2$. Now we can state the problem: given three linearly independent vectors $\mathbf{v}_1, \mathbf{v}_2, \mathbf{v}_3$, find a close *orthonormal* set of vectors $\mathbf{b}_1, \mathbf{b}_2, \mathbf{b}_3$.

Orthogonal projections and orthogonal components, as described in Section 2.8, are the foundation of this method. We will refer to the subspace formed by $\mathbf{v}_i$ as V_i and the subspace formed by $\mathbf{v}_1, \mathbf{v}_2$ as V_{12}. As we build an orthogonal frame, we normalize all the vectors. A notational shorthand that is helpful in making equations easier to read: if we normalize a vector $\mathbf{w}$, we will write $\mathbf{w}/\| \cdot \|$. Normalizing $\mathbf{v}_1$:

$$\mathbf{b}_1 = \frac{\mathbf{v}_1}{\| \cdot \|}. \tag{11.15}$$

Next, create $\mathbf{b}_2$ from the component of $\mathbf{v}_2$ that is orthogonal to the subspace V_1, which means we normalize $(\mathbf{v}_2 - \text{proj}_{V_1} \mathbf{v}_2)$:

$$\mathbf{b}_2 = \frac{\mathbf{v}_2 - (\mathbf{v}_2 \cdot \mathbf{b}_1)\mathbf{b}_1}{\| \cdot \|}. \tag{11.16}$$

As the last step, we create $\mathbf{b}_3$ from the component of $\mathbf{v}_3$ that is orthogonal to the subspace V_{12}, which means we normalize $(\mathbf{v}_3 - \text{proj}_{V_{12}} \mathbf{v}_3)$. This is done by separating the projection into the sum of a projection onto V_1 and onto V_2:

$$\mathbf{b}_3 = \frac{\mathbf{v}_3 - (\mathbf{v}_3 \cdot \mathbf{b}_1)\mathbf{b}_1 - (\mathbf{v}_3 \cdot \mathbf{b}_2)\mathbf{b}_2}{\| \cdot \|}. \tag{11.17}$$

Example 11.6

As illustrated in Sketch 11.16, we are given

$$\mathbf{v}_1 = \begin{bmatrix} 0 \\ -2 \\ 0 \end{bmatrix}, \qquad \mathbf{v}_2 = \begin{bmatrix} 2 \\ -1 \\ 0 \end{bmatrix}, \qquad \mathbf{v}_3 = \begin{bmatrix} 2 \\ -0.5 \\ 2 \end{bmatrix},$$

and in this example, unlike the scenario in the digitizing application, $\mathbf{v}_3$ is not the result of the cross product of $\mathbf{v}_1$ and $\mathbf{v}_2$, so the example can better illustrate each projection. Normalizing $\mathbf{v}_1$,

$$\mathbf{b}_1 = \frac{\begin{bmatrix} 0 \\ -2 \\ 0 \end{bmatrix}}{\| \cdot \|} = \begin{bmatrix} 0 \\ -1 \\ 0 \end{bmatrix}$$

The projection of $\mathbf{v}_2$ into the subspace V_1,

$$\mathbf{u} = \text{proj}_{V_1} \mathbf{v}_2 = \left(\begin{bmatrix} 2 \\ -1 \\ 0 \end{bmatrix} \cdot \begin{bmatrix} 0 \\ -1 \\ 0 \end{bmatrix} \right) \begin{bmatrix} 0 \\ -1 \\ 0 \end{bmatrix} = \begin{bmatrix} 0 \\ -1 \\ 0 \end{bmatrix},$$

thus $\mathbf{b}_2$ is the component of $\mathbf{v}_2$ that is orthogonal to $\mathbf{u}$, normalized

$$\mathbf{b}_2 = \frac{\begin{bmatrix} 2 \\ -1 \\ 0 \end{bmatrix} - \begin{bmatrix} 0 \\ -1 \\ 0 \end{bmatrix}}{\| \cdot \|} = \begin{bmatrix} 1 \\ 0 \\ 0 \end{bmatrix}.$$

The projection of $\mathbf{v}_3$ into the subspace V_{12} is computed as

$$\begin{aligned}
\mathbf{w} &= \text{proj}_{V_{12}} \mathbf{v}_3 \\
&= \text{proj}_{V_1} \mathbf{v}_3 + \text{proj}_{V_2} \mathbf{v}_3 \\
&= \left(\begin{bmatrix} 2 \\ -0.5 \\ 2 \end{bmatrix} \cdot \begin{bmatrix} 0 \\ -1 \\ 0 \end{bmatrix} \right) \begin{bmatrix} 0 \\ -1 \\ 0 \end{bmatrix} + \left(\begin{bmatrix} 2 \\ -0.5 \\ 2 \end{bmatrix} \cdot \begin{bmatrix} 1 \\ 0 \\ 0 \end{bmatrix} \right) \begin{bmatrix} 1 \\ 0 \\ 0 \end{bmatrix} \\
&= \begin{bmatrix} 2 \\ -0.5 \\ 0 \end{bmatrix}.
\end{aligned}$$

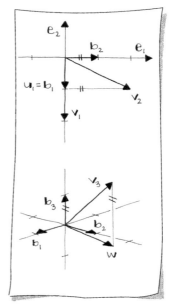

Sketch 11.16.
3D Gram-Schmidt orthonormalization.

Then $\mathbf{b}_3$ is the component of $\mathbf{v}_3$ orthogonal to $\mathbf{w}$, normalized

$$\mathbf{b}_3 = \frac{\begin{bmatrix} 2.0 \\ -0.5 \\ 2 \end{bmatrix} - \begin{bmatrix} 2 \\ -0.5 \\ 0 \end{bmatrix}}{\| \cdot \|} = \begin{bmatrix} 0 \\ 0 \\ 1 \end{bmatrix}.$$

As you might have observed, in 3D the Gram-Schmidt method is more work in terms of multiplications and additions than simply applying the cross product repeatedly. For example, if we form $\mathbf{b}_3 = \mathbf{b}_1 \wedge \mathbf{b}_2$, we get a normalized vector for free, and thus we save a square root, which is an expensive operation in terms of operating system cycles. The real advantage of the Gram-Schmidt method is for dimensions higher than three, where we don't have cross products. However, understanding the process in 3D makes the n-dimensional formulas easier to follow. A much more general version is discussed in Section 14.4.

- distance between a point and plane
- distance between two lines
- plane and line intersection
- triangle and line intersection

- reflection vector
- Householder matrix
- intersection of three planes
- intersection of two planes
- Gram-Schmidt orthonormalization

11.9   Exercises

For some of the exercises that follow, we will refer to the following two planes and line. $\mathbf{P}_1$ goes through a point $\mathbf{p}$ and has normal vector $\mathbf{n}$:

$$\mathbf{p} = \begin{bmatrix} 1 \\ 2 \\ 0 \end{bmatrix}, \quad \mathbf{n} = \begin{bmatrix} -1 \\ 0 \\ 0 \end{bmatrix}.$$

The plane $\mathbf{P}_2$ is given by its implicit form

$$x_1 + 2x_2 - 2x_3 - 1 = 0.$$

The line $\mathbf{l}$ goes through the point $\mathbf{q}$ and has direction $\mathbf{v}$,

$$\mathbf{q} = \begin{bmatrix} -1 \\ 2 \\ 0 \end{bmatrix}, \quad \mathbf{v} = \begin{bmatrix} 0 \\ 1 \\ 0 \end{bmatrix}.$$

1. Find the distance of the point

$$\mathbf{p} = \begin{bmatrix} 0 \\ 0 \\ 2 \end{bmatrix}$$

to the plane $\mathbf{P}_2$. Is this point on the same side as the normal direction?

2. Given the plane $(1/\sqrt{3})x_1 + (1/\sqrt{3})x_2 + (1/\sqrt{3})x_3 + (1/\sqrt{3}) = 0$ and point $\mathbf{p} = \begin{bmatrix} 1 & 1 & 4 \end{bmatrix}^{\mathrm{T}}$, what is the distance of $\mathbf{p}$ to the plane? What is the closest point to $\mathbf{p}$ in the plane?

3. Revisit Example 11.2, but set the point defining the line l_2 to be

$$\mathbf{p}_2 = \begin{bmatrix} 0 \\ 0 \\ 1 \end{bmatrix}.$$

The lines have not changed; how do you obtain the (unchanged) solutions $\mathbf{x}_1$ and $\mathbf{x}_2$?

4. Given two lines, $\mathbf{x}_i(s_i) = \mathbf{p}_i + s_i\mathbf{v}_i$, $i = 1, 2$, where

$$\mathbf{x}_1(s_1) = \begin{bmatrix} 1 \\ 1 \\ 0 \end{bmatrix} + s_1 \begin{bmatrix} -1 \\ -1 \\ 1 \end{bmatrix} \quad \text{and} \quad \mathbf{x}_2(s_2) = \begin{bmatrix} 1 \\ 1 \\ 1 \end{bmatrix} + s_2 \begin{bmatrix} -1 \\ -1 \\ -1 \end{bmatrix},$$

find the two points of closest proximity between the lines. What are the parameter values s_1 and s_2 for these two points?

5. Find the intersection of $\mathbf{P}_1$ with the line l.

6. Find the intersection of $\mathbf{P}_2$ with the line l.

7. Let $\mathbf{a}$ be an arbitrary vector. It may be projected along a direction $\mathbf{v}$ onto the plane $\mathbf{P}$ that passes through the origin with normal vector $\mathbf{n}$. What is its image $\mathbf{a}'$?

8. Does the ray $\mathbf{p} + t\mathbf{v}$ with

$$\mathbf{p} = \begin{bmatrix} -1 \\ -1 \\ 0 \end{bmatrix} \quad \text{and} \quad \mathbf{v} = \begin{bmatrix} 1 \\ 1 \\ 1 \end{bmatrix}$$

intersect the triangle with vertices

$$\begin{bmatrix} 3 \\ 0 \\ 0 \end{bmatrix}, \quad \begin{bmatrix} 0 \\ 2 \\ 1 \end{bmatrix}, \quad \begin{bmatrix} 2 \\ 2 \\ 3 \end{bmatrix}?$$

Solve this by setting up a linear system. *Hint:* $t = 3$.

9. Does the ray $\mathbf{p} + t\mathbf{v}$ with

$$\mathbf{p} = \begin{bmatrix} 2 \\ 2 \\ 2 \end{bmatrix} \quad \text{and} \quad \mathbf{v} = \begin{bmatrix} -1 \\ -1 \\ -2 \end{bmatrix}$$

intersect the triangle with vertices

$$\begin{bmatrix} 0 \\ 0 \\ 1 \end{bmatrix}, \quad \begin{bmatrix} 1 \\ 0 \\ 0 \end{bmatrix}, \quad \begin{bmatrix} 0 \\ 1 \\ 0 \end{bmatrix} ?$$

Hint: To solve the linear system, first express u and v in terms of t in order to solve for t first.

10. Given the point $\mathbf{p}$ in the plane P_1, what is the reflected direction of the vector

$$\mathbf{v} = \begin{bmatrix} 1/3 \\ 2/3 \\ -2/3 \end{bmatrix} ?$$

11. Given the ray defined by $\mathbf{q} + t\mathbf{v}$, where

$$\mathbf{q} = \begin{bmatrix} 0 \\ 0 \\ 1 \end{bmatrix} \quad \text{and} \quad \mathbf{v} = \begin{bmatrix} 0 \\ 0 \\ -1 \end{bmatrix},$$

find its reflection in the plane P_2 by defining the intersection point $\mathbf{p}$ in the plane and the reflection vector $\mathbf{v}'$.

12. Find the intersection of the three planes:

$$x_1 + x_2 = 1, \quad x_1 = 1, \quad x_3 = 4.$$

13. Find the intersection of the three planes

$$P_1(u, v) = \begin{bmatrix} 0 \\ 0 \\ 0 \end{bmatrix} + u \begin{bmatrix} 1 \\ 0 \\ 0 \end{bmatrix} + v \begin{bmatrix} 0 \\ 1 \\ 0 \end{bmatrix}$$

$$P_2(u, v) = \begin{bmatrix} 1 \\ 1 \\ 0 \end{bmatrix} + u \begin{bmatrix} -1 \\ 0 \\ 0 \end{bmatrix} + v \begin{bmatrix} 0 \\ 0 \\ 1 \end{bmatrix}$$

$$P_3(u, v) = \begin{bmatrix} 0 \\ 0 \\ 0 \end{bmatrix} + u \begin{bmatrix} 1 \\ 1 \\ 1 \end{bmatrix} + v \begin{bmatrix} 0 \\ 0 \\ 1 \end{bmatrix}.$$

14. Find the intersection of the planes:

$$x_1 = 1 \quad \text{and} \quad x_3 = 4.$$

15. Find the intersection of the planes:

$$x_1 - x_2 = 0 \quad \text{and} \quad x_1 + x_2 = 2.$$

16. Given vectors

$$\mathbf{v}_1 = \begin{bmatrix} 1 \\ 0 \\ 0 \end{bmatrix}, \quad \mathbf{v}_2 = \begin{bmatrix} 1 \\ 1 \\ 0 \end{bmatrix}, \quad \mathbf{v}_3 = \begin{bmatrix} -1 \\ -1 \\ 1 \end{bmatrix},$$

carry out the Gram-Schmidt orthonormalization.

17. Given the vectors $\mathbf{v}_i$ in Exercise 16, what are the cross products that will produce the orthonormal vectors $\mathbf{b}_i$?

12

Gauss for Linear Systems

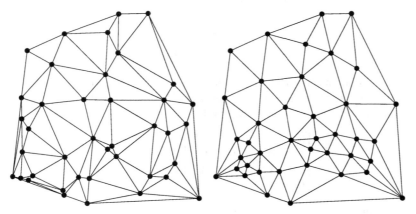

Figure 12.1.
Linear systems: the triangulation on the right was obtained from the left one by solving a linear system.

In Chapter 5, we studied linear systems of two equations in two unknowns. A whole chapter for such a humble task seems like a bit of overkill—its main purpose was really to lay the groundwork for this chapter.

Linear systems arise in virtually every area of science and engineering—some are as big as 1,000,000 equations in as many unknowns. Such huge systems require more sophisticated treatment than the methods introduced here. They *will* allow you to solve systems with several thousand equations without a problem.

Figure 12.1 illustrates the use of linear systems in the field of data smoothing. The left triangulation looks somewhat "rough"; after setting up an appropriate linear system, we compute the "smoother" triangulation on the right, in which the triangles are closer to being equilateral.

This chapter explains the basic ideas underlying linear systems. Readers eager for hands-on experience should get access to software such as Mathematica or MATLAB. Readers with advanced programming knowledge can download linear system solvers from the web. The most prominent collection of routines is LAPACK.

12.1 The Problem

A linear system is a set of equations like this:

$$
\begin{aligned}
3u_1 - 2u_2 - 10u_3 + u_4 &= 0 \\
u_1 - u_3 &= 4 \\
u_1 + u_2 - 2u_3 + 3u_4 &= 1 \\
u_2 + 2u_4 &= -4.
\end{aligned}
$$

The unknowns are the numbers u_1, u_2, u_3, u_4. There are as many equations as there are unknowns, four in this example. We rewrite this 4×4 linear system in matrix form:

$$
\begin{bmatrix}
3 & -2 & -10 & 1 \\
1 & 0 & -1 & 0 \\
1 & 1 & -2 & 3 \\
0 & 1 & 0 & 2
\end{bmatrix}
\begin{bmatrix}
u_1 \\ u_2 \\ u_3 \\ u_4
\end{bmatrix}
=
\begin{bmatrix}
0 \\ 4 \\ 1 \\ -4
\end{bmatrix}.
$$

A general $n \times n$ linear system looks like this:

$$
\begin{aligned}
a_{1,1}u_1 + a_{1,2}u_2 + \ldots + a_{1,n}u_n &= b_1 \\
a_{2,1}u_1 + a_{2,2}u_2 + \ldots + a_{2,n}u_n &= b_2 \\
&\vdots \\
a_{n,1}u_1 + a_{n,2}u_2 + \ldots + a_{n,n}u_n &= b_n.
\end{aligned}
$$

In matrix form, it becomes

$$
\begin{bmatrix}
a_{1,1} & a_{1,2} & \cdots & a_{1,n} \\
a_{2,1} & a_{2,2} & \cdots & a_{2,n} \\
& & \vdots & \\
a_{n,1} & a_{n,2} & \cdots & a_{n,n}
\end{bmatrix}
\begin{bmatrix}
u_1 \\ u_2 \\ \vdots \\ u_n
\end{bmatrix}
=
\begin{bmatrix}
b_1 \\ b_2 \\ \vdots \\ b_n
\end{bmatrix},
\tag{12.1}
$$

$$\begin{bmatrix} \mathbf{a}_1 & \mathbf{a}_2 & \ldots & \mathbf{a}_n \end{bmatrix} \mathbf{u} = \mathbf{b},$$

or even shorter

$$A\mathbf{u} = \mathbf{b}.$$

The *coefficient matrix* A has n rows and n columns. For example, the first row is

$$a_{1,1}, a_{1,2}, \ldots, a_{1,n},$$

and the second column is

$$a_{1,2}$$
$$a_{2,2}$$
$$\vdots$$
$$a_{n,2}.$$

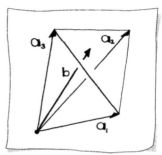

Sketch 12.1.
A solvable 3×3 system.

Equation (12.1) is a compact way of writing n equations for the n unknowns $u_1, \ldots, u_n$. In the 2×2 case, such systems had nice geometric interpretations; in the general case, that interpretation needs n-dimensional linear spaces, and is not very intuitive. Still, the methods that we developed for the 2×2 case can be gainfully employed here!

Some underlying principles with a geometric interpretation are best explained for the example $n = 3$. We are given a vector $\mathbf{b}$ and we try to write it as a linear combination of vectors $\mathbf{a}_1, \mathbf{a}_2, \mathbf{a}_3$,

$$\begin{bmatrix} \mathbf{a}_1 & \mathbf{a}_2 & \mathbf{a}_3 \end{bmatrix} \mathbf{u} = \mathbf{b}.$$

If the $\mathbf{a}_i$ are truly 3D, i.e., if they form a tetrahedron, then a *unique solution* may be found (see Sketch 12.1). But if the three $\mathbf{a}_i$ all lie in a plane (i.e., if the volume formed by them is zero), then you cannot write $\mathbf{b}$ as a linear combination of them, unless it is itself in that 2D plane. In this case, you cannot expect uniqueness for your answer. Sketch 12.2 covers these cases. In general, a linear system is uniquely solvable if the $\mathbf{a}_i$ have a nonzero n-dimensional volume. If they do not, they span a k-dimensional *subspace* (with $k < n$)—nonunique solutions exist only if $\mathbf{b}$ is itself in that subspace. A linear system is called *consistent* if at least one solution exists.

For 2D and 3D, we encountered many problems that lent themselves to constructing the linear system in terms of a linear combination of column vectors, the $\mathbf{a}_i$. However, in Section 5.11 we looked at how a linear system can be interpreted as a problem using equations built row by row. In n-dimensions, this commonly occurs. An example follows.

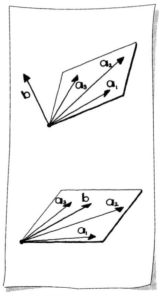

Sketch 12.2.
Top: no solution, bottom: nonunique solution.

Example 12.1

Suppose that at five time instances, say $t_i = 0, 0.25, 0.5, 0.75, 1$ seconds, we have associated observation data, $p(t_i) = 0, 1, 0.5, 0.5, 0$. We would like to fit a polynomial to these data so we can estimate values in between the observations. This is called *polynomial interpolation*. Five points require a degree four polynomial,

$$p(t) = c_0 + c_1 t + c_2 t^2 + c_3 t^3 + c_4 t^4,$$

which has five coefficients c_i. Immediately we see that we can write down five equations,

$$p(t_i) = c_0 + c_1 t_i + c_2 t_i^2 + c_3 t_i^3 + c_4 t_i^4, \qquad i = 0, \ldots, 4,$$

or in matrix form,

$$\begin{bmatrix} 1 & t_0 & t_0^2 & t_0^3 & t_0^4 \\ 1 & t_1 & t_1^2 & t_1^3 & t_1^4 \\ & & \vdots & & \\ 1 & t_4 & t_4^2 & t_4^3 & t_4^4 \end{bmatrix} \begin{bmatrix} c_0 \\ c_1 \\ \vdots \\ c_4 \end{bmatrix} = \begin{bmatrix} p(t_0) \\ p(t_1) \\ \vdots \\ p(t_4) \end{bmatrix}$$

Figure 12.2 illustrates the result, $p(t) = 12.667t - 50t^2 + 69.33t^3 - 32t^4$, for the given data.

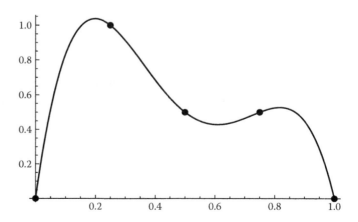

Figure 12.2.
Polynomial interpolation: a degree four polynomial fit to five data points.

12.2   The Solution via Gauss Elimination

The key to success in the 2×2 case was the application of a shear (forward elimination) so that the matrix A was transformed to *upper triangular*, meaning all entries below the diagonal are zero. Then it was possible to apply *back substitution* to solve for the unknowns. A shear was constructed to map the first column vector of the matrix onto the $\mathbf{e}_1$-axis. Revisiting an example from Chapter 5, we set

$$\begin{bmatrix} 2 & 4 \\ 1 & 6 \end{bmatrix} \begin{bmatrix} u_1 \\ u_2 \end{bmatrix} = \begin{bmatrix} 4 \\ 4 \end{bmatrix}. \tag{12.2}$$

The shear used was

$$S_1 = \begin{bmatrix} 1 & 0 \\ -1/2 & 1 \end{bmatrix},$$

which when applied to the system as

$$S_1 A \mathbf{u} = S_1 \mathbf{b}$$

produced the system

$$\begin{bmatrix} 2 & 4 \\ 0 & 4 \end{bmatrix} \begin{bmatrix} u_1 \\ u_2 \end{bmatrix} = \begin{bmatrix} 4 \\ 2 \end{bmatrix}.$$

Algebraically, what this shear did was to change the *rows* of the system in the following manner:

$$\text{row}_1 \leftarrow \text{row}_1 \quad \text{and} \quad \text{row}_2 \leftarrow \text{row}_2 - \frac{1}{2}\text{row}_1.$$

Each of these constitutes an *elementary row operation*. Back substitution came next, with

$$u_2 = \frac{1}{4} \times 2 = \frac{1}{2}$$

and then

$$u_1 = \frac{1}{2}(4 - 4u_2) = 1.$$

The divisions in the back substitution equations are actually scalings, thus they could be rewritten in terms of a scale matrix:

$$S_2 = \begin{bmatrix} 1/2 & 0 \\ 0 & 1/4 \end{bmatrix},$$

and then the system would be transformed via

$$S_2 S_1 A\mathbf{u} = S_2 S_1 \mathbf{b}.$$

The corresponding upper triangular matrix and system is

$$\begin{bmatrix} 1 & 2 \\ 0 & 1 \end{bmatrix} \begin{bmatrix} u_1 \\ u_2 \end{bmatrix} = \begin{bmatrix} 2 \\ 1/2 \end{bmatrix}.$$

Check for yourself that we get the same result.

Thus, we see the geometric steps for solving a linear system have methodical algebraic interpretations. We will be following this algebraic approach for the rest of the chapter. For general linear systems, the matrices, such as S_1 and S_2 above, are not actually constructed due to speed and storage expense. Notice that the shear to zero one element in the matrix, changed the elements in only one row, thus it is unnecessary to manipulate the other rows. This is an important observation for large systems.

In the general case, just as in the 2×2 case, *pivoting* will be used. Recall for the 2×2 case this meant that the equations were reordered such that the (pivot) matrix element $a_{1,1}$ is the largest one in the first column. A row exchange can be represented in terms of a *permutation matrix*. Suppose the 2×2 system in (12.2) was instead,

$$\begin{bmatrix} 1 & 6 \\ 2 & 4 \end{bmatrix} \begin{bmatrix} u_1 \\ u_2 \end{bmatrix} = \begin{bmatrix} 4 \\ 4 \end{bmatrix},$$

requiring pivoting as the first step. The permutation matrix that will exchange the two rows is

$$P_1 = \begin{bmatrix} 0 & 1 \\ 1 & 0 \end{bmatrix},$$

which is the identity matrix with the rows (columns) exchanged. After this row exchange, the steps for solving $P_1 A\mathbf{u} = P_1 \mathbf{b}$ are the same as for the system in (12.2): $S_2 S_1 P_1 A\mathbf{u} = S_2 S_1 P_1 \mathbf{b}$.

Example 12.2

Let's step through the necessary row exchanges and shears for a 3×3 linear system. The goal is to get it in upper triangular form so we may use back substitution to solve for the unknowns. The system is

$$\begin{bmatrix} 2 & -2 & 0 \\ 4 & 0 & -2 \\ 4 & 2 & -4 \end{bmatrix} \begin{bmatrix} u_1 \\ u_2 \\ u_3 \end{bmatrix} = \begin{bmatrix} 4 \\ -2 \\ 0 \end{bmatrix}.$$

The matrix element $a_{1,1}$ is not the largest in the first column, so we choose the 4 in the second row to be the *pivot element* and we reorder the rows:

$$\begin{bmatrix} 4 & 0 & -2 \\ 2 & -2 & 0 \\ 4 & 2 & -4 \end{bmatrix} \begin{bmatrix} u_1 \\ u_2 \\ u_3 \end{bmatrix} = \begin{bmatrix} -2 \\ 4 \\ 0 \end{bmatrix}.$$

The permutation matrix that achieves this row exchange is

$$P_1 = \begin{bmatrix} 0 & 1 & 0 \\ 1 & 0 & 0 \\ 0 & 0 & 1 \end{bmatrix}.$$

(The subscript 1 indicates that this matrix is designed to achieve the appropriate exchange for the first column.)

To zero entries in the first column apply:

$$\text{row}_2 \leftarrow \text{row}_2 - \frac{1}{2}\text{row}_1$$

$$\text{row}_3 \leftarrow \text{row}_3 - \text{row}_1,$$

and the system becomes

$$\begin{bmatrix} 4 & 0 & -2 \\ 0 & -2 & 1 \\ 0 & 2 & -2 \end{bmatrix} \begin{bmatrix} u_1 \\ u_2 \\ u_3 \end{bmatrix} = \begin{bmatrix} -2 \\ 5 \\ 2 \end{bmatrix}.$$

The shear matrix that achieves this is

$$G_1 = \begin{bmatrix} 1 & 0 & 0 \\ -1/2 & 1 & 0 \\ -1 & 0 & 1 \end{bmatrix}.$$

Now the first column consists of only zeroes except for $a_{1,1}$, meaning that it is lined up with the $\mathbf{e}_1$-axis.

Now work on the second column vector. First, check if pivoting is necessary; this means checking that $a_{2,2}$ is the largest in absolute value of all values in the second column that are below the diagonal. No pivoting is necessary. (We could say that the permutation matrix $P_2 = I$.) To zero the last element in this vector apply

$$\text{row}_3 \leftarrow \text{row}_3 + \text{row}_2,$$

which produces

$$\begin{bmatrix} 4 & 0 & -2 \\ 0 & -2 & 1 \\ 0 & 0 & -1 \end{bmatrix} \begin{bmatrix} u_1 \\ u_2 \\ u_3 \end{bmatrix} = \begin{bmatrix} -2 \\ 5 \\ 7 \end{bmatrix}.$$

The shear matrix that achieves this is

$$G_2 = \begin{bmatrix} 1 & 0 & 0 \\ 0 & 1 & 0 \\ 0 & 1 & 1 \end{bmatrix}.$$

By chance, the second column is aligned with $\mathbf{e}_2$ because $a_{1,2} = 0$. If this extra zero had not appeared, then the last operation would have mapped this 3D vector into the $[\mathbf{e}_1, \mathbf{e}_2]$-plane.

Now that we have the matrix in upper triangular form, we are ready for back substitution:

$$u_3 = \frac{1}{-1}(7)$$

$$u_2 = \frac{1}{-2}(5 - u_3)$$

$$u_1 = \frac{1}{4}(-2 + 2u_3).$$

This implicitly incorporates a scaling matrix. We obtain the solution

$$\begin{bmatrix} u_1 \\ u_2 \\ u_3 \end{bmatrix} = \begin{bmatrix} -4 \\ -6 \\ -7 \end{bmatrix}.$$

It is usually a good idea to insert the solution into the original equations:

$$\begin{bmatrix} 2 & -2 & 0 \\ 4 & 0 & -2 \\ 4 & 2 & -4 \end{bmatrix} \begin{bmatrix} -4 \\ -6 \\ -7 \end{bmatrix} = \begin{bmatrix} 4 \\ -2 \\ 0 \end{bmatrix}.$$

It works!

The example above illustrates each of the *elementary row operations* that take place during Gauss elimination:

- Pivoting results in the exchange of two rows.

- Shears result in adding a multiple of one row to another.

- Scaling results in multiplying a row by a scalar.

Gauss elimination for solving a linear system consists of two basic steps: forward elimination (pivoting and shears) and then back substitution (scaling). Here is the algorithm for solving a general $n \times n$ system of linear equations.

Gauss Elimination with Pivoting

Given: An $n \times n$ coefficient matrix A and a $n \times 1$ right-hand side **b** describing a linear system

$$A\mathbf{u} = \mathbf{b},$$

which is short for the more detailed (12.1).

Find: The unknowns $u_1, \ldots, u_n$.

Algorithm:

Initialize the $n \times n$ matrix $G = I$.
For $j = 1, \ldots, n-1$ (j counts columns)

Pivoting step:

Find the element with the largest absolute value in column j from $a_{j,j}$ to $a_{n,j}$; this is element $a_{r,j}$.
    If $r > j$, exchange equations r and j.

If $a_{j,j} = 0$, the system is not solvable.

Forward elimination step for column j:

For $i = j+1, \ldots, n$ (elements below diagonal of column j)
    Construct the *multiplier* $g_{i,j} = a_{i,j}/a_{j,j}$
    $a_{i,j} = 0$
    For $k = j+1, \ldots, n$ (each element in row i after
                                    column j)
        $a_{i,k} = a_{i,k} - g_{i,j}a_{j,k}$
    $b_i = b_i - g_{i,j}b_j$

At this point, all elements below the diagonal have been set to zero. The matrix is now in *upper triangular* form.

Back substitution:

$u_n = b_n/a_{n,n}$
For $j = n-1, \ldots, 1$
    $u_j = \frac{1}{a_{j,j}}[b_j - a_{j,j+1}u_{j+1} - \ldots - a_{j,n}u_n]$.

In a programming environment, it can be convenient to form an *augmented matrix*, which is the matrix A augmented with the vector **b**. Here is the idea for a 3×3 linear system:

$$\begin{bmatrix} a_{1,1} & a_{1,2} & a_{1,3} & b_1 \\ a_{2,1} & a_{2,2} & a_{2,3} & b_2 \\ a_{3,1} & a_{3,2} & a_{3,3} & b_3 \end{bmatrix}.$$

Then the k steps would run to $n+1$, and there would be no need for the extra line for the b_i element.

As demonstrated in Example 12.2, the operations in the elimination step above may be written in matrix form. If A is the current matrix, then at step j, we first check if a row exchange is necessary. This may be achieved with a permutation matrix, P_j. If no pivoting is necessary, $P_j = I$. To produce zeroes under $a_{j,j}$ the matrix product $G_j A$ is formed, where

$$G_j = \begin{bmatrix} 1 & & & & & & \\ & \ddots & & & & & \\ & & 1 & & & & \\ & & & 1 & & & \\ & & & -g_{j+1,j} & 1 & & \\ & & & \vdots & & \ddots & \\ & & & -g_{n,j} & & & 1 \end{bmatrix}. \tag{12.3}$$

The elements $-g_{i,j}$ of G_j are the *multipliers*. The matrix G_j is called a *Gauss matrix*. All entries except for the diagonal and the entries $-g_{i,j}$ below the diagonal of the jth column are zero. We could store all P_j and G_j in one matrix

$$G = G_{n-1} P_{n-1} \cdot \ldots \cdot G_2 \cdot P_2 \cdot G_1 \cdot P_1. \tag{12.4}$$

If no pivoting is required, then it is possible to store the $g_{i,j}$ in the zero elements of A rather than explicitly setting the $a_{i,j}$ element equal to zero. Regardless, it is more efficient with regard to speed and storage not to multiply A and **b** by the permutation and Gauss matrices because this would result in many unnecessary calculations.

To summarize, Gauss elimination with pivoting transforms the linear system $A\mathbf{u} = \mathbf{b}$ into the system

$$GA\mathbf{u} = G\mathbf{b},$$

which has the same solution as the original system. The matrix GA is upper triangular, and it is referred to as U. The diagonal elements of U are the pivots.

Example 12.3

We look at another example, taken from [2]. Let the system be given by

$$\begin{bmatrix} 2 & 2 & 0 \\ 1 & 1 & 2 \\ 2 & 1 & 1 \end{bmatrix} \begin{bmatrix} u_1 \\ u_2 \\ u_3 \end{bmatrix} = \begin{bmatrix} 6 \\ 9 \\ 7 \end{bmatrix}.$$

We start the algorithm with $j = 1$, and observe that no element in column 1 exceeds $a_{1,1}$ in absolute value, so no pivoting is necessary at this step, thus $P_1 = I$. Proceed with the elimination step for row 2 by constructing the multiplier

$$g_{2,1} = a_{2,1}/a_{1,1} = 1/2.$$

Change row 2 as follows:

$$\text{row}_2 \leftarrow \text{row}_2 - \frac{1}{2}\text{row}_1.$$

Remember, this includes changing the element b_2. Similarly for row 3,

$$g_{3,1} = a_{3,1}/a_{1,1} = 2/2 = 1$$

then

$$\text{row}_3 \leftarrow \text{row}_3 - \text{row}_1.$$

Step $j = 1$ is complete and the linear system is now

$$\begin{bmatrix} 2 & 2 & 0 \\ 0 & 0 & 2 \\ 0 & -1 & 1 \end{bmatrix} \begin{bmatrix} u_1 \\ u_2 \\ u_3 \end{bmatrix} = \begin{bmatrix} 6 \\ 6 \\ 1 \end{bmatrix}.$$

The Gauss matrix for $j = 1$,

$$G_1 = \begin{bmatrix} 1 & 0 & 0 \\ -1/2 & 1 & 0 \\ -1 & 0 & 1 \end{bmatrix}.$$

Next is column 2, so $j = 2$. Observe that $a_{2,2} = 0$, whereas $a_{3,2} = -1$. We exchange equations 2 and 3 and the system becomes

$$\begin{bmatrix} 2 & 2 & 0 \\ 0 & -1 & 1 \\ 0 & 0 & 2 \end{bmatrix} \begin{bmatrix} u_1 \\ u_2 \\ u_3 \end{bmatrix} = \begin{bmatrix} 6 \\ 1 \\ 6 \end{bmatrix}. \tag{12.5}$$

The permutation matrix for this exchange is

$$P_2 = \begin{bmatrix} 1 & 0 & 0 \\ 0 & 0 & 1 \\ 0 & 1 & 0 \end{bmatrix}.$$

If blindly following the algorithm above, we would proceed with the elimination for row 3 by forming the multiplier

$$g_{3,2} = \frac{a_{3,2}}{a_{2,2}} = \frac{0}{-1} = 0.$$

Then operate on the third row

$$\text{row}_3 \leftarrow \text{row}_3 - 0 \times \text{row}_2,$$

which doesn't change the row at all. (So we will record $G_2 = I$.) Due to numerical instabilities, $g_{3,2}$ might not be exactly zero. Without putting a special check for a zero multiplier, this unnecessary work takes place. Tolerances are very important here.

Apply back substitution by first solving for the last unknown:

$$u_3 = 3.$$

Start the back substitution loop with $j = 2$:

$$u_2 = \frac{1}{-1}[1 - u_3] = 2,$$

and finally

$$u_1 = \frac{1}{2}[6 - 2u_2] = 1.$$

It's a good idea to check the solution:

$$\begin{bmatrix} 2 & 2 & 0 \\ 1 & 1 & 2 \\ 2 & 1 & 1 \end{bmatrix} \begin{bmatrix} 1 \\ 2 \\ 3 \end{bmatrix} = \begin{bmatrix} 6 \\ 9 \\ 7 \end{bmatrix}.$$

The final matrix $G = G_2 P_2 G_1 P_1$ is

$$G = \begin{bmatrix} 1 & 0 & 0 \\ -1 & 0 & 1 \\ -1/2 & 1 & 0 \end{bmatrix}.$$

Check that $GA\mathbf{u} = G\mathbf{b}$ results in the linear system in (12.5).

Just before back substitution, we could scale to achieve ones along the diagonal of the matrix. Let's do precisely that to the linear system in Example 12.3. Multiply both sides of (12.5) by

$$\begin{bmatrix} 1/2 & 0 & 0 \\ 0 & -1 & 0 \\ 0 & 0 & 1/2 \end{bmatrix}.$$

This transforms the linear system to

$$\begin{bmatrix} 1 & 1 & 0 \\ 0 & 1 & -1 \\ 0 & 0 & 1 \end{bmatrix} \begin{bmatrix} u_1 \\ u_2 \\ u_3 \end{bmatrix} = \begin{bmatrix} 3 \\ -1 \\ 3 \end{bmatrix}.$$

This upper triangular matrix with rank $= n$ is said to be in *row echelon form*. If the matrix is *rank deficient*, rank $< n$, then the rows with all zeroes should be the last rows. Some definitions of row echelon do not require ones along the diagonal, as we have here; it is more efficient to do the scaling as part of back substitution.

Gauss elimination requires $O(n^3)$ operations.[1] Thus this algorithm is suitable for a system with thousands of equations, but not for a system with millions of equations. When the system is very large, often times many of the matrix elements are zero—a sparse linear system. Iterative methods, which are introduced in Section 13.6, are a better approach in this case.

12.3 Homogeneous Linear Systems

Let's revisit the topic of Section 5.8, *homogeneous linear systems*, which take the form

$$A\mathbf{u} = \mathbf{0}.$$

The trivial solution is always an option, but of little interest. How do we use Gauss elimination to find a nontrivial solution if it exists? Once we have one nontrivial solution, all multiples $c\mathbf{u}$ are solutions as well. The answer: slightly modify the back substitution step. An example will make this clear.

---

[1]Read this loosely as: an estimated number of n^3 operations.

Example 12.4

Start with the homogeneous system

$$\begin{bmatrix} 1 & 2 & 3 \\ 1 & 2 & 3 \\ 1 & 2 & 3 \end{bmatrix} \mathbf{u} = \begin{bmatrix} 0 \\ 0 \\ 0 \end{bmatrix}.$$

The matrix clearly has rank one. First perform forward elimination, arriving at

$$\begin{bmatrix} 1 & 2 & 3 \\ 0 & 0 & 0 \\ 0 & 0 & 0 \end{bmatrix} \mathbf{u} = \begin{bmatrix} 0 \\ 0 \\ 0 \end{bmatrix}.$$

For each zero row of the transformed system, set the corresponding u_i, the free variables, to one: $u_3 = 1$ and $u_2 = 1$. Back substituting these into the first equation gives $u_1 = -5$ for the pivot variable. Thus a final solution is

$$\mathbf{u} = \begin{bmatrix} -5 \\ 1 \\ 1 \end{bmatrix},$$

and all vectors $c\mathbf{u}$ are solutions as well.

Since the 3×3 matrix is rank one, it has a two dimensional null space. The number of free variables is equal to the dimension of the null space. We can systematically construct two vectors $\mathbf{u}_1, \mathbf{u}_2$ that span the null space by setting one of the free variables to one and the other to zero, resulting in

$$\mathbf{u}_1 = \begin{bmatrix} -3 \\ 0 \\ 1 \end{bmatrix} \quad \text{and} \quad \mathbf{u}_2 = \begin{bmatrix} -2 \\ 1 \\ 0 \end{bmatrix}.$$

All linear combinations of elements of the null space are also in the null space, for example, $\mathbf{u} = 1\mathbf{u}_1 + 1\mathbf{u}_2$.

Column pivoting might be required to ready the matrix for back substitution.

Example 12.5

The homogeneous system

$$\begin{bmatrix} 0 & 6 & 3 \\ 0 & 0 & 2 \\ 0 & 0 & 0 \end{bmatrix} \mathbf{u} = \begin{bmatrix} 0 \\ 0 \\ 0 \end{bmatrix}.$$

comes from an eigenvector problem similar to those in Section 7.3. (More on eigenvectors in higher dimensions in Chapter 15.)

The linear system in its existing form leads us to $0u_3 = 0$ and $2u_3 = 0$. To remedy this, we proceed with column exchanges:

$$\begin{bmatrix} 6 & 3 & 0 \\ 0 & 2 & 0 \\ 0 & 0 & 0 \end{bmatrix} \begin{bmatrix} u_2 \\ u_3 \\ u_1 \end{bmatrix} = \begin{bmatrix} 0 \\ 0 \\ 0 \end{bmatrix},$$

where column 1 was exchanged with column 2 and then column 2 was exchanged with column 3. Each exchange requires that the associated unknowns are exchanged as well. Set the free variable: $u_1 = 1$, then back substitution results in $u_3 = 0$ and $u_2 = 0$. All vectors

$$c \begin{bmatrix} 1 \\ 0 \\ 0 \end{bmatrix}$$

satisfy the original homogeneous system.

12.4   Inverse Matrices

The *inverse* of a square matrix A is the matrix that "undoes" A's action, i.e., the combined action of A and A^{-1} is the identity:

$$AA^{-1} = I. \tag{12.6}$$

We introduced the essentials of inverse matrices in Section 5.9 and reviewed properties of the inverse in Section 9.11. In this section, we introduce the inverse for $n \times n$ matrices and discuss inverses in the context of solving linear systems, Gauss elimination, and LU decomposition (covered in more detail in Section 12.5).

Example 12.6

The following scheme shows a matrix A multiplied by its inverse A^{-1}. The matrix A is on the left, A^{-1} is on top, and the result of the multiplication, the identity, is on the lower right:

$$
\begin{array}{ccc|ccc}
 & & & 1 & 0 & -1 \\
 & & & 3 & 1 & -3 \\
 & & & 1 & 2 & -2 \\
\hline
-4 & 2 & -1 & 1 & 0 & 0 \\
-3 & 1 & 0 & 0 & 1 & 0 \\
-5 & 2 & -1 & 0 & 0 & 1
\end{array}.
$$

How do we find the inverse of a matrix? In much the same way as we did in the 2×2 case in Section 5.9, we write

$$A \begin{bmatrix} \overline{\mathbf{a}}_1 & \ldots & \overline{\mathbf{a}}_n \end{bmatrix} = \begin{bmatrix} \mathbf{e}_1 & \ldots & \mathbf{e}_n \end{bmatrix}. \tag{12.7}$$

Here, the matrices are $n \times n$, and the vectors $\overline{\mathbf{a}}_i$ as well as the $\mathbf{e}_i$ are vectors with n components. The vector $\mathbf{e}_i$ has all zero entries except for its ith component; it equals 1.

We may now interpret (12.7) as n linear systems:

$$A\overline{\mathbf{a}}_1 = \mathbf{e}_1, \quad \ldots, \quad A\overline{\mathbf{a}}_n = \mathbf{e}_n. \tag{12.8}$$

In Example 5.8, we applied shears and a scaling to transform A into the identity matrix, and at the same time, the right-hand side ($\mathbf{e}_1$ and $\mathbf{e}_2$) was transformed into A^{-1}. Now, with Gauss elimination as per Section 12.2, we apply forward elimination to A and to each of the $\mathbf{e}_i$. Then with back substitution, we solve for each of the $\overline{\mathbf{a}}_i$ that form A^{-1}. However, as we will learn in Section 12.5, a more economical solution is found with LU decomposition. This method is tailored to solving multiple systems of equations that share the same matrix A, but have different right-hand sides.

Inverse matrices are primarily a theoretical concept. They suggest to solve a linear system $A\mathbf{v} = \mathbf{b}$ by computing A^{-1} and then to set $\mathbf{v} = A^{-1}\mathbf{b}$. Don't do that! It is a very expensive way to solve a linear system; simple Gauss elimination or LU decomposition is much cheaper. (Explicitly forming the inverse requires forward elimination, n back substitution algorithms, and then a matrix-vector multiplication. On the other hand, Gauss elimination requires forward elimination and just one back substitution algorithm.)

The inverse of a matrix A exists only if the matrix is square ($n \times n$) and the action of A does not reduce dimensionality, as in a projection. This means that all columns of A must be linearly independent. There is a simple way to see if a matrix A is invertible; just perform Gauss elimination for the first of the linear systems in (12.8). If you are able to transform A to upper triangular with all nonzero diagonal elements, then A is invertible. Otherwise, it is said to be *singular*. The term "nonzero" is to be taken with a grain of salt: real numbers are (almost) never zero, and tolerances must be employed.

Again we encounter the concept of matrix *rank*. An invertible matrix is said to have rank n or *full rank*. If a matrix reduces dimensionality by k, then it has rank $n - k$. The $n \times n$ identity matrix has rank n; the zero matrix has rank 0. An example of a matrix that does not have full rank is a projection. Review the structure of a 2D orthogonal projection in Section 4.8 and a 3D projection in Section 9.7 to confirm this statement about the rank.

Example 12.7

We apply forward elimination to three 4×4 matrices to achieve row echelon form:

$$M_1 = \begin{bmatrix} 1 & 3 & -3 & 0 \\ 0 & 3 & 3 & 1 \\ 0 & 0 & 0 & 0 \\ 0 & 0 & 0 & 0 \end{bmatrix},$$

$$M_2 = \begin{bmatrix} 1 & 3 & -3 & 0 \\ 0 & 3 & 3 & 1 \\ 0 & 0 & -1 & 0 \\ 0 & 0 & 0 & 0 \end{bmatrix},$$

$$M_3 = \begin{bmatrix} 1 & 3 & -3 & 0 \\ 0 & 3 & 3 & 1 \\ 0 & 0 & -1 & 0 \\ 0 & 0 & 0 & 2 \end{bmatrix}.$$

M_1 has rank 2, M_2 has rank 3, and M_3 has rank 4 or full rank.

Example 12.8

Let us compute the inverse of the $n \times n$ matrix G_j as defined in (12.3):

$$G_j^{-1} = \begin{bmatrix} 1 & & & & & & \\ & \ddots & & & & & \\ & & 1 & & & & \\ & & & 1 & & & \\ & & & g_{j+1,j} & 1 & & \\ & & & \vdots & & \ddots & \\ & & & g_{n,j} & & & 1 \end{bmatrix}.$$

That's simple! To make some geometric sense of this, you should realize that G_j is a shear, and so is G_j^{-1}, and it "undoes" G_j.

Here is another interesting property of the inverse of a matrix. Suppose $k \neq 0$ and kA is an invertible matrix, then

$$(kA)^{-1} = \frac{1}{k}A^{-1}.$$

And yet another: If two matrices, A and B, are invertible, then the product AB is invertible, too.

12.5 LU Decomposition

Gauss elimination has two major parts: transforming the system to upper triangular form with forward elimination and back substitution. The creation of the upper triangular matrix may be written in terms of matrix multiplications using Gauss matrices G_j. For now, assume that no pivoting is necessary. If we denote the final upper triangular matrix by U, then we have

$$G_{n-1} \cdot \ldots \cdot G_1 \cdot A = U. \tag{12.9}$$

It follows that

$$A = G_1^{-1} \cdot \ldots \cdot G_{n-1}^{-1} U.$$

The neat thing about the product $G_1^{-1} \cdot \ldots \cdot G_{n-1}^{-1}$ is that it is a *lower triangular* matrix with elements $g_{i,j}$ below the diagonal and zeroes above the diagonal:

$$G_1^{-1} \cdot \ldots \cdot G_{n-1}^{-1} = \begin{bmatrix} 1 & & & \\ g_{2,1} & 1 & & \\ \vdots & \ddots & \ddots & \\ g_{n,1} & \cdots & g_{n,n-1} & 1 \end{bmatrix}.$$

We denote this product by L (for lower triangular). Thus,

$$A = LU, \tag{12.10}$$

which is known as the *LU decomposition* of A. It is also called the *triangular factorization* of A. Every invertible matrix A has such a decomposition, although it may be necessary to employ pivoting.

Denote the elements of L by $l_{i,j}$ (keeping in mind that $l_{i,i} = 1$) and those of U by $u_{i,j}$. A simple 3×3 example will help illustrate the idea.

$$\begin{array}{ccc|ccc} & & & u_{1,1} & u_{1,2} & u_{1,3} \\ & & & 0 & u_{2,2} & u_{2,3} \\ & & & 0 & 0 & u_{3,3} \\ \hline 1 & 0 & 0 & a_{1,1} & a_{1,2} & a_{1,3} \\ l_{2,1} & 1 & 0 & a_{2,1} & a_{2,2} & a_{2,3} \\ l_{3,1} & l_{3,2} & 1 & a_{3,1} & a_{3,2} & a_{3,3} \end{array}.$$

In this scheme, we are given the $a_{i,j}$ and we want the $l_{i,j}$ and $u_{i,j}$. This is systematically achieved using the following.

Observe that elements of A below the diagonal may be rewritten as

$$a_{i,j} = l_{i,1}u_{1,j} + \ldots + l_{i,j-1}u_{j-1,j} + l_{i,j}u_{j,j}; \quad j < i,$$

For the elements of A that are on or above the diagonal, we get

$$a_{i,j} = l_{i,1}u_{1,j} + \ldots + l_{i,i-1}u_{i-1,j} + l_{i,i}u_{i,j}; \quad j \geq i.$$

This leads to

$$l_{i,j} = \frac{1}{u_{j,j}}(a_{i,j} - l_{i,1}u_{1,j} - \ldots - l_{i,j-1}u_{j-1,j}); \quad j < i \tag{12.11}$$

and

$$u_{i,j} = a_{i,j} - l_{i,1}u_{1,j} - \ldots - l_{i,i-1}u_{i-1,j}; \quad j \geq i. \tag{12.12}$$

If A has a decomposition $A = LU$, then the system can be written as

$$LU\mathbf{u} = \mathbf{b}. \tag{12.13}$$

The matrix vector product $U\mathbf{u}$ results in a vector; call this $\mathbf{y}$. Reexamining (12.13), it becomes a two-step problem. First solve

$$L\mathbf{y} = \mathbf{b}, \qquad (12.14)$$

then solve

$$U\mathbf{u} = \mathbf{y}. \qquad (12.15)$$

If $U\mathbf{u} = \mathbf{y}$, then $LU\mathbf{u} = L\mathbf{y} = \mathbf{b}$. The two systems in (12.14) and (12.15) are triangular and easy to solve. *Forward substitution* is applied to the matrix L. (See Exercise 21 and its solution for an algorithm.) Back substitution is applied to the matrix U. An algorithm is provided in Section 12.2.

A more direct method for forming L and U is achieved with (12.11) and (12.12), rather than through Gauss elimination. This then is the method of LU decomposition.

LU Decomposition

Given: A coefficient matrix A and a right-hand side $\mathbf{b}$ describing a linear system
$$A\mathbf{u} = \mathbf{b}.$$

Find: The unknowns $u_1, \ldots, u_n$.

Algorithm:

> Initialize L as the identity matrix and U as the zero matrix.
> Calculate the nonzero elements of L and U:
> For $k = 1, \ldots, n$
> $\quad u_{k,k} = a_{k,k} - l_{k,1}u_{1,k} - \ldots - l_{k,k-1}u_{k-1,k}$
> $\quad$ For $i = k+1, \ldots, n$
> $\quad\quad l_{i,k} = \dfrac{1}{u_{k,k}}[a_{i,k} - l_{i,1}u_{1,k} - \ldots - l_{i,k-1}u_{k-1,k}]$
> $\quad$ For $j = k+1, \ldots, n$
> $\quad\quad u_{k,j} = a_{k,j} - l_{k,1}u_{1,j} - \ldots - l_{k,k-1}u_{k-1,j}$
>
> Using forward substitution solve $L\mathbf{y} = \mathbf{b}$.
> Using back substitution solve $U\mathbf{u} = \mathbf{y}$.

The $u_{k,k}$ term must not be zero; we had a similar situation with Gauss elimination. This situation either requires pivoting or the matrix might be singular.

The construction of the LU decomposition takes advantage of the triangular structure of L and U combined with a particular computation order. The matrix L is being filled column by column and the matrix U is being filled row by row.

Example 12.9

Let's use LU decomposition to solve the linear system

$$A = \begin{bmatrix} 2 & 2 & 4 \\ -1 & 2 & -3 \\ 1 & 2 & 2 \end{bmatrix} \mathbf{u} = \begin{bmatrix} 1 \\ 1 \\ 1 \end{bmatrix}.$$

The first step is to decompose A. Following the steps in the algorithm above, we calculate the matrix entries:

$$k = 1: \quad u_{1,1} = a_{1,1} = 2,$$
$$l_{2,1} = \frac{a_{2,1}}{u_{1,1}} = \frac{-1}{2},$$
$$l_{3,1} = \frac{a_{3,1}}{u_{1,1}} = \frac{1}{2},$$
$$u_{1,2} = a_{1,2} = 2,$$
$$u_{1,3} = a_{1,3} = 4,$$

$$k = 2: \quad u_{2,2} = a_{2,2} - l_{2,1}u_{1,2} = 2 + 1 = 3,$$
$$l_{3,2} = \frac{1}{u_{2,2}}[a_{3,2} - l_{3,1}u_{1,2}] = \frac{1}{3}[2 - 1] = \frac{1}{3},$$
$$u_{2,3} = a_{2,3} - l_{2,1}u_{1,3} = -3 + 2 = -1,$$

$$k = 3: \quad u_{3,3} = a_{3,3} - l_{3,1}u_{1,3} - l_{3,2}u_{2,3} = 2 - 2 + \frac{1}{3} = \frac{1}{3}.$$

Check that this produces valid entries for L and U:

$$
\begin{array}{ccc|ccc}
 & & & 2 & 2 & 4 \\
 & & & 0 & 3 & -1 \\
 & & & 0 & 0 & 1/3 \\
\hline
1 & 0 & 0 & 2 & 2 & 4 \\
-1/2 & 1 & 0 & -1 & 2 & -3 \\
1/2 & 1/3 & 1 & 1 & 2 & 2
\end{array}.
$$

Next, we solve $L\mathbf{y} = \mathbf{b}$ with forward substitution—solving for y_1, then y_2, and then y_3—and find that

$$\mathbf{y} = \begin{bmatrix} 1 \\ 3/2 \\ 0 \end{bmatrix}.$$

The last step is to solve $U\mathbf{u} = \mathbf{y}$ with back substitution—as we did in Gauss elimination,

$$\mathbf{u} = \begin{bmatrix} 0 \\ 1/2 \\ 0 \end{bmatrix}.$$

It is simple to check that this solution is correct since clearly, the column vector $\mathbf{a}_2$ is a multiple of $\mathbf{b}$, and that is reflected in $\mathbf{u}$.

Suppose A is nonsingular, but in need of pivoting. Then a permutation matrix P is used to exchange (possibly multiple) rows so it is possible to create the LU decomposition. The system is now $PA\mathbf{u} = P\mathbf{b}$ and we find $PA = LU$.

Finally, the major benefit of the LU decomposition: speed. For cases in which we have to solve multiple linear systems with the same coefficient matrix, LU decomposition is a big timesaver. We perform it once, and then perform the forward and backward substitutions (12.14) and (12.15) for each right-hand side. This is significantly less work than performing a complete Gauss elimination every time! Finding the inverse of a matrix, as described in (12.8), is an example of a problem that requires solving multiple linear systems with the same coefficient matrix.

12.6   Determinants

With the introduction of the scalar triple product, Section 8.5 provided a geometric derivation of 3×3 determinants; they measure volume. And then in Section 9.8 we learned more about determinants from the perspective of linear maps. Let's revisit that approach for $n \times n$ determinants.

When we apply forward elimination to A, transforming it to upper triangular form U, we apply a sequence of shears and row exchanges. Shears do not change the volumes. As we learned in Section 9.8, a row exchange will change the sign of the determinant. Thus the column vectors of U span the same volume as did those of A, however the sign might change. This volume is given by the signed product of the diagonal entries of U and is called the *determinant* of A:

$$\det A = (-1)^k(u_{1,1} \times \ldots \times u_{n,n}), \tag{12.16}$$

where k is the number of row exchanges. In general, this is the best (and most stable) method for finding the determinant (but also see the method in Section 16.4).

Example 12.10

Let's revisit Example 12.3 to illustrate how to calculate the determinant with the upper triangular form, and how row exchanges influence the sign of the determinant.

Use the technique of cofactor expansion, as defined by (9.14) to find the determinant of the given 3×3 matrix A:

$$\det A = 2\begin{vmatrix} 1 & 2 \\ 1 & 1 \end{vmatrix} - 2\begin{vmatrix} 1 & 2 \\ 2 & 1 \end{vmatrix} = 4.$$

Now, apply (12.16) to the upper triangular form U (12.5) from the example, and notice that we did one row exchange, $k = 1$:

$$\det A = (-1)^1[2 \times -1 \times 2] = 4.$$

So the shears of Gauss elimination have not changed the absolute value of the determinant, and by modifying the sign based on the number of row exchanges, we can determine $\det A$ from $\det U$.

The technique of cofactor expansion that was used for the 3×3 matrix in Example 12.10 may be generalized to $n \times n$ matrices. Choose any column or row of the matrix, for example entries $a_{1,j}$ as above, and then

$$\det A = a_{1,1}C_{1,1} + a_{1,2}C_{1,2} + \ldots + a_{1,n}C_{1,n},$$

where each cofactor is defined as

$$C_{i,j} = (-1)^{i+j}M_{i,j},$$

and the $M_{i,j}$ are called the *minors*; each is the determinant of the matrix with the ith row and jth column removed. The $M_{i,j}$ are $(n-1) \times (n-1)$ determinants, and they are computed by yet another cofactor expansion. This process is repeated until we have 2×2 determinants. This technique is also known as *expansion by minors*.

Example 12.11

Let's look at repeated application of cofactor expansion to find the determinant. Suppose we are given the following matrix,

$$A = \begin{bmatrix} 2 & 2 & 0 & 4 \\ 0 & -1 & 1 & 3 \\ 0 & 0 & 2 & 0 \\ 0 & 0 & 0 & 5 \end{bmatrix}.$$

We may choose any row or column from which to form the cofactors, so in this example, we will have less work to do if we choose the first column. The cofactor expansion is

$$\det A = 2 \begin{vmatrix} -1 & 1 & 3 \\ 0 & 2 & 0 \\ 0 & 0 & 5 \end{vmatrix} = 2(-1) \begin{vmatrix} 2 & 0 \\ 0 & 5 \end{vmatrix} = 2(-1)(10) = -20.$$

Since the matrix is in upper triangular form, we could use (12.16) and immediately see that this is in fact the correct determinant.

Cofactor expansion is more a theoretical tool than a computational one. This method of calculating the determinant plays an important theoretical role in the analysis of linear systems, and there are advanced theorems involving cofactor expansion and the inverse of a matrix. Computationally, Gauss elimination and the calculation of $\det U$ is superior.

In our first encounter with solving linear systems via Cramer's rule in Section 5.3, we learned that the solution to a linear system may be found by simply forming quotients of areas. Now with our knowledge of $n \times n$ determinants, let's revisit *Cramer's rule*. If $A\mathbf{u} = \mathbf{b}$ is an $n \times n$ linear system such that $\det A \neq 0$, then the system has the following unique solution:

$$u_1 = \frac{\det A_1}{\det A}, \quad u_2 = \frac{\det A_2}{\det A}, \quad \ldots, \quad u_n = \frac{\det A_n}{\det A}, \qquad (12.17)$$

where A_i is the matrix obtained by replacing the entries in the ith column by **b**. Cramer's rule is an important theoretical tool; however, use it *only* for 2×2 or 3×3 linear systems.

Example 12.12

Let's solve the linear system from Example 12.3 using Cramer's rule. Following (12.17), we have

$$u_1 = \frac{\begin{vmatrix} 6 & 2 & 0 \\ 9 & 1 & 2 \\ 7 & 1 & 1 \end{vmatrix}}{\begin{vmatrix} 2 & 2 & 0 \\ 1 & 1 & 2 \\ 2 & 1 & 1 \end{vmatrix}}, \quad u_2 = \frac{\begin{vmatrix} 2 & 6 & 0 \\ 1 & 9 & 2 \\ 2 & 7 & 1 \end{vmatrix}}{\begin{vmatrix} 2 & 2 & 0 \\ 1 & 1 & 2 \\ 2 & 1 & 1 \end{vmatrix}}, \quad u_3 = \frac{\begin{vmatrix} 2 & 2 & 6 \\ 1 & 1 & 9 \\ 2 & 1 & 7 \end{vmatrix}}{\begin{vmatrix} 2 & 2 & 0 \\ 1 & 1 & 2 \\ 2 & 1 & 1 \end{vmatrix}}.$$

We have computed the determinant of the coefficient matrix A in Example 12.10, $\det A = 4$. With the application of cofactor expansion for each numerator, we find that

$$u_1 = \frac{4}{4} = 1, \quad u_2 = \frac{8}{4} = 2, \quad u_3 = \frac{12}{4} = 3,$$

which is identical to the solution found with Gauss elimination.

The determinant of a positive definite matrix is always positive, and therefore the matrix is always nonsingular. The *upper-left submatrices* of an $n \times n$ matrix A are

$$A_1 = \begin{bmatrix} a_{1,1} \end{bmatrix}, \quad A_2 = \begin{bmatrix} a_{1,1} & a_{1,2} \\ a_{2,1} & a2,2 \end{bmatrix}, \quad \ldots, \quad A_n = A.$$

If A is positive definite, then the determinants of all A_i are positive.

Rules for working with determinants are given in Section 9.8.

12.7   Least Squares

When presented with large amounts of data, we often look for methods to create a simpler view or synopsis of the data. For example, Figure 12.3 is a graph of AIG's monthly average stock price over twelve years. We see a lot of activity in the price, but there is a clear

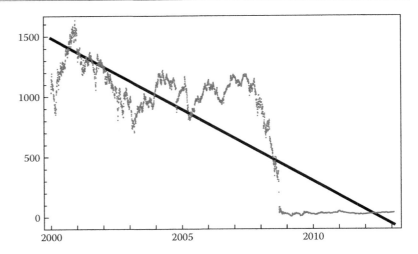

Figure 12.3.
Least squares: fitting a straight line to stock price data for AIG from 2000 to 2013.

declining trend. A mathematical tool to capture this, which works when the trend is not as clear as it is here, is *linear least squares approximation.* The line illustrated in Figure 12.3 is the "best fit" line or best approximating line.

Linear least squares approximation is also useful when analyzing experimental data, which can be "noisy," either from the data capture or observation method or from round-off from computations that generated the data. We might want to make summary statements about the data, estimate values where data is missing, or predict future values.

As a concrete (simple) example, suppose our experimental data are temperature (Celsius) over time (seconds):

$$\begin{bmatrix} \text{time} \\ \text{temperature} \end{bmatrix} \quad \begin{bmatrix} 0 \\ 30 \end{bmatrix} \begin{bmatrix} 10 \\ 25 \end{bmatrix} \begin{bmatrix} 20 \\ 40 \end{bmatrix} \begin{bmatrix} 30 \\ 40 \end{bmatrix} \begin{bmatrix} 40 \\ 30 \end{bmatrix} \begin{bmatrix} 50 \\ 5 \end{bmatrix} \begin{bmatrix} 60 \\ 25 \end{bmatrix},$$

which are plotted in Figure 12.4. We want to establish a simple linear relationship between the variables,

$$\text{temperature} = a \times \text{time} + b,$$

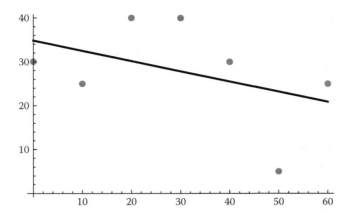

Figure 12.4.
Least squares: a linear approximation to experimental data of time and temperature pairs.

Writing down all relationships between knowns and unknowns, we obtain linear equations of the form

$$
\begin{bmatrix} 0 & 1 \\ 10 & 1 \\ 20 & 1 \\ 30 & 1 \\ 40 & 1 \\ 50 & 1 \\ 60 & 1 \end{bmatrix} \begin{bmatrix} a \\ b \end{bmatrix} = \begin{bmatrix} 30 \\ 25 \\ 40 \\ 40 \\ 30 \\ 5 \\ 25 \end{bmatrix}. \tag{12.18}
$$

We write the system as

$$
A\mathbf{u} = \mathbf{b}, \qquad \text{where} \quad \mathbf{u} = \begin{bmatrix} a \\ b \end{bmatrix}.
$$

This system of seven equations in two unknowns is *overdetermined* and in general it will not have solutions; it is inconsistent. After all, it is not very likely that **b** lives in the subspace $\mathcal{V}$ formed by the columns of A. (As an analogy, consider the likelihood of a randomly selected 3D vector living in the $[\mathbf{e}_1, \mathbf{e}_2]$-plane.) But there is a recipe for finding an *approximate solution*.

Denoting by $\mathbf{b}'$ a vector in $\mathcal{V}$, the system

$$
A\mathbf{u} = \mathbf{b}' \tag{12.19}
$$

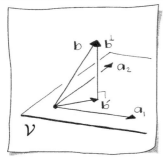

Sketch 12.3.

Least squares.

is solvable (consistent), but it is still overdetermined since we have seven equations in two unknowns. Recall from Section 2.8 that we can write $\mathbf{b}$ as the sum of its orthogonal projection into $\mathcal{V}$ and the component of $\mathbf{b}$ orthogonal to $\mathcal{V}$,

$$\mathbf{b} = \mathbf{b}' + \mathbf{b}^{\perp}. \tag{12.20}$$

Also recall that $\mathbf{b}'$ is closest to $\mathbf{b}$ and in $\mathcal{V}$. Sketch 12.3 illustrates this idea in 3D.

Since $\mathbf{b}^{\perp}$ is orthogonal to $\mathcal{V}$, we can use matrix notation to formalize this relationship,

$$\mathbf{a}_1^{\mathrm{T}}\mathbf{b}^{\perp} = 0 \qquad \text{and} \qquad \mathbf{a}_2^{\mathrm{T}}\mathbf{b}^{\perp} = 0,$$

which is equivalent to

$$A^{\mathrm{T}}\mathbf{b}^{\perp} = \mathbf{0}.$$

Based on (12.20), we can substitute $\mathbf{b} - \mathbf{b}'$ for $\mathbf{b}^{\perp}$,

$$A^{\mathrm{T}}(\mathbf{b} - \mathbf{b}') = \mathbf{0}$$
$$A^{\mathrm{T}}(\mathbf{b} - A\mathbf{u}) = \mathbf{0}$$
$$A^{\mathrm{T}}\mathbf{b} - A^{\mathrm{T}}A\mathbf{u} = \mathbf{0}.$$

Rearranging this last equation, we have the *normal equations*

$$A^{\mathrm{T}}A\mathbf{u} = A^{\mathrm{T}}\mathbf{b}. \tag{12.21}$$

This is a linear system with a square matrix $A^{\mathrm{T}}A$! Even more, that matrix is symmetric. The solution to the new system (12.21), when it has one, is the one that minimizes the *error*

$$\|A\mathbf{u} - \mathbf{b}\|^2,$$

and for this reason, it is called the *least squares solution* of the original system. Recall that $\mathbf{b}'$ is closest to $\mathbf{b}$ in $\mathcal{V}$ and since we solved (12.19), we have in effect minimized $\|\mathbf{b}' - \mathbf{b}\|$.

It seems pretty amazing that by simply multiplying both sides by A^{T}, we have a "best" solution to the original problem!

Example 12.13

Returning to the system in (12.18), we form the normal equations,

$$\begin{bmatrix} 9100 & 210 \\ 210 & 7 \end{bmatrix} \begin{bmatrix} a \\ b \end{bmatrix} = \begin{bmatrix} 5200 \\ 195 \end{bmatrix}.$$

(Notice that the matrix is symmetric.)

The least squares solution is the solution of this linear system,

$$\begin{bmatrix} a \\ b \end{bmatrix} = \begin{bmatrix} -0.23 \\ 34.8 \end{bmatrix},$$

which corresponds to the line $x_2 = -0.23x_1 + 34.8$. Figure 12.4 illustrates this line with negative slope and $\mathbf{e}_2$ intercept of 34.8.

Imagine a scenario where our data capture method failed due to some environmental condition. We might want to remove data points if they seem outside the norm. These are called *outliers*. Point six in Figure 12.4 looks to be an outlier. The least squares approximating line provides a means for determining that this point is something of an exception.

Linear least squares approximation can also serve as a method for data compression.

Numerical problems can creep into the normal equations of the linear system (12.21). This is particularly so when the $n \times m$ matrix A has many more equations than unknowns, $n \gg m$. In Section 13.1, we will examine the Householder method for finding the least squares solution to the linear system $A\mathbf{u} = \mathbf{b}$ directly, without forming the normal equations. Example 12.13 will be revisited in Example 13.3. And yet another look at the least squares solution is possible with the singular value decomposition of A in Section 16.6.

12.8   Application: Fitting Data to a Femoral Head

In prosthetic surgery, a common task is that of *hip bone replacement*. This involves removing an existing femoral head and replacing it by a transplant, consisting of a new head and a shaft for attaching it into the existing femur. The transplant is typically made from titanium or ceramic; the part that is critical for perfect fit and thus function is the spherical head as shown in Figure 12.5. Data points are collected from the existing femoral head by means of MRI or PET scans, then a spherical fit is obtained, and finally the transplant is manufactured. The fitting process is explained next.

We are given a set of 3D vectors $\mathbf{v}_1, \ldots, \mathbf{v}_L$ that are of approximately equal length, $\rho_1, \ldots, \rho_L$. We would like to find a sphere (centered at the origin) with radius r that closely fits the $\mathbf{v}_i$.

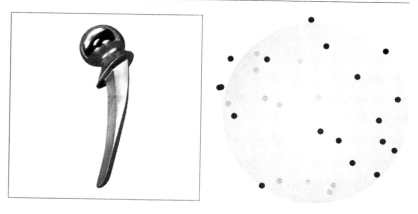

Figure 12.5.
Femur transplant: left, a titanium femoral head with shaft. Right, an example of a sphere fit. Black points are "in front," gray points are occluded.

If all $\mathbf{v}_i$ were on that sphere, we would have

$$r \quad = \rho_1 \tag{12.22}$$

$$\vdots \tag{12.23}$$

$$r \quad = \rho_L. \tag{12.24}$$

This is a very overdetermined linear system—L equations in only one unknown, r!

In matrix form:

$$
\begin{bmatrix} 1 \\ \vdots \\ 1 \end{bmatrix} [r] = \begin{bmatrix} \rho_1 \\ \vdots \\ \rho_L \end{bmatrix}.
$$

Be sure to verify that the matrix dimensions work out!

Multiplying both sides by $[1 \quad \ldots \quad 1]$ gives

$$Lr = \rho_1 + \ldots + \rho_L$$

with the final result

$$r = \frac{\rho_1 + \ldots + \rho_L}{L}.$$

Thus the least squares solution is simply the average of the given radii—just as our intuition would have suggested in the first place. Things are not that simple if it comes to more unknowns, see Section 16.6.

- $n \times n$ linear system
- coefficient matrix
- consistent system
- subspace
- solvable system
- unsolvable system
- Gauss elimination
- upper triangular matrix
- forward elimination
- back substitution
- elementary row operation
- permutation matrix
- row echelon form
- pivoting
- Gauss matrix
- multiplier
- augmented matrix
- singular matrix
- matrix rank
- full rank
- rank deficient
- homogeneous linear system
- inverse matrix
- LU decomposition
- factorization
- forward substitution
- lower triangular matrix
- determinant
- cofactor expansion
- expansion by minors
- Cramer's rule
- overdetermined system
- least squares solution
- normal equations

12.9 Exercises

1. Does the linear system

$$\begin{bmatrix} 1 & 2 & 0 \\ 0 & 0 & 0 \\ 1 & 2 & 1 \end{bmatrix} \mathbf{u} = \begin{bmatrix} 1 \\ 2 \\ 3 \end{bmatrix}$$

   have a unique solution? Is it consistent?

2. Does the linear system

$$\begin{bmatrix} 1 & 1 & 5 \\ 1 & -1 & 1 \\ 1 & 2 & 7 \end{bmatrix} \mathbf{u} = \begin{bmatrix} 3 \\ 3 \\ 3 \end{bmatrix}$$

   have a unique solution? Is it consistent?

3. Examine the linear system in Example 12.1. What restriction on the t_i is required to guarantee a unique solution?

4. Solve the linear system $A\mathbf{v} = \mathbf{b}$ where

$$A = \begin{bmatrix} 1 & 0 & -1 & 2 \\ 0 & 0 & 1 & -2 \\ 2 & 0 & 0 & 1 \\ 1 & 1 & 1 & 0 \end{bmatrix}, \quad \text{and} \quad \mathbf{b} = \begin{bmatrix} -1 \\ 2 \\ 1 \\ -3 \end{bmatrix}.$$

   Show all the steps from the Gauss elimination algorithm.

5. Solve the linear system $A\mathbf{v} = \mathbf{b}$ where

$$A = \begin{bmatrix} 0 & 0 & 1 \\ 1 & 0 & 0 \\ 1 & 1 & 1 \end{bmatrix}, \quad \text{and} \quad \mathbf{b} = \begin{bmatrix} -1 \\ 0 \\ -1 \end{bmatrix}.$$

Show all the steps from the Gauss elimination algorithm.

6. Solve the linear system $A\mathbf{v} = \mathbf{b}$ where

$$A = \begin{bmatrix} 4 & 2 & 1 \\ 2 & 2 & 0 \\ 4 & 2 & 3 \end{bmatrix}, \quad \text{and} \quad \mathbf{b} = \begin{bmatrix} 7 \\ 2 \\ 9 \end{bmatrix}.$$

7. Transform the following linear system to row echelon form.

$$\begin{bmatrix} 3 & 2 & 0 \\ 3 & 1 & 2 \\ 0 & 2 & 0 \end{bmatrix} \mathbf{u} = \begin{bmatrix} 1 \\ 1 \\ 1 \end{bmatrix}.$$

8. What is the rank of the following matrix?

$$\begin{bmatrix} 3 & 2 & 0 & 1 \\ 0 & 0 & 0 & 1 \\ 0 & 1 & 2 & 0 \\ 0 & 0 & 0 & 1 \end{bmatrix}$$

9. What is the permutation matrix that will exchange rows 3 and 4 in a 5×5 matrix?

10. What is the permutation matrix that will exchange rows 2 and 4 in a 4×4 matrix?

11. What is the matrix G as defined in (12.4) for Example 12.2?

12. What is the matrix G as defined in (12.4) for Exercise 6?

13. Solve the linear system

$$\begin{bmatrix} 4 & 1 & 2 \\ 2 & 1 & 1 \\ 2 & 1 & 1 \end{bmatrix} \mathbf{u} = \begin{bmatrix} 0 \\ 0 \\ 0 \end{bmatrix}$$

with Gauss elimination with pivoting.

14. Solve the linear system

$$\begin{bmatrix} 3 & 6 & 1 \\ 6 & 12 & 2 \\ 9 & 18 & 3 \end{bmatrix} \mathbf{u} = \begin{bmatrix} 0 \\ 0 \\ 0 \end{bmatrix}$$

with Gauss elimination with pivoting.

15. Find the inverse of the matrix from Exercise 5.

16. Find the inverse of
$$\begin{bmatrix} 3 & 2 & 1 \\ 0 & 2 & 1 \\ 0 & 2 & 1 \end{bmatrix}.$$

17. Find the inverse of
$$\begin{bmatrix} \cos\theta & 0 & -\sin\theta \\ 0 & 1 & 0 \\ \sin\theta & 0 & \cos\theta \end{bmatrix}.$$

18. Find the inverse of
$$\begin{bmatrix} 5 & 0 & 0 & 0 & 0 \\ 0 & 4 & 0 & 0 & 0 \\ 0 & 0 & 3 & 0 & 0 \\ 0 & 0 & 0 & 2 & 0 \\ 0 & 0 & 0 & 0 & 1 \end{bmatrix}.$$

19. Find the inverse of
$$\begin{bmatrix} 3 & 0 \\ 2 & 1 \\ 1 & 4 \\ 3 & 2 \end{bmatrix}.$$

20. Calculate the LU decomposition of the matrix
$$A = \begin{bmatrix} 3 & 0 & 1 \\ 1 & 2 & 0 \\ 1 & 1 & 1 \end{bmatrix}.$$

21. Write a forward substitution algorithm for solving the lower triangular system (12.14).

22. Use the LU decomposition of A from Exercise 20 to solve the linear system $A\mathbf{u} = \mathbf{b}$, where
$$\mathbf{b} = \begin{bmatrix} 4 \\ 0 \\ 4 \end{bmatrix}.$$

23. Calculate the determinant of
$$A = \begin{bmatrix} 3 & 0 & 1 \\ 1 & 2 & 0 \\ 1 & 1 & 1 \end{bmatrix}.$$

24. What is the rank of the matrix in Exercise 23?

25. Calculate the determinant of the matrix
$$A = \begin{bmatrix} 2 & 4 & 3 & 6 \\ 1 & 0 & 0 & 0 \\ 2 & 1 & 0 & 1 \\ 1 & 1 & 2 & 0 \end{bmatrix}$$

using expansion by minors. Show all steps.

26. Apply Cramer's rule to solve the following linear system:

$$\begin{bmatrix} 3 & 0 & 1 \\ 1 & 2 & 0 \\ 1 & 1 & 1 \end{bmatrix} \mathbf{u} = \begin{bmatrix} 8 \\ 6 \\ 6 \end{bmatrix}.$$

Hint: Reuse your work from Exercise 23.

27. Apply Cramer's rule to solve the following linear system:

$$\begin{bmatrix} 3 & 0 & 1 \\ 0 & 2 & 0 \\ 0 & 2 & 1 \end{bmatrix} \mathbf{u} = \begin{bmatrix} 6 \\ 4 \\ 7 \end{bmatrix}.$$

28. Set up and solve the linear system for solving the intersection of the three planes,

$$x_1 + x_3 = 1, \quad x_3 = 1, \quad x_2 = 2.$$

29. Find the intersection of the plane

$$\mathbf{x}(u_1, u_2) = \begin{bmatrix} 0 \\ 0 \\ 1 \end{bmatrix} + u_1 \begin{bmatrix} 1 \\ 0 \\ -1 \end{bmatrix} u_2 \begin{bmatrix} 0 \\ 1 \\ -1 \end{bmatrix}$$

and the line

$$\mathbf{p}(t) = \begin{bmatrix} 1 \\ 1 \\ 2 \end{bmatrix} + t \begin{bmatrix} 0 \\ 0 \\ 1 \end{bmatrix}$$

by setting up the problem as a linear system and solving it.

30. Let five points be given by

$$\mathbf{p}_1 = \begin{bmatrix} -2 \\ 2 \end{bmatrix}, \quad \mathbf{p}_2 = \begin{bmatrix} -1 \\ 1 \end{bmatrix}, \quad \mathbf{p}_3 = \begin{bmatrix} 0 \\ 0 \end{bmatrix}, \quad \mathbf{p}_4 = \begin{bmatrix} 1 \\ 1 \end{bmatrix}, \quad \mathbf{p}_5 = \begin{bmatrix} 2 \\ 2 \end{bmatrix}.$$

Find the linear least squares approximation.

31. Let five points be given by

$$\mathbf{p}_1 = \begin{bmatrix} 0 \\ 0 \end{bmatrix}, \quad \mathbf{p}_2 = \begin{bmatrix} 1 \\ 1 \end{bmatrix}, \quad \mathbf{p}_3 = \begin{bmatrix} 2 \\ 2 \end{bmatrix}, \quad \mathbf{p}_4 = \begin{bmatrix} 1 \\ -1 \end{bmatrix}, \quad \mathbf{p}_5 = \begin{bmatrix} 2 \\ -2 \end{bmatrix}.$$

Find the linear least squares approximation.

32. Let two points be given by

$$\mathbf{p}_1 = \begin{bmatrix} -4 \\ 1 \end{bmatrix}, \quad \mathbf{p}_2 = \begin{bmatrix} 4 \\ 3 \end{bmatrix}.$$

Find the linear least squares approximation.

13

Alternative System Solvers

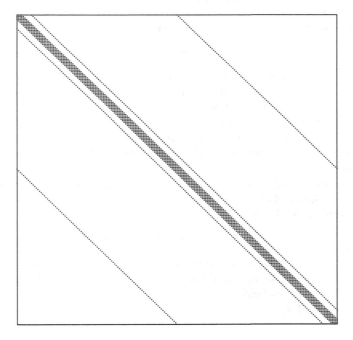

Figure 13.1.
A sparse matrix: all nonzero entries are marked.

We have encountered Gauss elimination methods for solving linear systems. These work well for moderately sized systems (up to a few thousand equations), in particular if they do not pose numerical problems. Ill-conditioned problems are more efficiently attacked using the Householder method below. Huge linear systems (up to a million equations) are more successfully solved with iterative methods. Often times, these linear systems are defined by a sparse matrix, one that has many zero entries, as is illustrated in Figure 13.1. All these alternative methods are the topics of this chapter.

13.1   The Householder Method

Let's revisit the problem of solving the linear system $A\mathbf{u} = \mathbf{b}$, where A is an $n \times n$ matrix. We may think of A as n column vectors, each with n elements,

$$[\mathbf{a}_1 \ldots \mathbf{a}_n]\mathbf{u} = \mathbf{b}.$$

In Section 12.2 we examined the classical method of Gauss elimination, or the process of applying shears G_i to the vectors $\mathbf{a}_1 \ldots \mathbf{a}_n$ and $\mathbf{b}$ in order to convert A to upper triangular form, or

$$G_{n-1} \ldots G_1 A\mathbf{u} = G_{n-1} \ldots G_1 \mathbf{b},$$

and then we were able to solve for $\mathbf{u}$ with back substitution. Each G_i is a shear matrix, constructed to transform the ith column vector $G_{i-1} \ldots G_1 \mathbf{a}_i$ to a vector with zeroes below the diagonal element, $a_{i,i}$.

As it turns out, Gauss elimination is not the most robust method for solving a linear system. A more numerically stable method may be found by replacing shears with *reflections*. This is the Householder method.

The Householder method applied to a linear system takes the same form as Gauss elimination. A series of reflections H_i is constructed and applied to the system,

$$H_{n-1} \ldots H_1 A\mathbf{u} = H_{n-1} \ldots H_1 \mathbf{b},$$

where each H_i transforms the column vector $H_{i-1} \ldots H_1 \mathbf{a}_i$ to a vector with zeroes below the diagonal element. Let's examine how we construct a *Householder transformation* H_i.

A simple 2×2 matrix

$$\begin{bmatrix} 1 & -2 \\ 1 & 0 \end{bmatrix}$$

will help to illustrate the construction of a Householder transformation. The column vectors of this matrix are illustrated in Sketch 13.1. The first transformation, $H_1 A$, reflects $\mathbf{a}_1$ onto the $\mathbf{e}_1$ axis to the vector $\mathbf{a}'_1 = \|\mathbf{a}_1\|\mathbf{e}_1$, or

$$\begin{bmatrix} 1 \\ 1 \end{bmatrix} \quad \rightarrow \quad \begin{bmatrix} \sqrt{2} \\ 0 \end{bmatrix}.$$

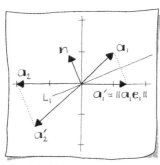

Sketch 13.1.

Householder reflection of vector $\mathbf{a}_1$ to $\|\mathbf{a}_1\|\mathbf{e}_1$.

We will reflect about the line L_1 illustrated in Sketch 13.1, so we must construct a normal $\mathbf{n}_1$ to this line:

$$\mathbf{n}_1 = \frac{\mathbf{a}_1 - \|\mathbf{a}_1\|\mathbf{e}_1}{\|\mathbf{a}_1 - \|\mathbf{a}_1\|\mathbf{e}_1\|},$$

which is simply the normalized difference between the original vector and the target vector (after the reflection). The implicit equation of the line L_1 is

$$\mathbf{n}_1^{\mathrm{T}}\mathbf{x} = 0,$$

and $\mathbf{n}_1^{\mathrm{T}}\mathbf{a}_1$ is the distance of the point $(\mathbf{o} + \mathbf{a}_1)$ to L_1. Therefore, the reflection constitutes moving twice the distance to the line in the direction of the normal, so

$$\mathbf{a}_1' = \mathbf{a}_1 - \left[2\mathbf{n}_1^{\mathrm{T}}\mathbf{a}_1\right]\mathbf{n}_1.$$

(Notice that $2\mathbf{n}_1^{\mathrm{T}}\mathbf{a}_1$ is a scalar.) To write this reflection in matrix form, we rearrange the terms,

$$\mathbf{a}_1' = \mathbf{a}_1 - 2\mathbf{n}_1\left[\mathbf{n}_1^{\mathrm{T}}\mathbf{a}_1\right]$$
$$= \left(I - 2\mathbf{n}_1\mathbf{n}_1^{\mathrm{T}}\right)\mathbf{a}_1$$

Notice that $2\mathbf{n}_1\mathbf{n}_1^{\mathrm{T}}$ is a dyadic matrix, as introduced in Section 11.5. We now have the matrix H_1 defining one Householder transformation:

$$H_1 = I - 2\mathbf{n}_1\mathbf{n}_1^{\mathrm{T}}.$$

This is precisely the reflection we constructed in (11.13)!

Example 13.1

The Householder matrix H_1 for our 2×2 example is formed with

$$\mathbf{n}_1 = \begin{bmatrix} -0.382 \\ 0.923 \end{bmatrix},$$

then

$$H_1 = I - 2\begin{bmatrix} 0.146 & -0.353 \\ -0.353 & 0.853 \end{bmatrix} = \begin{bmatrix} 0.707 & 0.707 \\ 0.707 & -0.707 \end{bmatrix}.$$

The transformed matrix is formed from the column vectors

$$H_1\mathbf{a}_1 = \begin{bmatrix} \sqrt{2} \\ 0 \end{bmatrix} \quad \text{and} \quad H_1\mathbf{a}_2 = \begin{bmatrix} -\sqrt{2} \\ -\sqrt{2} \end{bmatrix}.$$

The 2×2 example serves only to illustrate the underlying geometry of a reflection matrix. The construction of a general Householder transformation H_i is a little more complicated. Suppose now that we have the following matrix

$$A = \begin{bmatrix} a_{1,1} & a_{1,2} & a_{1,3} & a_{1,4} \\ 0 & a_{2,2} & a_{2,3} & a_{2,4} \\ 0 & 0 & a_{3,3} & a_{3,4} \\ 0 & 0 & a_{4,3} & a_{4,4} \end{bmatrix}.$$

We need to construct H_3 to zero the element $a_{4,3}$ while preserving A's upper triangular nature. In other words, we want to preserve the elimination done by previous transformations, H_1 and H_2. To achieve this, construct

$$\bar{\mathbf{a}}_3 = \begin{bmatrix} 0 \\ 0 \\ a_{3,3} \\ a_{4,3} \end{bmatrix},$$

and then construct H_3 so that

$$H_3 \bar{\mathbf{a}}_3 = \gamma \mathbf{e}_3 = \begin{bmatrix} 0 \\ 0 \\ \gamma \\ 0 \end{bmatrix},$$

where $\gamma = \pm \|\bar{\mathbf{a}}_3\|$. The matrix H_3 has been designed so that $H_3 \mathbf{a}_3$ will modify only elements $a_{3,3}$ and $a_{4,3}$ and the length of $\mathbf{a}_3$ will be preserved.

That is the idea, so let's develop it for $n \times n$ matrices. Suppose we have

$$\bar{\mathbf{a}}_i = \begin{bmatrix} 0 \\ \vdots \\ 0 \\ a_{i,i} \\ \vdots \\ a_{n,i} \end{bmatrix}.$$

We want to construct the Householder matrix H_i for the following transformation

$$H_i \bar{\mathbf{a}}_i = \gamma \mathbf{e}_i = \begin{bmatrix} 0 \\ \vdots \\ \gamma \\ \vdots \\ 0 \end{bmatrix},$$

where $\gamma = \pm \|\bar{\mathbf{a}}_i\|$. Just as we developed for the 2×2 example,

$$H_i = I - 2\mathbf{n}_i \mathbf{n}_i^{\mathrm{T}}, \tag{13.1}$$

but we make a slight modification of the normal,

$$\mathbf{n}_i = \frac{\bar{\mathbf{a}}_i - \gamma \mathbf{e}_i}{\| \cdot \|},$$

If $\bar{\mathbf{a}}_i$ is nearly parallel to $\mathbf{e}_i$ then numerical problems can creep into our construction due to loss of *significant digits* in subtraction of nearly equal numbers. Therefore, it is better to reflect onto the direction of the $\mathbf{e}_i$-axis that represents the largest reflection.

Let $N_i = \mathbf{n}_i \mathbf{n}_i^{\mathrm{T}}$ and note that it is symmetric and idempotent. Properties of H_i include being:

- symmetric: $H_i = H_i^{\mathrm{T}}$, since N_i is symmetric;

- involutory: $H_i H_i = I$ and thus $H_i = H_i^{-1}$,

- unitary (orthogonal): $H_i^{\mathrm{T}} H_i = I$, and thus $H_i \mathbf{v}$ has the same length as $\mathbf{v}$.

Implementation of Householder transformations doesn't involve explicit construction and multiplication by the Householder matrix in (13.1). A numerically and computationally more efficient algorithm is easy to achieve since we know quite a bit about how each H_i acts on the column vectors. So the following describes some of the variables in the algorithm that aid optimization.

Some optimization of the algorithm can be done by rewriting $\mathbf{n}$; let

$$\mathbf{v}_i = \bar{\mathbf{a}}_i - \gamma \mathbf{e}_i, \qquad \text{where} \quad \gamma = \begin{cases} -\operatorname{sign} a_{i,i} \|\bar{\mathbf{a}}_i\| & \text{if } a_{i,i} \neq 0 \\ -\|\bar{\mathbf{a}}_i\| & \text{otherwise.} \end{cases}$$

Then we have that

$$2\mathbf{n}\mathbf{n}^{\mathrm{T}} = \frac{\mathbf{v}\mathbf{v}^{\mathrm{T}}}{\frac{1}{2}\mathbf{v}^{\mathrm{T}}\mathbf{v}} = \frac{\mathbf{v}\mathbf{v}^{\mathrm{T}}}{\alpha}$$

thus

$$\alpha = \gamma^2 - a_{i,i}\gamma.$$

When H_i is applied to column vector $\mathbf{c}$,

$$H_i\mathbf{c} = \left(I - \frac{\mathbf{v}\mathbf{v}^{\mathrm{T}}}{\alpha}\right)\mathbf{c} = \mathbf{c} - s\mathbf{v}.$$

In the Householder algorithm that follows, as we work on the jth column vector, we call

$$\hat{\mathbf{a}}_k = \begin{bmatrix} a_{j,k} \\ \vdots \\ a_{n,k} \end{bmatrix}$$

to indicate that only elements $j, \ldots, n$ of the kth column vector $\mathbf{a}_k$ (with $k \geq j$) are involved in a calculation. Thus application of H_j results in changes in the sub-block of A with $a_{j,j}$ at the upper-left corner. Hence, the vector $\mathbf{a}_j$ and $H_j\mathbf{a}_j$ coincide in the first $j - 1$ components.

For a more detailed discussion, see [2] or [11].

The Householder Method

Algorithm:

> *Input:*
> An $n \times m$ matrix A, where $n \geq m$ and rank of A is m;
> n vector $\mathbf{b}$, augmented to A as the $(m + 1)$st column.
> *Output:*
> Upper triangular matrix HA written over A;
> $H\mathbf{b}$ written over $\mathbf{b}$ in the augmented $(m + 1)$st
> column of A.
> $(H = H_{n-1} \ldots H_1)$

If $n = m$ then $p = n - 1$; else $p = m$ (p is last column to
transform)

For $j = 1, 2, \ldots, p$
    $a = \hat{\mathbf{a}}_j \cdot \hat{\mathbf{a}}_j$
    $\gamma = -\mathrm{sign}(a_{j,j})\sqrt{a}$
    $\alpha = a - a_{j,j}\gamma$
    Temporarily set $a_{j,j} = a_{j,j} - \gamma$
    For $k = j + 1, \ldots, m + 1$
        $s = \frac{1}{\alpha}(\hat{\mathbf{a}}_j \cdot \hat{\mathbf{a}}_k)$
        $\hat{\mathbf{a}}_k = \hat{\mathbf{a}}_k - s\hat{\mathbf{a}}_j$
    Set $\hat{\mathbf{a}}_j = \begin{bmatrix} \gamma & 0 & \ldots & 0 \end{bmatrix}^{\mathrm{T}}$

Example 13.2

Let's apply the Householder algorithm to the linear system

$$\begin{bmatrix} 1 & 1 & 0 \\ 1 & -1 & 0 \\ 0 & 0 & 1 \end{bmatrix} \mathbf{u} = \begin{bmatrix} -1 \\ 0 \\ 1 \end{bmatrix}.$$

For $j = 1$ in the algorithm, we calculate the following values: $\gamma = -\sqrt{2}$, $\alpha = 2 + \sqrt{2}$, and then we temporarily set

$$\hat{\mathbf{a}}_1 = \begin{bmatrix} 1 + \sqrt{2} \\ 1 \\ 0 \end{bmatrix}.$$

For $k = 2$, $s = \sqrt{2}/(2 + \sqrt{2})$. This results in

$$\hat{\mathbf{a}}_2 = \begin{bmatrix} 0 \\ -\sqrt{2} \\ 0 \end{bmatrix}.$$

For $k = 3$, $s = 0$ and $\hat{\mathbf{a}}_3$ remains unchanged. For $k = 4$, $s = -\sqrt{2}/2$ and then the right-hand side vector becomes

$$\hat{\mathbf{a}}_4 = \begin{bmatrix} \sqrt{2}/2 \\ \sqrt{2}/2 \\ 0 \end{bmatrix}.$$

Now we set $\hat{\mathbf{a}}_1$, and the reflection H_1 results in the linear system

$$\begin{bmatrix} -\sqrt{2} & 0 & 0 \\ 0 & -\sqrt{2} & 0 \\ 0 & 0 & 1 \end{bmatrix} \mathbf{u} = \begin{bmatrix} \sqrt{2}/2 \\ \sqrt{2}/2 \\ 1 \end{bmatrix}.$$

Although not explicitly computed,

$$\mathbf{n}_1 = \frac{\begin{bmatrix} 1 + \sqrt{2} \\ 1 \\ 0 \end{bmatrix}}{\| \cdot \|}$$

and the Householder matrix

$$H_1 = \begin{bmatrix} -\sqrt{2}/2 & -\sqrt{2}/2 & 0 \\ -\sqrt{2}/2 & \sqrt{2}/2 & 0 \\ 0 & 0 & 1 \end{bmatrix}.$$

Notice that $\mathbf{a}_3$ was not affected because it is in the plane about which we were reflecting, and this is a result of the involutary property of the Householder matrix. The length of each column vector was not changed as a result of the orthogonal property. Since the matrix is upper triangular, we may now use back substitution to find the solution vector

$$\mathbf{u} = \begin{bmatrix} -1/2 \\ -1/2 \\ 1 \end{bmatrix}.$$

The Householder algorithm is the method of choice if one is dealing with ill-conditioned systems. An example: for some data sets, the least squares solution to an overdetermined linear system can produce unreliable results because of the nature of $A^{\mathrm{T}}A$. In Section 13.4, we will examine this point in more detail. The Householder algorithm above is easy to use for such overdetermined problems: The input matrix A is of dimension $n \times m$ where $n \geq m$. The following example illustrates that the Householder method will result in the least squares solution to an overdetermined system.

Example 13.3

Let's revisit the least squares line-fitting problem from Example 12.13. See that example for a problem description, and see Figure 12.4 for an illustration. The overdetermined linear system for this problem is

$$\begin{bmatrix} 0 & 1 \\ 10 & 1 \\ 20 & 1 \\ 30 & 1 \\ 40 & 1 \\ 50 & 1 \\ 60 & 1 \end{bmatrix} \begin{bmatrix} a \\ b \end{bmatrix} = \begin{bmatrix} 30 \\ 25 \\ 40 \\ 40 \\ 30 \\ 5 \\ 25 \end{bmatrix}.$$

After the first Householder reflection ($j = 1$), the linear system becomes

$$\begin{bmatrix} -95.39 & -2.20 \\ 0 & 0.66 \\ 0 & 0.33 \\ 0 & -0.0068 \\ 0 & -0.34 \\ 0 & -0.68 \\ 0 & -1.01 \end{bmatrix} \begin{bmatrix} a \\ b \end{bmatrix} = \begin{bmatrix} -54.51 \\ 16.14 \\ 22.28 \\ 13.45 \\ -5.44 \\ -39.29 \\ -28.15 \end{bmatrix}.$$

For the second Householder reflection $(j = 2)$, the linear system becomes

$$\begin{bmatrix} -95.39 & -2.20 \\ 0 & -1.47 \\ 0 & 0 \\ 0 & 0 \\ 0 & 0 \\ 0 & 0 \\ 0 & 0 \end{bmatrix} \begin{bmatrix} a \\ b \end{bmatrix} = \begin{bmatrix} -54.51 \\ -51.10 \\ 11.91 \\ 13.64 \\ 5.36 \\ -17.91 \\ 3.81 \end{bmatrix}.$$

We can now solve the system with back substitution, starting with the first nonzero row, and the solution is

$$\begin{bmatrix} a \\ b \end{bmatrix} = \begin{bmatrix} -0.23 \\ 34.82 \end{bmatrix}.$$

Excluding numerical round-off, this is the same solution found using the normal equations in Example 12.13.

13.2 Vector Norms

Finding the magnitude or length of a 3D vector is fundamental to many geometric operations. It is also fundamental for n-dimensional vectors, even though such vectors might no longer have geometric meaning. For example, in Section 13.6, we examine iterative methods for solving linear systems, and vector length is key for monitoring improvements in the solution.

The magnitude or length of an $n-$dimensional vector $\mathbf{v}$ is typically measured by

$$\|\mathbf{v}\|_2 = \sqrt{v_1^2 + \ldots + v_n^2}. \tag{13.2}$$

The (nonnegative) scalar $\|\mathbf{v}\|_2$ is also referred to as the *Euclidean norm* because in $\mathbb{R}^3$ it is Euclidean length. Since this is the "usual" way to measure length, the subscript 2 often is omitted.

More generally, (13.2) is a *vector norm*. Other vector norms are conceivable, for instance the 1-norm

$$\|\mathbf{v}\|_1 = |v_1| + |v_2| + \ldots + |v_n| \tag{13.3}$$

or the ∞-norm

$$\|\mathbf{v}\|_\infty = \max_i |v_i|. \tag{13.4}$$

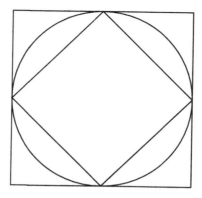

Figure 13.2.

Vector norms: outline of the unit vectors for the 2-norm is a circle, ∞-norm is a square, 1-norm is a diamond.

We can gain an understanding of these norms by studying the familiar case $n = 2$. Recall that a *unit vector* is one whose norm equals one. For the 2-norm $\|\mathbf{v}\|_2$, all unit vectors form a circle of radius one. With a bit of thinking, we can see that for the 1-norm all unit vectors form a diamond, while in the ∞-norm, all unit vectors form a square of edge length two. This geometric interpretation is shown in Figure 13.2.

All three of these norms may be viewed as members of a whole family of norms, referred to as *p*-norms. The *p*-norm of $\mathbf{v}$ is given by

$$\|\mathbf{v}\|_p = (v_1^p + v_2^p + \ldots + v_n^p)^{1/p}.$$

It is easy to see that the 1- and 2-norm fit this mold. That $p \to \infty$ gives the ∞-norm requires a little more thought; give it a try!

Example 13.4

Let

$$\mathbf{v} = \begin{bmatrix} 1 \\ 0 \\ -2 \end{bmatrix}.$$

Then

$$\|\mathbf{v}\|_1 = 3, \qquad \|\mathbf{v}\|_2 = \sqrt{5} \approx 2.24, \qquad \|\mathbf{v}\|_\infty = 2.$$

This example and Figure 13.2 demonstrate that

$$\|\mathbf{v}\|_1 \geq \|\mathbf{v}\|_2 \geq \|\mathbf{v}\|_\infty; \tag{13.5}$$

this is also true in n dimensions.

All vector norms share some basic properties:

$$\|\mathbf{v}\| \geq 0, \tag{13.6}$$

$$\|\mathbf{v}\| = 0 \quad \text{if and only if} \quad \mathbf{v} = \mathbf{0}, \tag{13.7}$$

$$\|c\mathbf{v}\| = |c|\|\mathbf{v}\| \quad \text{for} \quad c \in \mathbb{R}, \tag{13.8}$$

$$\|\mathbf{v} + \mathbf{w}\| \leq \|\mathbf{v}\| + \|\mathbf{w}\|. \tag{13.9}$$

The last of these is the familiar *triangle inequality*.

By this time, we have developed a familiarity with the 2-norm, and these properties are easily verified. Recall the steps for the triangle inequality were demonstrated in (2.24).

Let's show that the vector norm properties hold for the ∞-norm. Examining (13.4), for each $\mathbf{v}$ in $\mathbb{R}^n$, by definition, $\max_i |v_i| \geq 0$. If $\max_i |v_i| = 0$ then $v_i = 0$ for each $i = 1, \ldots, n$ and $\mathbf{v} = \mathbf{0}$. Conversely, if $\mathbf{v} = \mathbf{0}$ then $\max_i |v_i| = 0$. Thus we have established that the first two properties hold. The third property is easily shown by the properties of the absolute value function:

$$\|c\mathbf{v}\|_\infty = \max_i |cv_i| = |c| \max_i |v_i| = |c|\|\mathbf{v}\|_\infty.$$

For the triangle inequality:

$$\begin{aligned}
\|\mathbf{v} + \mathbf{w}\|_\infty &= \max_i |v_i + w_i| \\
&\leq \max_i \{|v_i| + |w_i|\} \\
&\leq \max_i |v_i| + \max_i |w_i| \\
&= \|\mathbf{v}\|_\infty + \|\mathbf{w}\|_\infty.
\end{aligned}$$

From a 2D geometric perspective, consider the 1-norm: starting at the origin, move v_1 units along the $\mathbf{e}_1$-axis and then move v_2 units along the $\mathbf{e}_2$-axis. The sum of these travels determines $\mathbf{v}$'s length. This is analogous to driving in a city with rectilinear streets, and thus this norm is also called the Manhattan or taxicab norm.

The ∞-norm is also called the max-norm. Our standard measure of length, the 2-norm, can take more computing power than we might be willing to spend on a proximity problem, where we want to determine whether two points are closer than a given tolerance. Problems of this sort are often repeated many times and against many points. Instead, the max norm might be more suitable from a computing point of view and sufficient given the relationship in (13.5). This method of using an inexpensive measure to exclude many possibilities is called *trivial reject* in some disciplines such as computer graphics.

13.3 Matrix Norms

A vector norm measures the magnitude of a vector; does it also make sense to talk about the magnitude of a matrix?

Instead of exploring the general case right away, we will first give some 2D insight. Let A be a 2×2 matrix. It maps the unit circle to an ellipse—the action ellipse. In Figure 13.3 we see points on the unit circle at the end of unit vectors $\mathbf{v}_i$ and their images $A\mathbf{v}_i$, forming this action ellipse. We can learn more about this ellipse by looking at $A^{\mathrm{T}}A$. It is *symmetric* and *positive definite*, and thus has real and positive eigenvalues, which are λ_1' and λ_2', in decreasing value. We then define

$$\sigma_i = \sqrt{\lambda_i'}, \tag{13.10}$$

to be the *singular values* of A. (In Chapter 16, we will explore singular values in more detail.)

The singular value σ_1 is the length of the action ellipse's semi-major axis and σ_2 is the length of the action ellipse's semi-minor axis. If A

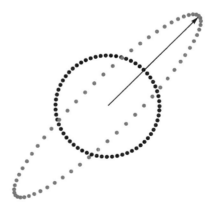

Figure 13.3.

Action of a 2×2 matrix: points on the unit circle (black) are mapped to points on an ellipse (gray) This is the *action ellipse.* The vector is the semi-major axis of the ellipse, which has length σ_1.

is a symmetric positive definite matrix, then its singular values are equal to its eigenvalues.

How much does A distort the unit circle? This is measured by its 2-norm, $\|A\|_2$. If we find the largest $\|A\mathbf{v}_i\|_2$, then we have an indication of how much A distorts. Assuming that we have k unit vectors $\mathbf{v}_i$, we can compute

$$\|A\|_2 \approx \max_i \|A\mathbf{v}_i\|_2. \tag{13.11}$$

As we increase k, (13.11) gives a better and better approximation to $\|A\|_2$. We then have

$$\|A\|_2 = \max_{\|\mathbf{v}\|_2 = 1} \|A\mathbf{v}\|_2. \tag{13.12}$$

It should be clear by now that this development is not restricted to 2×2 matrices, but holds for $n \times n$ ones as well. From now on, we will discuss this general case, where the matrix 2-norm is more complicated to compute than its companion vector norm. It is given by

$$\|A\|_2 = \sigma_1,$$

where σ_1 is A's largest singular value.

Recall that the inverse matrix A^{-1} "undoes" the action of A. Let the singular values of A^{-1} be called $\hat{\sigma}_i$, then

$$\hat{\sigma}_1 = \frac{1}{\sigma_n}, \quad \ldots, \quad \hat{\sigma}_n = \frac{1}{\sigma_1}$$

and

$$\|A^{-1}\|_2 = \frac{1}{\sigma_n}.$$

Singular values are frequently used in the numerical analysis of matrix operations. If you do not have access to software for singular values, then (13.11) will give a decent approximation. Singular values are typically computed using a method called *singular value decomposition*, or *SVD*, and they are the focus of Chapter 16.

Analogous to vector norms, there are other matrix norms. We give two important examples:

$$\|A\|_1 \; : \; \text{maximum absolute column sum},$$
$$\|A\|_\infty \; : \; \text{maximum absolute row sum}.$$

Because the notation for matrix and vector norms is identical, it is important to note what is enclosed in the norm symbols—a vector or a matrix.

Another norm that is sometimes used is the *Frobenius norm*, which is given by

$$\|A\|_F = \sqrt{\sigma_1^2 + \ldots + \sigma_n^2}, \tag{13.13}$$

where the σ_i are A's singular values.

The following is not too obvious. The *Euclidean norm*: add up the squares of all matrix elements, and take the square root,

$$\|A\|_F = \sqrt{a_{1,1}^2 + a_{1,2}^2 + \ldots + a_{n,n}^2}, \tag{13.14}$$

is identical to the Frobenius norm.

A matrix A maps all unit vectors to an ellipsoid whose semi-axis lengths are given by $\sigma_1, \ldots, \sigma_n$. Then the Frobenius norm gives the total distortion caused by A.

Example 13.5

Let's examine matrix norms for

$$A = \begin{bmatrix} 1 & 2 & 3 \\ 3 & 4 & 5 \\ 5 & 6 & -7 \end{bmatrix}.$$

Its singular values are given by 10.5, 7.97, 0.334 resulting in

$$\|A\|_2 = \max\{10.5, 7.97, 0.334\} = 10.5,$$
$$\|A\|_1 = \max\{9, 12, 15\} = 15,$$
$$\|A\|_\infty = \max\{6, 12, 18\} = 18$$
$$\|A\|_F = \sqrt{1^2 + 3^2 + \ldots (-7)^2} = \sqrt{10.5^2 + 7.97^2 + 0.334} = 13.2.$$

From the examples above, we see that matrix norms are real-valued functions of the linear space defined over all $n \times n$ matrices.[1] Matrix norms satisfy conditions very similar to the vector norm conditions (13.6)–(13.9):

$$\|A\| > 0 \quad \text{for} \quad A \neq Z, \tag{13.15}$$
$$\|A\| = 0 \quad \text{for} \quad A = Z, \tag{13.16}$$
$$\|cA\| = |c| \|A\| \quad c \in \mathbb{R}, \tag{13.17}$$
$$\|A + B\| \leq \|A\| + \|B\|, \tag{13.18}$$
$$\|AB\| \leq \|A\| \|B\|, \tag{13.19}$$

Z being the zero matrix.

---

[1]More on this type of linear space in Chapter 14.

How to choose a matrix norm? Computational expense and properties of the norm are the deciders. For example, the Frobenius and 2-norms are invariant with respect to orthogonal transformations. Many numerical methods texts provide a wealth of information on matrix norms and their relationships.

13.4   The Condition Number

We would like to know how sensitive the solution to $A\mathbf{u} = \mathbf{b}$ is to changes in A and $\mathbf{b}$. For this problem, the matrix 2-norm helps us define the condition of the map.

In most of our figures describing linear maps, we have mapped a circle, formed by many unit vectors $\mathbf{v}_i$, to an ellipse, formed by vectors $A\mathbf{v}_i$. This ellipse is evidently closely related to the geometry of the map, and indeed its semi-major and semi-minor axis lengths correspond to A's singular values. For $n = 2$, the eigenvalues of $A^{\mathrm{T}}A$ are λ'_1 and λ'_2, in decreasing value. The singular values of A are defined in (13.10). If σ_1 is very large and σ_2 is very small, then the ellipse will be very elongated, as illustrated by the example in Figure 13.4. The ratio

$$\kappa(A) = \|A\|_2 \|A^{-1}\|_2 = \sigma_1/\sigma_n$$

is called the *condition number* of A. In Figure 13.4, we picked the (symmetric, positive definite) matrix

$$A = \begin{bmatrix} 1.5 & 0 \\ 0 & 0.05 \end{bmatrix},$$

which has condition number $\kappa(A) = 1.5/0.05 = 30$.

Since $A^{\mathrm{T}}A$ is symmetric and positive definite, $\kappa(A) \geq 1$. If a matrix has a condition number close to one, it is called *well-conditioned*. If the condition number is 1 (such as for the identity matrix), then no distortion happens. The larger the $\kappa(A)$, the more A distorts, and if $\kappa(A)$ is "large," the matrix is called *ill-conditioned*.

Figure 13.4.

Condition number: The action of A with $\kappa(A) = 30$.

Example 13.6

Let

$$A = \begin{bmatrix} \cos\alpha & -\sin\alpha \\ \sin\alpha & \cos\alpha \end{bmatrix},$$

meaning that A is an α degree rotation. Clearly, $A^{\mathrm{T}}A = I$, where I is the identity matrix. Thus, $\sigma_1 = \sigma_2 = 1$. Hence the condition number of a rotation matrix is 1. Since a rotation does not distort, this is quite intuitive.

Example 13.7

Now let

$$A = \begin{bmatrix} 100 & 0 \\ 0 & 0.01 \end{bmatrix},$$

a matrix that scales by 100 in the $\mathbf{e}_1$-direction and by 0.01 in the $\mathbf{e}_2$-direction. This matrix is severely distorting! We quickly find $\sigma_1 = 100$ and $\sigma_2 = 0.01$ and thus the condition number is $100/0.01 = 10,000$. The fact that A distorts is clearly reflected in the magnitude of its condition number.

Back to the initial problem: solving $A\mathbf{u} = \mathbf{b}$. We will always have round-off error to deal with, but we want to avoid creating a poorly designed linear system, which would mean a matrix A with a "large" condition number. The definition of large is subjective and problem-specific. A guideline: $\kappa(A) \approx 10^k$ can result in a loss of k digits of accuracy. If the condition number of a matrix is large, then irrespective of round-off, the solution cannot be depended upon. Practically speaking, a large condition number means that the solution to the linear system is numerically very sensitive to small changes in A or $\mathbf{b}$. Alternatively, we can say that we can confidently calculate the inverse of a well-conditioned matrix.

The condition number is a better measure of singularity than the determinant because it is a scale and size n invariant measure. If s is a scalar, then $\kappa(sA) = \kappa(A)$.

Example 13.8

Let $n = 100$. Form the identity matrix I and $J = 0.1I$.

$$\det I = 1, \quad \kappa(I) = 1, \quad \det J = 10^{-100}, \quad \kappa(J) = 1.$$

The determinant of J is small, indicating that there is a problem with this matrix. However, in solving a linear system with such a coefficient matrix, the scale of J poses no problem.

In Section 12.7, we examined overdetermined linear systems $A\mathbf{u} = \mathbf{b}$ in m equations and n unknowns, where $m > n$. (Thus A is an $m \times n$ matrix.) Our approach to this problem depended on the approximation method of least squares, and we solved the system $A^\mathrm{T} A\mathbf{u} = A^\mathrm{T}\mathbf{b}$. The condition number for this new matrix, $\kappa(A^\mathrm{T} A) = \kappa(A)^2$, so if A has a high condition number to start with, this process will create an ill-posed problem. In this situation, a method such as the Householder method (see Section 13.1), is preferred. We will revisit this topic in more detail in Section 16.6.

An advanced linear algebra or numerical analysis text will provide an in-depth study of the condition number and error analysis.

13.5   Vector Sequences

You are familiar with sequences of real numbers such as

$$1, \frac{1}{2}, \frac{1}{4}, \frac{1}{8}, \ldots$$

or

$$1, 2, 4, 8, \ldots$$

The first of these has the *limit* 0, whereas the second one does not have a limit. One way of saying a sequence of real numbers a_i has a limit a is that beyond some index i, all a_i differ from the limit a by an arbitrarily small amount ϵ.

Vector sequences in $\mathbb{R}^n$,

$$\mathbf{v}^{(0)}, \mathbf{v}^{(1)}, \mathbf{v}^{(2)}, \ldots,$$

are not all that different. A vector sequence has a limit if each component has a limit.

Example 13.9

Let a vector sequence be given by

$$\mathbf{v}^{(i)} = \begin{bmatrix} 1/i \\ 1/i^2 \\ 1/i^3 \end{bmatrix}.$$

This sequence has the limit

$$\mathbf{v} = \begin{bmatrix} 0 \\ 0 \\ 0 \end{bmatrix}.$$

Now take the sequence

$$\mathbf{v}^{(i)} = \begin{bmatrix} i \\ 1/i^2 \\ 1/i^3 \end{bmatrix}.$$

It does not have a limit: even though the last two components each have a limit, the first component diverges.

We say that a vector sequence converges to $\mathbf{v}$ with respect to a norm if for any tolerance $\epsilon > 0$, there exists an integer m such that

$$\|\mathbf{v}^{(i)} - \mathbf{v}\| < \epsilon \qquad \text{for all } i > m. \tag{13.20}$$

In other words, from some index i on, the distance of any $\mathbf{v}^{(i)}$ from $\mathbf{v}$ is smaller than an arbitrarily small amount ϵ. See Figure 13.5 for an example. If the sequence converges with respect to one norm, it will converge with respect to all norms. Our focus will be on the (usual) Euclidean or 2-norm, and the subscript will be omitted.

Vector sequences are key to iterative methods, such as the two methods for solving $A\mathbf{u} = \mathbf{b}$ in Section 13.6 and the power method for finding the dominant eigenvalue in Section 15.2. In practical applications, the limit vector $\mathbf{v}$ is not known. For some special problems, we can say whether a limit exists; however, we will not know it *a priori*. So we will modify our theoretical convergence measure (13.20) to examine the distance between iterations. This can take

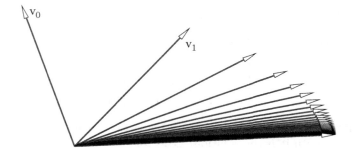

Figure 13.5.
Vector sequences: a sequence that converges.

on different forms depending on the problem at hand. In general, it will lead to testing the condition

$$\|\mathbf{v}^{(i)} - \mathbf{v}^{(i+1)}\| < \epsilon,$$

which measures the change from one iteration to the next. In the case of an iterative solution to a linear system, $\mathbf{u}^{(i)}$, we will test for the condition that $\|A\mathbf{u}^{(i)} - \mathbf{b}\| < \epsilon$, which indicates that this iteration has provided an acceptable solution.

13.6 Iterative System Solvers: Gauss-Jacobi and Gauss-Seidel

In applications such as *finite element methods* (*FEM*) in the context of the solution of fluid flow problems, scientists are faced with linear systems with many thousands of equations. Gauss elimination would work, but would be far too slow. Typically, huge linear systems have one advantage: the coefficient matrix is *sparse*, meaning it has only very few (such as ten) nonzero entries per row. Thus, a 100,000 × 100,000 system would only have 1,000,000 nonzero entries, compared to 10,000,000,000 matrix elements! In these cases, one does not store the whole matrix, but only its nonzero entries, together with their i, j location. An example of a sparse matrix is shown in Figure 13.1. The solution to such systems is typically obtained by *iterative methods*, which we will discuss next.

Example 13.10

Let a system[2] be given by

$$\begin{bmatrix} 4 & 1 & 0 \\ 2 & 5 & 1 \\ -1 & 2 & 4 \end{bmatrix} \begin{bmatrix} u_1 \\ u_2 \\ u_3 \end{bmatrix} = \begin{bmatrix} 1 \\ 0 \\ 3 \end{bmatrix}.$$

An iterative method starts from a guess for the solution and then refines it until it *is* the solution. Let's take

$$\mathbf{u}^{(1)} = \begin{bmatrix} u_1^{(1)} \\ u_2^{(1)} \\ u_3^{(1)} \end{bmatrix} = \begin{bmatrix} 1 \\ 1 \\ 1 \end{bmatrix}$$

for our first guess, and note that it clearly is not the solution to our system: $A\mathbf{u}^{(1)} \neq \mathbf{b}$.

A better guess ought to be obtained by using the current guess and solving the first equation for a new $u_1^{(2)}$, the second for a new $u_2^{(2)}$, and so on. This gives us

$$4u_1^{(2)} + 1 = 1,$$
$$2 + 5u_2^{(2)} + 1 = 0,$$
$$-1 + 2 + 4u_3^{(2)} = 3,$$

and thus

$$\mathbf{u}^{(2)} = \begin{bmatrix} 0 \\ -0.6 \\ 0.5 \end{bmatrix}.$$

The next iteration becomes

$$4u_1^{(3)} - 0.6 = 1,$$
$$5u_2^{(3)} + 0.5 = 0,$$
$$-1.2 + 4u_3^{(3)} = 3,$$

and thus

$$\mathbf{u}^{(3)} = \begin{bmatrix} 0.4 \\ -0.1 \\ 1.05 \end{bmatrix}.$$

[2]This example was taken from Johnson and Riess [11].

After a few more iterations, we will be close enough to the true solution

$$\mathbf{u} = \begin{bmatrix} 0.333 \\ -0.333 \\ 1.0 \end{bmatrix}.$$

Try one more iteration for yourself.

This iterative method is known as *Gauss-Jacobi iteration*. Let us now formulate this process for the general case. We are given a linear system with n equations and n unknowns u_i, written in matrix form as

$$A\mathbf{u} = \mathbf{b}.$$

Let us also assume that we have an initial guess $\mathbf{u}^{(1)}$ for the solution vector $\mathbf{u}$.

We now define two matrices D and R as follows: D is the diagonal matrix whose diagonal elements are those of A, and R is the matrix obtained from A by setting all its diagonal elements to zero. Clearly then

$$A = D + R$$

and our linear system becomes

$$D\mathbf{u} + R\mathbf{u} = \mathbf{b}$$

or

$$\mathbf{u} = D^{-1}[\mathbf{b} - R\mathbf{u}].$$

In the spirit of our previous development, we now write this as

$$\mathbf{u}^{(k+1)} = D^{-1}\left[\mathbf{b} - R\mathbf{u}^{(k)}\right],$$

meaning that we attempt to compute a new estimate $\mathbf{u}^{(k+1)}$ from an existing one $\mathbf{u}^{(k)}$. Note that D must not contain zeroes on the diagonal; this can be achieved by row or column interchanges.

Example 13.11

With this new framework, let us reconsider our last example. We have

$$A = \begin{bmatrix} 4 & 1 & 0 \\ 2 & 5 & 1 \\ -1 & 2 & 4 \end{bmatrix}, \quad R = \begin{bmatrix} 0 & 1 & 0 \\ 2 & 0 & 1 \\ -1 & 2 & 0 \end{bmatrix}, \quad D^{-1} = \begin{bmatrix} 0.25 & 0 & 0 \\ 0 & 0.2 & 0 \\ 0 & 0 & 0.25 \end{bmatrix}.$$

Then

$$\mathbf{u}^{(2)} = \begin{bmatrix} 0.25 & 0 & 0 \\ 0 & 0.2 & 0 \\ 0 & 0 & 0.25 \end{bmatrix} \left(\begin{bmatrix} 1 \\ 0 \\ 3 \end{bmatrix} - \begin{bmatrix} 0 & 1 & 0 \\ 2 & 0 & 1 \\ -1 & 2 & 0 \end{bmatrix} \begin{bmatrix} 1 \\ 1 \\ 1 \end{bmatrix} \right) = \begin{bmatrix} 0 \\ -0.6 \\ 0.5 \end{bmatrix}$$

Will the Gauss-Jacobi method succeed, i.e., will the sequence of vectors $\mathbf{u}^{(k)}$ converge? The answer is: sometimes yes, and sometimes no. It will *always* succeed if A is diagonally dominant, which means that if, for every row, the absolute value of its diagonal element is larger than the sum of the absolute values of its remaining elements. In this case, it will succeed no matter what our initial guess $\mathbf{u}^{(1)}$ was. Strictly diagonally dominant matrices are nonsingular. Many practical problems, for example, finite element ones, result in diagonally dominant systems.

In a practical setting, how do we determine if convergence is taking place? Ideally, we would like $\mathbf{u}^{(k)} = \mathbf{u}$, the true solution, after a number of iterations. Equality will most likely not happen, but the length of the *residual vector*

$$\|A\mathbf{u}^{(k)} - \mathbf{b}\|$$

should become small (i.e., less than some preset tolerance). Thus, we check the length of the residual vector after each iteration, and stop once it is smaller than our tolerance.

A modification of the Gauss-Jacobi method is known as *Gauss-Seidel* iteration. When we compute $\mathbf{u}^{(k+1)}$ in the Gauss-Jacobi method, we can observe the following: the second element, $u_2^{(k+1)}$, is computed using $u_1^{(k)}, u_3^{(k)}, \ldots, u_n^{(k)}$. We had just computed $u_1^{(k+1)}$. It stands to reason that using it instead of $u_1^{(k)}$ would be advantageous. This idea gives rise to the Gauss-Seidel method: as soon as a new element $u_i^{(k+1)}$ is computed, the estimate vector $\mathbf{u}^{(k+1)}$ is updated.

In summary, Gauss-Jacobi updates the new estimate vector once all of its elements are computed, Gauss-Seidel updates as soon as a new element is computed. Typically, Gauss-Seidel iteration converges faster than Gauss-Jacobi iteration.

- reflection matrix
- Householder method
- overdetermined system
- symmetric matrix
- involutary matrix
- orthogonal matrix
- unitary matrix
- vector norm
- vector norm properties
- Euclidean norm
- L^2 norm
- ∞ norm
- max norm
- Manhattan norm
- matrix norm
- matrix norm properties
- Frobenius norm
- action ellipse axes
- singular values
- condition number
- well-conditioned matrix
- ill-conditioned matrix
- vector sequence
- convergence
- iterative method
- sparse matrix
- Gauss-Jacobi method
- Gauss-Seidel method
- residual vector

13.7 Exercises

1. Use the Householder method to solve the following linear system

$$\begin{bmatrix} 1 & 1.1 & 1.1 \\ 1 & 0.9 & 0.9 \\ 0 & -0.1 & 0.2 \end{bmatrix} \mathbf{u} = \begin{bmatrix} 1 \\ 1 \\ 0.3 \end{bmatrix}.$$

   Notice that the columns are almost linearly dependent.

2. Show that the Householder matrix is involuntary.

3. What is the Euclidean norm of

$$\mathbf{v} = \begin{bmatrix} 1 \\ 1 \\ 1 \\ 1 \end{bmatrix}.$$

4. Examining the 1- and 2-norms defined in Section 13.2, how would you define a 3-norm?

5. Show that the $\|\mathbf{v}\|_1$ satisfies the properties (13.6)–(13.9) of a vector norm.

6. Define a new vector norm to be $\max\{2|v_1|, |v_2|, \ldots, |v_n|\}$. Show that this is indeed a vector norm. For vectors in $\mathbb{R}^2$, sketch the outline of all unit vectors with respect to this norm.

7. Let
$$A = \begin{bmatrix} 1 & 0 & 1 \\ 0 & 1 & 2 \\ 0 & 0 & 1 \end{bmatrix}.$$

What is A's 2-norm?

8. Let four unit vectors be given by
$$\mathbf{v}_1 = \begin{bmatrix} 1 \\ 0 \end{bmatrix}, \quad \mathbf{v}_2 = \begin{bmatrix} 0 \\ 1 \end{bmatrix}, \quad \mathbf{v}_3 = \begin{bmatrix} -1 \\ 0 \end{bmatrix}, \quad \mathbf{v}_4 = \begin{bmatrix} 0 \\ -1 \end{bmatrix}.$$

Using the matrix
$$A = \begin{bmatrix} 1 & 1 \\ 0 & 1 \end{bmatrix},$$

compute the four image vectors $A\mathbf{v}_i$. Use these to find an approximation to $\|A\|_2$.

9. Is the determinant of a matrix a norm?

10. Why is 1 the smallest possible condition number for a nonsingular matrix?

11. What is the condition number of the matrix
$$A = \begin{bmatrix} 0.7 & 0.3 \\ -1 & 1 \end{bmatrix}$$

that generated Figure 7.11?

12. What can you say about the condition number of a rotation matrix?

13. What is the condition number of the matrix
$$\begin{bmatrix} 1 & 0 \\ 0 & 0 \end{bmatrix}?$$

What type of matrix is it, and is it invertible?

14. Is the condition number of a matrix a norm?

15. Define a vector sequence by
$$\mathbf{u}^i = \begin{bmatrix} 0 & 0 & 1/i \\ 0 & 1 & 0 \\ -1/i & 0 & 0 \end{bmatrix} \begin{bmatrix} 1 \\ 1 \\ 1 \end{bmatrix}; \quad i = 1, 2, \ldots.$$

Does it have a limit? If so, what is it?

16. Let a vector sequence be given by
$$\mathbf{u}^{(1)} = \begin{bmatrix} 1 \\ -1 \end{bmatrix} \quad \text{and} \quad \mathbf{u}^{(i+1)} = \begin{bmatrix} 1/2 & 1/4 \\ 0 & 3/4 \end{bmatrix} \mathbf{u}^{(i)}.$$

Will this sequence converge? If so, to what vector?

17. Let a linear system be given by

$$A = \begin{bmatrix} 4 & 0 & -1 \\ 2 & 8 & 2 \\ 1 & 0 & 2 \end{bmatrix} \quad \text{and} \quad \mathbf{b} = \begin{bmatrix} 2 \\ -2 \\ 0 \end{bmatrix}.$$

Carry out three iterations of the Gauss-Jacobi iteration starting with an initial guess

$$\mathbf{u}^{(1)} = \begin{bmatrix} 0 \\ 0 \\ 0 \end{bmatrix}.$$

18. Let a linear system be given by

$$A = \begin{bmatrix} 4 & 0 & 1 \\ 2 & -8 & 2 \\ 1 & 0 & 2 \end{bmatrix} \quad \text{and} \quad \mathbf{b} = \begin{bmatrix} 0 \\ 2 \\ 0 \end{bmatrix}.$$

Carry out three iterations of the Gauss-Jacobi iteration starting with an initial guess

$$\mathbf{u}^{(1)} = \begin{bmatrix} 0 \\ 0 \\ 0 \end{bmatrix}.$$

19. Carry out three Gauss-Jacobi iterations for the linear system

$$A = \begin{bmatrix} 4 & 1 & 0 & 1 \\ 1 & 4 & 1 & 0 \\ 0 & 1 & 4 & 1 \\ 1 & 0 & 1 & 4 \end{bmatrix} \quad \text{and} \quad \mathbf{b} = \begin{bmatrix} 0 \\ 1 \\ 1 \\ 0 \end{bmatrix},$$

starting with the initial guess

$$\mathbf{u}^{(1)} = \begin{bmatrix} 1 \\ 1 \\ 1 \\ 1 \end{bmatrix}.$$

20. Carry out three iterations of Gauss-Seidel for Example 13.10. Which method, Gauss-Jacobi or Gauss-Seidel, is converging to the solution faster? Why?

14

General Linear Spaces

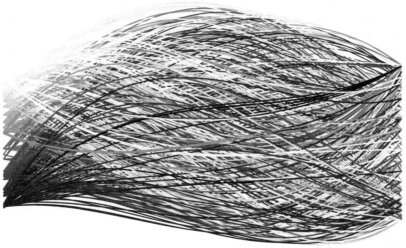

Figure 14.1.
General linear spaces: all cubic polynomials over the interval $[0, 1]$ form a linear space. Some elements of this space are shown.

In Sections 4.3 and 9.2, we had a first look at the concept of linear spaces, also called vector spaces, by examining the properties of the spaces of all 2D and 3D vectors. In this chapter, we will provide a framework for linear spaces that are not only of dimension two or three, but of possibly much higher dimension. These spaces tend to be somewhat abstract, but they are a powerful concept in dealing with many real-life problems, such as car crash simulations, weather

303

forecasts, or computer games. The linear space of cubic polynomials over $[0, 1]$ is important for many applications. Figure 14.1 provides an artistic illustration of some of the elements of this linear space. Hence the term "general" in the chapter title refers to the dimension and abstraction that we will study.

14.1 Basic Properties of Linear Spaces

We denote a *linear space* of dimension n by $\mathcal{L}_n$. The elements of $\mathcal{L}_n$ are vectors, denoted (as before) by boldface letters such as $\mathbf{u}$ or $\mathbf{v}$. We need two operations to be defined on the elements of $\mathcal{L}_n$: addition and multiplication by a scalar such as s or t. With these operations in place, the defining property for a linear space is that any *linear combination* of vectors,

$$\mathbf{w} = s\mathbf{u} + t\mathbf{v}, \tag{14.1}$$

results in a vector in the same space. This is called the *linearity property*. Note that both s and t may be zero, asserting that every linear space has a zero vector in it. This equation is familiar to us by now, as we first encountered it in Section 2.6, and linear combinations have been central to working in the linear spaces of $\mathbb{R}^2$ and $\mathbb{R}^3$.[1]

But in this chapter, we want to generalize linear spaces; we want to include new kinds of vectors. For example, our linear space could be the quadratic polynomials or all 3×3 matrices. Because the objects in the linear space are not always in the format of what we traditionally call a vector, we choose to call these linear spaces, which focuses on the linearity property rather than vector spaces. Both terms are accepted. Thus, we will expand our notation for objects of a linear space—we will not always use boldface vector notation. In this section, we will look at just a few examples of linear spaces, more examples of linear spaces will appear in Section 14.5.

Example 14.1

Let's start with a very familiar example of a linear space: $\mathbb{R}^2$. Suppose

$$\mathbf{u} = \begin{bmatrix} 1 \\ 1 \end{bmatrix} \quad \text{and} \quad \mathbf{v} = \begin{bmatrix} -2 \\ 3 \end{bmatrix}$$

are elements of this space; we know that

$$\mathbf{w} = 2\mathbf{u} + \mathbf{v} = \begin{bmatrix} 0 \\ 5 \end{bmatrix}$$

is also in $\mathbb{R}^2$.

---

[1]Note that we will not always use the $\mathcal{L}$ notation, but rather the standard name for the space when one exists.

Example 14.2

Let the linear space $\mathcal{M}_{2\times2}$ be the set of all 2×2 matrices. We know that the linearity property (14.1) holds because of the rules of matrix arithmetic from Section 4.12.

Example 14.3

Let's define $\mathcal{V}_2$ to be all vectors $\mathbf{w}$ in $\mathbb{R}^2$ that satisfy $w_2 \geq 0$. For example, $\mathbf{e}_1$ and $\mathbf{e}_2$ live in $\mathcal{V}_2$. Is this a linear space?

It is not: we can form a linear combination as in (14.1) and produce

$$\mathbf{v} = 0 \times \mathbf{e}_1 + -1 \times \mathbf{e}_2 = \begin{bmatrix} 0 \\ -1 \end{bmatrix},$$

which is not in $\mathcal{V}_2$.

In a linear space $\mathcal{L}_n$, let's look at a set of r vectors, where $1 \leq r \leq n$. A set of vectors $\mathbf{v}_1, \ldots, \mathbf{v}_r$ is called *linearly independent* if it is impossible to express one of them as a linear combination of the others. For example, the equation

$$\mathbf{v}_1 = s_2\mathbf{v}_2 + s_3\mathbf{v}_3 + \ldots + s_r\mathbf{v}_r$$

will not have a solution set $s_2, \ldots, s_r$ in case the vectors $\mathbf{v}_1, \ldots, \mathbf{v}_r$ are linearly independent. As a simple consequence, the zero vector can be expressed only in a trivial manner in terms of linearly independent vectors, namely if

$$\mathbf{0} = s_1\mathbf{v}_1 + \ldots + s_r\mathbf{v}_r$$

then $s_1 = \ldots = s_r = 0$. If the zero vector *can* be expressed as a nontrivial combination of r vectors, then we say these vectors are *linearly dependent*.

If the vectors $\mathbf{v}_1, \ldots, \mathbf{v}_r$ are linearly independent, then the set of all vectors that may be expressed as a linear combination of them forms a *subspace* of $\mathcal{L}_n$ of dimension r. We also say this subspace is *spanned* by $\mathbf{v}_1, \ldots, \mathbf{v}_r$. If this subspace equals the whole space $\mathcal{L}_n$, then we call $\mathbf{v}_1, \ldots, \mathbf{v}_n$ a *basis* for $\mathcal{L}_n$. If $\mathcal{L}_n$ is a linear space of dimension n, then any $n + 1$ vectors in it are linearly dependent.

Example 14.4

Let's consider one very familiar linear space: $\mathbb{R}^3$. A commonly used basis for $\mathbb{R}^3$ are the linearly independent vectors $\mathbf{e}_1, \mathbf{e}_2, \mathbf{e}_3$. A linear combination of these basis vectors, for example,

$$\mathbf{v} = \begin{bmatrix} 3 \\ 4 \\ 7 \end{bmatrix} = 3 \begin{bmatrix} 1 \\ 0 \\ 0 \end{bmatrix} + 4 \begin{bmatrix} 0 \\ 1 \\ 0 \end{bmatrix} + 7 \begin{bmatrix} 0 \\ 0 \\ 1 \end{bmatrix}$$

is also in $\mathbb{R}^3$. Then the four vectors $\mathbf{v}, \mathbf{e}_1, \mathbf{e}_2, \mathbf{e}_3$ are linearly dependent.

Any one of the four vectors forms a one-dimensional subspace of $\mathbb{R}^3$. Since $\mathbf{v}$ and each of the $\mathbf{e}_i$ are linearly independent, any two vectors here form a two-dimensional subspace of $\mathbb{R}^3$.

Example 14.5

In a 4D space $\mathbb{R}^4$, let three vectors be given by

$$\mathbf{v}_1 = \begin{bmatrix} -1 \\ 0 \\ 0 \\ 1 \end{bmatrix}, \qquad \mathbf{v}_2 = \begin{bmatrix} 5 \\ 0 \\ -3 \\ 1 \end{bmatrix}, \qquad \mathbf{v}_3 = \begin{bmatrix} 3 \\ 0 \\ -3 \\ 0 \end{bmatrix}.$$

These three vectors are linearly dependent since

$$\mathbf{v}_2 = \mathbf{v}_1 + 2\mathbf{v}_3 \quad \text{or} \quad \mathbf{0} = \mathbf{v}_1 - \mathbf{v}_2 + 2\mathbf{v}_3.$$

Our set $\{\mathbf{v}_1, \mathbf{v}_2, \mathbf{v}_3\}$ contains only two linearly independent vectors, hence any two of them spans a subspace of $\mathbb{R}^4$ of dimension two.

Example 14.6

In a 3D space $\mathbb{R}^3$, let four vectors be given by

$$\mathbf{v}_1 = \begin{bmatrix} -1 \\ 0 \\ 0 \end{bmatrix}, \qquad \mathbf{v}_2 = \begin{bmatrix} 1 \\ 2 \\ 0 \end{bmatrix}, \qquad \mathbf{v}_3 = \begin{bmatrix} 1 \\ 2 \\ -3 \end{bmatrix}, \qquad \mathbf{v}_4 = \begin{bmatrix} 0 \\ 0 \\ -3 \end{bmatrix}.$$

These four vectors are linearly dependent since

$$\mathbf{v}_3 = -\mathbf{v}_1 + 2\mathbf{v}_2 + \mathbf{v}_4.$$

However, any set of three of these vectors is a basis for $\mathbb{R}^3$.

14.2   Linear Maps

The *linear map* A that transforms the linear space $\mathcal{L}_n$ to the linear space $\mathcal{L}_m$, written as $A : \mathcal{L}_n \to \mathcal{L}_m$, preserves linear relationships. Let three *preimage* vectors $\mathbf{v}_1, \mathbf{v}_2, \mathbf{v}_3$ in $\mathcal{L}_n$ be mapped to three *image* vectors $A\mathbf{v}_1, A\mathbf{v}_2, A\mathbf{v}_3$ in $\mathcal{L}_m$. If there is a linear relationship among the preimages, $\mathbf{v}_i$, then the same relationship will hold for the images:

$$\mathbf{v}_1 = \alpha\mathbf{v}_2 + \beta\mathbf{v}_3 \quad \text{implies} \quad A\mathbf{v}_1 = \alpha A\mathbf{v}_2 + \beta A\mathbf{v}_3. \tag{14.2}$$

Maps that do not have this property are called *nonlinear maps* and are much harder to deal with.

Linear maps are conveniently written in terms of *matrices*. A vector $\mathbf{v}$ in $\mathcal{L}_n$ is mapped to $\mathbf{v}'$ in $\mathcal{L}_m$,

$$\mathbf{v}' = A\mathbf{v},$$

where A has m rows and n columns. The matrix A describes the map from the $[\mathbf{e}_1, \ldots, \mathbf{e}_n]$-system to the $[\mathbf{a}_1, \ldots, \mathbf{a}_n]$-system, where the $\mathbf{a}_i$ are in $\mathcal{L}_m$. This means that $\mathbf{v}'$ is a linear combination $\mathbf{v}$ of the $\mathbf{a}_i$,

$$\mathbf{v}' = v_1\mathbf{a}_1 + v_2\mathbf{a}_2 + \ldots v_n\mathbf{a}_n,$$

and therefore it is in the column space of A.

Example 14.7

Suppose we have a map $A : \mathbb{R}^2 \to \mathbb{R}^3$, defined by

$$A = \begin{bmatrix} 1 & 0 \\ 0 & 1 \\ 2 & 2 \end{bmatrix}.$$

And suppose we have three vectors in $\mathbb{R}^2$,

$$\mathbf{v}_1 = \begin{bmatrix} 1 \\ 0 \end{bmatrix}, \qquad \mathbf{v}_2 = \begin{bmatrix} 0 \\ 1 \end{bmatrix}, \qquad \mathbf{v}_3 = \begin{bmatrix} 2 \\ 1 \end{bmatrix},$$

which are mapped to

$$\hat{\mathbf{v}}_1 = \begin{bmatrix} 1 \\ 0 \\ 2 \end{bmatrix}, \qquad \hat{\mathbf{v}}_2 = \begin{bmatrix} 0 \\ 1 \\ 2 \end{bmatrix}, \qquad \hat{\mathbf{v}}_3 = \begin{bmatrix} 2 \\ 1 \\ 6 \end{bmatrix},$$

in $\mathbb{R}^3$ by A.

The $\mathbf{v}_i$ are linearly dependent since $\mathbf{v}_3 = 2\mathbf{v}_1 + \mathbf{v}_2$. This same linear combination holds for the $\hat{\mathbf{v}}_i$, asserting that the map preserves linear relationships.

The matrix A has a certain rank k—how can we infer this rank from the matrix? First of all, a matrix of size $m \times n$, can be at most of rank $k = \min\{m, n\}$. This is called *full rank*. In other words, a linear map can never *increase* dimension. It is possible to map $\mathcal{L}_n$ to a higher-dimensional space $\mathcal{L}_m$. However, the images of $\mathcal{L}_n$'s n basis vectors will span a subspace of $\mathcal{L}_m$ of dimension at most n. Example 14.7 demonstrates this idea: the matrix A has rank 2, thus the $\hat{\mathbf{v}}_i$ live in a dimension 2 subspace of $\mathbb{R}^3$. A matrix with rank less than this $\min\{m, n\}$ is called *rank deficient*. We perform forward elimination (possibly with row exchanges) until the matrix is in upper triangular form. If after forward elimination there are k nonzero rows, then the rank of A is k. This is equivalent to our definition in Section 4.2 that the rank is equal to the number of linearly independent column vectors. Figure 14.2 gives an illustration of some possible scenarios.

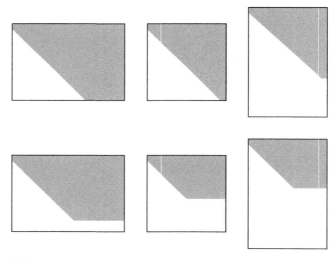

Figure 14.2.
The three types of matrices: from left to right, $m < n$, $m = n$, $m > n$. Examples of full rank matrices are on the top row, and examples of rank deficient matrices are on the bottom row. In each, gray indicates nonzero entries and white indicates zero entries after forward elimination was performed.

Example 14.8

Let us determine the rank of the matrix

$$\begin{bmatrix} 1 & 3 & 4 \\ 0 & 1 & 2 \\ 1 & 2 & 2 \\ -1 & 1 & 1 \end{bmatrix}.$$

We perform forward elimination to obtain

$$\begin{bmatrix} 1 & 3 & 4 \\ 0 & 1 & 2 \\ 0 & 0 & -3 \\ 0 & 0 & 0 \end{bmatrix}.$$

There is one row of zeroes, and we conclude that the matrix has rank 3, which is full rank since $\min\{4, 3\} = 3$.

Next, let us take the matrix

$$\begin{bmatrix} 1 & 3 & 4 \\ 0 & 1 & 2 \\ 1 & 2 & 2 \\ 0 & 1 & 2 \end{bmatrix}.$$

Forward elimination yields

$$\begin{bmatrix} 1 & 3 & 4 \\ 0 & 1 & 2 \\ 0 & 0 & 0 \\ 0 & 0 & 0 \end{bmatrix},$$

and we conclude that this matrix has rank 2, which is rank deficient.

Let's review some other features of linear maps that we have encountered in earlier chapters. A square, $n \times n$ matrix A of rank n is *invertible*, which means that there is a matrix that undoes A's action. This is the inverse matrix, denoted by A^{-1}. See Section 12.4 on how to compute the inverse.

If a matrix is invertible, then it does not reduce dimension and its *determinant* is nonzero. The determinant of a square matrix measures the volume of the n-dimensional parallelepiped, which is defined by its column vectors. The determinant of a matrix is computed by subjecting its column vectors to a sequence of shears until it is of upper triangular form (forward elimination). The value is then the product of the diagonal elements. (If pivoting takes place, the row exchanges need to be documented in order to determine the correct sign of the determinant.)

14.3   Inner Products

Let $\mathbf{u}, \mathbf{v}, \mathbf{w}$ be elements of a linear space $\mathcal{L}_n$ and let α and β be scalars. A map from $\mathcal{L}_n$ to the reals $\mathbb{R}$ is called an *inner product* if it assigns a real number $\langle \mathbf{v}, \mathbf{w} \rangle$ to the pair $\mathbf{v}, \mathbf{w}$ such that:

$$\langle \mathbf{v}, \mathbf{w} \rangle = \langle \mathbf{w}, \mathbf{v} \rangle \qquad \text{(symmetry)}, \qquad (14.3)$$

$$\langle \alpha\mathbf{v}, \mathbf{w} \rangle = \alpha\langle \mathbf{w}, \mathbf{v} \rangle \qquad \text{(homogeneity)}, \qquad (14.4)$$

$$\langle \mathbf{u} + \mathbf{v}, \mathbf{w} \rangle = \langle \mathbf{u}, \mathbf{w} \rangle + \langle \mathbf{v}, \mathbf{w} \rangle \qquad \text{(additivity)}, \qquad (14.5)$$

$$\text{for all } \mathbf{v} \quad \langle \mathbf{v}, \mathbf{v} \rangle \geq 0 \quad \text{ and}$$

$$\langle \mathbf{v}, \mathbf{v} \rangle = 0 \text{ if and only if } \mathbf{v} = \mathbf{0} \quad \text{(positivity)}. \qquad (14.6)$$

The homogeneity and additivity properties can be nicely combined into

$$\langle \alpha\mathbf{u} + \beta\mathbf{v}, \mathbf{w} \rangle = \alpha\langle \mathbf{u}, \mathbf{w} \rangle + \beta\langle \mathbf{v}, \mathbf{w} \rangle.$$

A linear space with an inner product is called an *inner product space*.

We can consider the inner product to be a generalization of the dot product. For the dot product in $\mathbb{R}^n$, we write

$$\langle \mathbf{v}, \mathbf{w} \rangle = \mathbf{v} \cdot \mathbf{w} = v_1 w_1 + v_2 w_2 + \ldots + v_n w_n.$$

From our experience with 2D and 3D, we can easily show that the dot product satisfies the inner product properties.

Example 14.9

Suppose $\mathbf{u}, \mathbf{v}, \mathbf{w}$ live in $\mathbb{R}^2$. Let's define the following "test" inner product in $\mathbb{R}^2$,

$$\langle \mathbf{v}, \mathbf{w} \rangle = 4v_1 w_1 + 2v_2 w_2, \qquad (14.7)$$

which is an odd variation of the standard dot product. To get a feel for it, consider the three unit vectors, $\mathbf{e}_1$, $\mathbf{e}_2$, and $\mathbf{r} = [1/\sqrt{2} \ \ 1/\sqrt{2}]^{\mathrm{T}}$, then

$$\langle \mathbf{e}_1, \mathbf{e}_2 \rangle = 4(1)(0) + 2(0)(1) = 0,$$

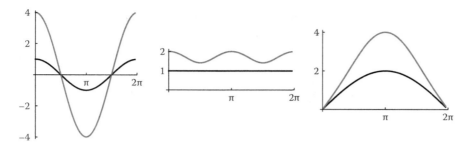

Figure 14.3.

Inner product: comparing the dot product (black) with the test inner product from Example 14.9 (gray). For each plot, the unit vector **r** is rotated in the range $[0, 2\pi]$. Left: the graph of the inner products, $\mathbf{e}_1 \cdot \mathbf{r}$ and $\langle \mathbf{e}_1, \mathbf{r} \rangle$. Middle: length for the inner products, $\sqrt{\mathbf{r} \cdot \mathbf{r}}$ and $\sqrt{\langle \mathbf{r}, \mathbf{r} \rangle}$. Right: distance for the inner products, $\sqrt{(\mathbf{e}_1 - \mathbf{r}) \cdot (\mathbf{e}_1 - \mathbf{r})}$ and $\sqrt{\langle (\mathbf{e}_1 - \mathbf{r}), (\mathbf{e}_1 - \mathbf{r}) \rangle}$.

which is the same result as the dot product, but

$$\langle \mathbf{e}_1, \mathbf{r} \rangle = 4(1) \left(\frac{1}{\sqrt{2}} \right) + 2(0) \left(\frac{1}{\sqrt{2}} \right) = \frac{4}{\sqrt{2}}$$

differs from $\mathbf{e}_1 \cdot \mathbf{r} = 1/\sqrt{2}$. In Figure 14.3 (left), this test inner product and the dot product are graphed for **r** that is rotated in the range $[0, 2\pi]$. Notice that the graph of the dot product is the graph of $\cos(\theta)$.

We will now show that this definition satisfies the inner product properties.

Symmetry:
$$\begin{aligned} \langle \mathbf{v}, \mathbf{w} \rangle &= 4v_1 w_1 + 2v_2 w_2 \\ &= 4w_1 v_1 + 2w_2 v_2 \\ &= \langle \mathbf{w}, \mathbf{v} \rangle. \end{aligned}$$

Homogeneity:
$$\begin{aligned} \langle \alpha \mathbf{v}, \mathbf{w} \rangle &= 4(\alpha v_1) w_1 + 2(\alpha v_2) w_2 \\ &= \alpha(4v_1 w_1 + 2v_2 w_2) \\ &= \alpha \langle \mathbf{v}, \mathbf{w} \rangle. \end{aligned}$$

Additivity:
$$\begin{aligned} \langle \mathbf{u} + \mathbf{v}, \mathbf{w} \rangle &= 4(u_1 + v_1) w_1 + 2(u_2 + v_2) w_2 \\ &= (4u_1 w_1 + 2u_2 w_2) + (4v_1 w_1 + 2v_2 w_2) \\ &= \langle \mathbf{u}, \mathbf{w} \rangle + \langle \mathbf{v}, \mathbf{w} \rangle. \end{aligned}$$

Positivity:
$$\begin{aligned} \langle \mathbf{v}, \mathbf{v} \rangle &= 4v_1^2 + 2v_2^2 \geq 0, \\ \langle \mathbf{v}, \mathbf{v} \rangle &= 0 \text{ if and only if } \mathbf{v} = \mathbf{0}. \end{aligned}$$

We can question the usefulness of this inner product, but it does satisfy the necessary properties.

Inner product spaces offer the concept of *length*,

$$\|\mathbf{v}\| = \sqrt{\langle \mathbf{v}, \mathbf{v} \rangle}.$$

This is also called the 2-norm or Euclidean norm and is denoted as $\|\mathbf{v}\|_2$. Since the 2-norm is the most commonly used norm, the subscript is typically omitted. (More vector norms were introduced in Section 13.2.) Then the distance between two vectors,

$$\text{dist}(\mathbf{u}, \mathbf{v}) = \sqrt{\langle \mathbf{u} - \mathbf{v}, \mathbf{u} - \mathbf{v} \rangle} = \|\mathbf{u} - \mathbf{v}\|.$$

For vectors in $\mathbb{R}^n$ and the dot product, we have the Euclidean norm

$$\|\mathbf{v}\| = \sqrt{v_1^2 + v_2^2 + \ldots + v_n^2}$$

and

$$\text{dist}(\mathbf{u}, \mathbf{v}) = \sqrt{(u_1 - v_1)^2 + (u_2 - v_2)^2 + \ldots + (u_n - v_n)^2}.$$

Example 14.10

Let's get a feel for the norm and distance concept for the test inner product in (14.7). We have

$$\|\mathbf{e}_1\| = \sqrt{\langle \mathbf{e}_1, \mathbf{e}_1 \rangle} = 4(1)^2 + 2(0)^2 = 4$$

$$\text{dist}(\mathbf{e}_1, \mathbf{e}_2) = \sqrt{4(1-0)^2 + 2(0-1)^2} = \sqrt{6},$$

The dot product produces $\|\mathbf{e}_1\| = 1$ and $\text{dist}(\mathbf{e}_1, \mathbf{e}_2) = \sqrt{2}$.

Figure 14.3 illustrates the difference between the dot product and the test inner product. Again, $\mathbf{r}$ is a unit vector, rotated in the range $[0, 2\pi]$. The middle plot shows that unit length vectors with respect to the dot product are not unit length with respect to the test inner product. The right plot shows that the distance between two vectors differs too.

Orthogonality is important to establish as well. If two vectors $\mathbf{v}$ and $\mathbf{w}$ in a linear space $\mathcal{L}_n$ satisfy

$$\langle \mathbf{v}, \mathbf{w} \rangle = 0,$$

then they are called *orthogonal*. If $\mathbf{v}_1, \ldots, \mathbf{v}_n$ form a basis for $\mathcal{L}_n$ and all $\mathbf{v}_i$ are mutually orthogonal, $\langle \mathbf{v}_i, \mathbf{v}_j \rangle = 0$ for $i \neq j$, then the $\mathbf{v}_i$ are said to form an *orthogonal basis*. If in addition they are also of unit length, $\|\mathbf{v}_i\| = 1$, they form an *orthonormal basis*. Any basis of a linear space may be transformed to an orthonormal basis by the Gram-Schmidt process, described in Section 14.4.

The Cauchy-Schwartz inequality was introduced in (2.23) for $\mathbb{R}^2$, and we repeat it here in the context of inner product spaces,

$$\langle \mathbf{v}, \mathbf{w} \rangle^2 \leq \langle \mathbf{v}, \mathbf{v} \rangle \langle \mathbf{w}, \mathbf{w} \rangle.$$

Equality holds if and only if $\mathbf{v}$ and $\mathbf{w}$ are linearly dependent.

If we restate the Cauchy-Schwartz inequality

$$\langle \mathbf{v}, \mathbf{w} \rangle^2 \leq \|\mathbf{v}\|^2 \|\mathbf{w}\|^2$$

and rearrange,

$$\left(\frac{\langle \mathbf{v}, \mathbf{w} \rangle}{\|\mathbf{v}\| \|\mathbf{w}\|} \right)^2 \leq 1,$$

we obtain

$$-1 \leq \frac{\langle \mathbf{v}, \mathbf{w} \rangle}{\|\mathbf{v}\| \|\mathbf{w}\|} \leq 1.$$

Now we can express the angle θ between $\mathbf{v}$ and $\mathbf{w}$ by

$$\cos \theta = \frac{\langle \mathbf{v}, \mathbf{w} \rangle}{\|\mathbf{v}\| \|\mathbf{w}\|}.$$

The properties defining an inner product (14.3)–(14.6) suggest

$$\|\mathbf{v}\| \geq 0,$$
$$\|\mathbf{v}\| = 0 \text{ if and only if } \mathbf{v} = 0,$$
$$\|\alpha \mathbf{v}\| = |\alpha| \|\mathbf{v}\|.$$

A fourth property is the triangle inequality,

$$\|\mathbf{v} + \mathbf{w}\| \leq \|\mathbf{v}\| + \|\mathbf{w}\|,$$

which we derived from the Cauchy-Schwartz inequality in Section 2.9. See Sketch 2.22 for an illustration of this with respect to a triangle formed from two vectors in $\mathbb{R}^2$. See the exercises for the generalized Pythagorean theorem.

The tools associated with the inner product are key for orthogonal decomposition and best approximation in a linear space. Recall that these concepts were introduced in Section 2.8 and have served as the building blocks for the 3D Gram-Schmidt method for construction of an orthonormal coordinate frame (Section 11.8) and least squares approximation (Section 12.7).

We finish with the general definition of a projection. Let $\mathbf{u}_1, \ldots, \mathbf{u}_k$ span a subspace $\mathcal{L}_k$ of $\mathcal{L}$. If $\mathbf{v}$ is a vector not in that space, then

$$P\mathbf{v} = \langle \mathbf{v}, \mathbf{u}_1 \rangle \mathbf{u}_1 + \ldots + \langle \mathbf{v}, \mathbf{u}_k \rangle \mathbf{u}_k$$

is $\mathbf{v}$'s orthogonal projection into $\mathcal{L}_k$.

14.4   Gram-Schmidt Orthonormalization

In Section 11.8 we built an orthonormal coordinate frame in $\mathbb{R}^3$ using the Gram-Schmidt method. Every inner product space has an orthonormal basis. The key elements of its construction are projections and vector decomposition.

Let $\mathbf{b}_1, \ldots, \mathbf{b}_r$ be a set of orthonormal vectors, forming the basis of an r-dimensional subspace $\mathcal{S}_r$ of $\mathcal{L}_n$, where $n > r$. We want to find $\mathbf{b}_{r+1}$ orthogonal to the given $\mathbf{b}_i$. Let $\mathbf{u}$ be an arbitrary vector in $\mathcal{L}_n$, but not in $\mathcal{S}_r$. Define a vector $\hat{\mathbf{u}}$ by

$$\hat{\mathbf{u}} = \text{proj}_{\mathcal{S}_r} \mathbf{u} = \langle \mathbf{u}, \mathbf{b}_1 \rangle \mathbf{b}_1 + \ldots + \langle \mathbf{u}, \mathbf{b}_r \rangle \mathbf{b}_r.$$

This vector is $\mathbf{u}$'s *orthogonal projection* into $\mathcal{S}_r$. To see this, we check that the difference vector $\mathbf{u} - \hat{\mathbf{u}}$ is orthogonal to each of the $\mathbf{b}_i$. We first check that it is orthogonal to $\mathbf{b}_1$ and observe

$$\langle \mathbf{u} - \hat{\mathbf{u}}, \mathbf{b}_1 \rangle = \langle \mathbf{u}, \mathbf{b}_1 \rangle - \langle \mathbf{u}, \mathbf{b}_1 \rangle \langle \mathbf{b}_1, \mathbf{b}_1 \rangle - \ldots - \langle \mathbf{u}, \mathbf{b}_r \rangle \langle \mathbf{b}_1, \mathbf{b}_r \rangle.$$

All terms $\langle \mathbf{b}_1, \mathbf{b}_2 \rangle$, $\langle \mathbf{b}_1, \mathbf{b}_3 \rangle$, etc. vanish since the $\mathbf{b}_i$ are mutually orthogonal and $\langle \mathbf{b}_1, \mathbf{b}_1 \rangle = 1$. Thus, $\langle \mathbf{u} - \hat{\mathbf{u}}, \mathbf{b}_1 \rangle = 0$. In the same manner, we show that $\mathbf{u} - \hat{\mathbf{u}}$ is orthogonal to the remaining $\mathbf{b}_i$. Thus

$$\mathbf{b}_{r+1} = \frac{\mathbf{u} - \text{proj}_{\mathcal{S}_r} \mathbf{u}}{\| \cdot \|},$$

and the set $\mathbf{b}_1, \ldots, \mathbf{b}_{r+1}$ forms an orthonormal basis for the subspace $\mathcal{S}_{r+1}$ of $\mathcal{L}_n$. We may repeat this process until we have found an orthonormal basis for all of $\mathcal{L}_n$.

Sketch 14.1 provides an illustration in which $\mathcal{S}_2$ is depicted as $\mathbb{R}^2$. This process is known as *Gram-Schmidt* orthonormalization. Given any basis $\mathbf{v}_1, \ldots, \mathbf{v}_n$ of $\mathcal{L}_n$, we can find an orthonormal basis by setting $\mathbf{b}_1 = \mathbf{v}_1/\|\cdot\|$ and continuing to construct, one by one, vectors $\mathbf{b}_2, \ldots, \mathbf{b}_n$ using the above procedure. For instance,

$$\mathbf{b}_2 = \frac{\mathbf{v}_2 - \text{proj}_{\mathcal{S}_1}\mathbf{v}_2}{\|\cdot\|} = \frac{\mathbf{v}_2 - \langle \mathbf{v}_2, \mathbf{b}_1 \rangle \mathbf{b}_1}{\|\cdot\|},$$

$$\mathbf{b}_3 = \frac{\mathbf{v}_3 - \text{proj}_{\mathcal{S}_2}\mathbf{v}_3}{\|\cdot\|} = \frac{\mathbf{v}_3 - \langle \mathbf{v}_3, \mathbf{b}_1 \rangle \mathbf{b}_1 - \langle \mathbf{v}_3, \mathbf{b}_2 \rangle \mathbf{b}_2}{\|\cdot\|}.$$

Consult Section 11.8 for an example in $\mathbb{R}^3$.

Sketch 14.1.
Gram-Schmidt orthonormalization.

Example 14.11

Suppose we are given the following vectors in $\mathbb{R}^4$,

$$\mathbf{v}_1 = \begin{bmatrix} 1 \\ 0 \\ 0 \\ 0 \end{bmatrix}, \quad \mathbf{v}_2 = \begin{bmatrix} 1 \\ 1 \\ 1 \\ 1 \end{bmatrix}, \quad \mathbf{v}_3 = \begin{bmatrix} 1 \\ 1 \\ 0 \\ 0 \end{bmatrix}, \quad \mathbf{v}_4 = \begin{bmatrix} 0 \\ 0 \\ 1 \\ 0 \end{bmatrix},$$

and we wish to form an orthonormal basis, $\mathbf{b}_1, \mathbf{b}_2, \mathbf{b}_3, \mathbf{b}_4$.

The Gram-Schmidt method, as defined in the displayed equations above, produces

$$\mathbf{b}_1 = \begin{bmatrix} 1 \\ 0 \\ 0 \\ 0 \end{bmatrix}, \quad \mathbf{b}_2 = \begin{bmatrix} 0 \\ 1/\sqrt{3} \\ 1/\sqrt{3} \\ 1/\sqrt{3} \end{bmatrix}, \quad \mathbf{b}_3 = \begin{bmatrix} 0 \\ 2/\sqrt{6} \\ -1/\sqrt{6} \\ -1/\sqrt{6} \end{bmatrix}.$$

The final vector, $\mathbf{b}_4$, is defined as

$$\mathbf{b}_4 = \mathbf{v}_4 - \langle \mathbf{v}_4, \mathbf{b}_1 \rangle \mathbf{b}_1 - \langle \mathbf{v}_4, \mathbf{b}_2 \rangle \mathbf{b}_2 - \langle \mathbf{v}_4, \mathbf{b}_3 \rangle \mathbf{b}_3 = \begin{bmatrix} 0 \\ 0 \\ 1/\sqrt{2} \\ -1/\sqrt{2} \end{bmatrix}$$

Knowing that the $\mathbf{b}_i$ are normalized and checking that $\mathbf{b}_i \cdot \mathbf{b}_j = 0$, we can be confident that this is an orthonormal basis. Another tool we have is the determinant, which will be one,

$$\begin{vmatrix} \mathbf{b}_1 & \mathbf{b}_2 & \mathbf{b}_3 & \mathbf{b}_4 \end{vmatrix} = 1.$$

14.5   A Gallery of Spaces

In this section, we highlight some special linear spaces—but there are many more!

For a first example, consider all polynomials of a fixed degree n. These are functions of the form

$$p(t) = a_0 + a_1 t + a_2 t^2 + \ldots + a_n t^n$$

where t is the independent variable of $p(t)$. Thus we can construct a linear space $\mathcal{P}_n$ whose elements are all polynomials of a fixed degree. Addition in this space is addition of polynomials, i.e., coefficient by coefficient; multiplication is multiplication of a polynomial by a real number. It is easy to check that these polynomials have the linearity property (14.1). For example, if $p(t) = 3 - 2t + 3t^2$ and $q(t) = -1 + t + 2t^2$, then $2p(t) + 3q(t) = 3 - t + 12t^2$ is yet another polynomial of the same degree.

This example also serves to introduce a not-so-obvious linear map: the operation of forming derivatives! The derivative p' of a degree n polynomial p is a polynomial of degree $n - 1$, given by

$$p'(t) = a_1 + 2a_2 t + \ldots + n a_n t^{n-1}.$$

The linear map of forming derivatives thus maps the space of all degree n polynomials into that of all degree $n - 1$ polynomials. The rank of this map is thus $n - 1$.

Example 14.12

Let us consider the two cubic polynomials

$$p(t) = 3 - t + 2t^2 + 3t^3 \quad \text{and} \quad q(t) = 1 + t - t^3$$

in the linear space of cubic polynomials, $\mathcal{P}_3$. Let

$$r(t) = 2p(t) - q(t) = 5 - 3t + 4t^2 + 7t^3.$$

Now

$$r'(t) = -3 + 8t + 21t^2,$$
$$p'(t) = -1 + 4t + 9t^2,$$
$$q'(t) = 1 - 3t^2.$$

It is now trivial to check that $r'(t) = 2p'(t) - q'(t)$, thus asserting the linearity of the derivative map.

For a second example, a linear space is given by the set of all real-valued continuous functions over the interval $[0, 1]$. This space is typically named $C[0, 1]$. Clearly the linearity condition is met: if f and g are elements of $C[0, 1]$, then $\alpha f + \beta g$ is also in $C[0, 1]$. Here we have an example of a linear space that is infinite-dimensional, meaning that no finite set of functions forms a basis for $C[0, 1]$.

For a third example, consider the set of all 3×3 matrices. They form a linear space; this space consists of matrices. In this space, linear combinations are formed using standard matrix addition and multiplication with a scalar as summarized in Section 9.11.

And, finally, a more abstract example. The set of all linear maps from a linear space $\mathcal{L}_n$ into the reals forms a linear space itself and it is called the *dual space* $\mathcal{L}_n^*$ of $\mathcal{L}_n$. As indicated by the notation, its dimension equals that of $\mathcal{L}_n$. The linear maps in $\mathcal{L}_n^*$ are known as *linear functionals*.

For an example, let a fixed vector $\mathbf{v}$ and a variable vector $\mathbf{u}$ be in $\mathcal{L}_n$. The linear functionals defined by $\Phi_{\mathbf{v}}(\mathbf{u}) = \langle \mathbf{u}, \mathbf{v} \rangle$ are in $\mathcal{L}_n^*$. Then, for any basis $\mathbf{b}_1, \ldots, \mathbf{b}_n$ of $\mathcal{L}_n$, we can define linear functionals

$$\Phi_{\mathbf{b}_i}(\mathbf{u}) = \langle \mathbf{u}, \mathbf{b}_i \rangle \quad \text{for } i = 1, \ldots, n.$$

These functionals form a basis for $\mathcal{L}_n^*$.

Example 14.13

In $\mathbb{R}^2$, consider the fixed vector

$$\mathbf{v} = \begin{bmatrix} 1 \\ -2 \end{bmatrix}.$$

Then $\Phi_{\mathbf{v}}(\mathbf{u}) = \langle \mathbf{u}, \mathbf{v} \rangle = u_1 - 2u_2$ for all vectors $\mathbf{u}$, where $\langle \cdot, \cdot \rangle$ is the dot product.

Example 14.14

Pick $\mathbf{e}_1, \mathbf{e}_2$ for a basis in $\mathbb{R}^2$. The associated linear functionals are

$$\Phi_{\mathbf{e}_1}(\mathbf{u}) = u_1, \quad \Phi_{\mathbf{e}_2}(\mathbf{u}) = u_2.$$

Any linear functional Φ can now be defined as

$$\Phi(\mathbf{u}) = r_1 \Phi_{\mathbf{e}_1}(\mathbf{u}) + r_2 \Phi_{\mathbf{e}_2}(\mathbf{u}),$$

where r_1 and r_2 are scalars.

- linear space
- vector space
- dimension
- linear combination
- linearity property
- linearly independent
- subspace
- span
- linear map
- image
- preimage
- domain
- range
- rank
- full rank
- rank deficient
- inverse

- determinant
- inner product
- inner product space
- distance in an inner product space
- length in an inner product space
- orthogonal
- Gram-Schmidt method
- projection
- basis
- orthonormal
- orthogonal decomposition
- best approximation
- dual space
- linear functional

14.6   Exercises

1. Given elements of $\mathbb{R}^4$,

$$\mathbf{u} = \begin{bmatrix} 1 \\ 2 \\ 0 \\ 4 \end{bmatrix}, \quad \mathbf{v} = \begin{bmatrix} 0 \\ 0 \\ 2 \\ 7 \end{bmatrix}, \quad \mathbf{w} = \begin{bmatrix} 3 \\ 1 \\ 2 \\ 0 \end{bmatrix},$$

   is $\mathbf{r} = 3\mathbf{u} + 6\mathbf{v} + 2\mathbf{w}$ also in $\mathbb{R}^4$?

2. Given matrices that are elements of $\mathcal{M}_{3 \times 3}$,

$$A = \begin{bmatrix} 1 & 0 & 2 \\ 2 & 0 & 1 \\ 1 & 1 & 3 \end{bmatrix} \quad \text{and} \quad B = \begin{bmatrix} 3 & 0 & 0 \\ 0 & 3 & 1 \\ 4 & 1 & 7 \end{bmatrix},$$

   is $C = 4A + B$ an element of $\mathcal{M}_{3 \times 3}$?

3. Does the set of all polynomials with $a_n = 1$ form a linear space?

4. Does the set of all 3D vectors with nonnegative components form a subspace of $\mathbb{R}^3$?

5. Are

$$\mathbf{u} = \begin{bmatrix} 1 \\ 0 \\ 1 \\ 0 \\ 1 \end{bmatrix}, \quad \mathbf{v} = \begin{bmatrix} 1 \\ 1 \\ 1 \\ 1 \\ 1 \end{bmatrix}, \quad \mathbf{w} = \begin{bmatrix} 0 \\ 1 \\ 0 \\ 1 \\ 0 \end{bmatrix}$$

linearly independent?

6. Are

$$\mathbf{u} = \begin{bmatrix} 1 \\ 2 \\ 1 \end{bmatrix}, \quad \mathbf{v} = \begin{bmatrix} 3 \\ 6 \\ 1 \end{bmatrix}, \quad \mathbf{w} = \begin{bmatrix} 2 \\ 2 \\ 1 \end{bmatrix}$$

linearly independent?

7. Is the vector

$$\mathbf{r} = \begin{bmatrix} 2 \\ 3 \\ 2 \\ 3 \\ 2 \end{bmatrix}$$

in the subspace defined by $\mathbf{u}, \mathbf{v}, \mathbf{w}$ defined in Exercise 5?

8. Is the vector

$$\mathbf{r} = \begin{bmatrix} 2 \\ 3 \\ 2 \\ 3 \\ 2 \end{bmatrix}$$

in the subspace defined by $\mathbf{u}, \mathbf{v}, \mathbf{w}$ defined in Exercise 6?

9. What is the dimension of the linear space formed by all $n \times n$ matrices?

10. Suppose we are given a linear map $A : \mathbb{R}^4 \to \mathbb{R}^2$, preimage vectors $\mathbf{v}_i$, and corresponding image vectors $\mathbf{w}_i$. What are the dimensions of the matrix A? The following linear relationship exists among the preimages,
$$\mathbf{v}_4 = 3\mathbf{v}_1 + 6\mathbf{v}_2 + 9\mathbf{v}_3.$$
What relationship holds for $\mathbf{w}_4$ with respect to the $\mathbf{w}_i$?

11. What is the rank of the matrix
$$\begin{bmatrix} 1 & 2 & 0 \\ -1 & -2 & 1 \\ 0 & 0 & 1 \\ 2 & 4 & -1 \end{bmatrix}?$$

12. What is the rank of the matrix
$$\begin{bmatrix} 1 & 2 & 0 & 0 & 0 \\ -1 & 0 & 0 & -2 & 1 \\ 0 & 0 & 1 & 0 & 1 \end{bmatrix}?$$

13. Given the vectors

$$\mathbf{w} = \begin{bmatrix} 1/\sqrt{10} \\ 3/\sqrt{10} \end{bmatrix} \quad \text{and} \quad \mathbf{r} = \begin{bmatrix} -3/\sqrt{10} \\ 1/\sqrt{10} \end{bmatrix},$$

find $\langle \mathbf{w}, \mathbf{r} \rangle$, $\|\mathbf{w}\|$, $\|\mathbf{r}\|$, and $\text{dist}(\mathbf{w}, \mathbf{r})$ with respect to the dot product and then for the test inner product in (14.7).

14. For $\mathbf{v}, \mathbf{w}$ in $\mathbb{R}^3$, does

$$\langle \mathbf{v}, \mathbf{w} \rangle = v_1^2 w_1^2 + v_2^2 w_2^2 + v_3^2 w_3^2$$

satisfy the requirements of an inner product?

15. For $\mathbf{v}, \mathbf{w}$ in $\mathbb{R}^3$, does

$$\langle \mathbf{v}, \mathbf{w} \rangle = 4v_1 w_1 + v_2 w_2 + 2v_3 w_3$$

satisfy the requirements of an inner product?

16. In the space of all 3×3 matrices, is

$$\langle A, B \rangle = a_{1,1}b_{1,1} + a_{1,2}b_{1,2} + a_{1,3}b_{1,3} + a_{2,1}b_{2,1} + \ldots + a_{3,3}b_{3,3}$$

an inner product?

17. Let $p(t) = p_0 + p_1 t + p_2 t^2$ and $q(t) = q_0 + q_1 t + q_2 t^2$ be two quadratic polynomials. Define

$$\langle p, q \rangle = p_0 q_0 + p_1 q_1 + p_2 q_2.$$

Is this an inner product for the space of all quadratic polynomials?

18. Show that the Pythagorean theorem

$$\|\mathbf{v} + \mathbf{w}\|^2 = \|\mathbf{v}\|^2 + \|\mathbf{w}\|^2$$

holds for orthogonal vectors $\mathbf{v}$ and $\mathbf{w}$ in an inner product space.

19. Given the following vectors in $\mathbb{R}^3$,

$$\mathbf{v}_1 = \begin{bmatrix} 1 \\ 1 \\ 0 \end{bmatrix}, \quad \mathbf{v}_2 = \begin{bmatrix} 0 \\ 1 \\ 0 \end{bmatrix}, \quad \mathbf{v}_3 = \begin{bmatrix} -1 \\ -1 \\ -1 \end{bmatrix},$$

form an orthonormal basis, $\mathbf{b}_1, \mathbf{b}_2, \mathbf{b}_3$.

20. Given the following vectors in $\mathbb{R}^4$,

$$\mathbf{v}_1 = \begin{bmatrix} 0 \\ 0 \\ 1 \\ 0 \end{bmatrix}, \quad \mathbf{v}_2 = \begin{bmatrix} 1 \\ 1 \\ 1 \\ 1 \end{bmatrix}, \quad \mathbf{v}_3 = \begin{bmatrix} 1 \\ 1 \\ 0 \\ 0 \end{bmatrix}, \quad \mathbf{v}_4 = \begin{bmatrix} 1 \\ 0 \\ 0 \\ 0 \end{bmatrix},$$

form an orthonormal basis, $\mathbf{b}_1 = \mathbf{v}_1, \mathbf{b}_2, \mathbf{b}_3, \mathbf{b}_4$.

21. Find a basis for the linear space formed by all 2×2 matrices.

22. Does the set of all monotonically increasing functions over $[0, 1]$ form a linear space?

23. Let $\mathcal{L}$ be a linear space. Is the map $\Phi(\mathbf{u}) = \|\mathbf{u}\|$ an element of the dual space $\mathcal{L}^*$?

24. Show the linearity of the derivative map on the linear space of quadratic polynomials $\mathcal{P}_2$.

15

Eigen Things Revisited

Figure 15.1.
Google matrix: part of the connectivity matrix for Wikipedia pages in 2009, which is used to find the webpage ranking. (Source: Wikipedia, Google matrix.)

Chapter 7 "Eigen Things" introduced the basics of eigenvalues and eigenvectors in terms of 2×2 matrices. But also for any $n \times n$ matrix A, we may ask if it has fixed directions and what are the corresponding eigenvalues.

In this chapter we go a little further and examine the *power method* for finding the eigenvector that corresponds to the dominant eigenvalue. This method is paired with an application section describing how a search engine might rank webpages based on this special eigenvector, given a fun, slang name—the Google eigenvector.

We explore "Eigen Things" of function spaces that are even more general than those in the gallery in Section 14.5.

"Eigen Things" characterize a map by revealing its action and geometry. This is key to understanding the behavior of any system. A great example of this interplay is provided by the collapse of the Tacoma Narrows Bridge in Figures 7.1 and 7.2. But "Eigen Things" are important in many other areas: characterizing harmonics of musical instruments, moderating movement of fuel in a ship, and analysis of large data sets, such as the Google matrix in Figure 15.1.

15.1 The Basics Revisited

If an $n \times n$ matrix A has fixed directions, then there are vectors $\mathbf{r}$ that are mapped to multiples λ of themselves by A. Such vectors are characterized by

$$A\mathbf{r} = \lambda\mathbf{r}$$

or

$$[A - \lambda I]\mathbf{r} = \mathbf{0}. \tag{15.1}$$

Since $\mathbf{r} = \mathbf{0}$ trivially satisfies this equation, we will not consider it from now on. In (15.1), we see that the matrix $[A - \lambda I]$ maps a nonzero vector $\mathbf{r}$ to the zero vector. Thus, its determinant must vanish (see Section 4.9):

$$\det[A - \lambda I] = 0. \tag{15.2}$$

The term $\det[A - \lambda I]$ is called the *characteristic polynomial* of A. It is a polynomial of degree n in λ, and its zeroes are A's eigenvalues.

Example 15.1

Let

$$A = \begin{bmatrix} 1 & 1 & 0 & 0 \\ 0 & 3 & 1 & 0 \\ 0 & 0 & 4 & 1 \\ 0 & 0 & 0 & 2 \end{bmatrix}.$$

We find the degree four characteristic polynomial $p(\lambda)$:

$$p(\lambda) = \det[A - \lambda I] = \begin{vmatrix} 1 - \lambda & 1 & 0 & 0 \\ 0 & 3 - \lambda & 1 & 0 \\ 0 & 0 & 4 - \lambda & 1 \\ 0 & 0 & 0 & 2 - \lambda \end{vmatrix},$$

resulting in

$$p(\lambda) = (1 - \lambda)(3 - \lambda)(4 - \lambda)(2 - \lambda).$$

The zeroes of this polynomial are found by solving $p(\lambda) = 0$. In our slightly contrived example, we find $\lambda_1 = 4, \lambda_2 = 3, \lambda_3 = 2, \lambda_4 = 1$. Convention orders the eigenvalues in decreasing order. Recall that the largest eigenvalue in absolute value is called the *dominant eigenvalue*.

Observe that this matrix is upper triangular. Forming $p(\lambda)$ with expansion by minors (see Section 9.8), it is clear that the eigenvalues for this type of matrix are on the main diagonal.

We learned that Gauss elimination or LU decomposition allows us to transform a matrix into upper triangular form. However, elementary row operations change the eigenvalues, so this is not a means for finding eigenvalues. Instead, diagonalization is used when possible, and this is the topic of Chapter 16.

Example 15.2

Here is a simple example showing that elementary row operations change the eigenvalues. Start with the matrix

$$A = \begin{bmatrix} 2 & 2 \\ 1 & 2 \end{bmatrix}$$

that has $\det A = 2$ and eigenvalues $\lambda_1 = 2 + \sqrt{2}$ and $\lambda_2 = 2 - \sqrt{2}$. After one step of forward elimination, we have an upper triangular matrix

$$A' = \begin{bmatrix} 2 & 2 \\ 0 & 1 \end{bmatrix}.$$

The determinant is invariant under forward elimination, $\det A' = 2$; however, the eigenvalues are not: A' has eigenvalues $\lambda_1 = 2$ and $\lambda_2 = 1$.

The bad news is that one is not always dealing with upper triangular matrices like the one in Example 15.1. A general $n \times n$ matrix has a degree n characteristic polynomial

$$p(\lambda) = \det[A - \lambda I] = (\lambda_1 - \lambda)(\lambda_2 - \lambda) \cdot \ldots \cdot (\lambda_n - \lambda), \qquad (15.3)$$

and the eigenvalues are the zeroes of this polynomial. Finding the zeroes of an nth degree polynomial is a nontrivial numerical task. In fact, for $n \geq 5$, it is not certain that we can algebraically find the factorization in (15.3) because there is no general formula like we have for $n = 2$, the quadratic equation. An iterative method for finding the dominant eigenvalue is described in Section 15.2.

For 2×2 matrices, in (7.3) and (7.4), we observed that the characteristic polynomial easily reveals that the determinant is the product of the eigenvalues. For $n \times n$ matrices, we have the same situation. Consider $\lambda = 0$ in (15.3), then $p(0) = \det A = \lambda_1 \lambda_2 \cdot \ldots \cdot \lambda_n$.

Needless to say, not all eigenvalues of a matrix are real in general. But the important class of *symmetric* matrices always does have real eigenvalues.

Two more properties of eigenvalues:

- The matrices A and A^T have the same eigenvalues.

- Suppose A is invertible and has eigenvalues λ_i, then A^{-1} has eigenvalues $1/\lambda_i$.

Having found the λ_i, we can now solve homogeneous linear systems

$$[A - \lambda_i I]\mathbf{r}_i = \mathbf{0}$$

in order to find the eigenvectors $\mathbf{r}_i$ for $i = 1, n$. The $\mathbf{r}_i$ are in the null space of $[A - \lambda_i I]$. These are homogeneous systems, and thus have no unique solution. Oftentimes, one will normalize all eigenvectors in order to eliminate this ambiguity.

Example 15.3

Now we find the eigenvectors for the matrix in Example 15.1. Starting with $\lambda_1 = 4$, the corresponding homogeneous linear system is

$$\begin{bmatrix} -3 & 1 & 0 & 0 \\ 0 & -1 & 1 & 0 \\ 0 & 0 & 0 & 1 \\ 0 & 0 & 0 & -2 \end{bmatrix} \mathbf{r}_1 = \mathbf{0},$$

and we solve it using the homogeneous linear system techniques from Section 12.3. Repeating for all eigenvalues, we find eigenvectors

$$\mathbf{r}_1 = \begin{bmatrix} 1/3 \\ 1 \\ 1 \\ 0 \end{bmatrix}, \quad \mathbf{r}_2 = \begin{bmatrix} 1/2 \\ 1 \\ 0 \\ 0 \end{bmatrix}, \quad \mathbf{r}_3 = \begin{bmatrix} 1/2 \\ 1/2 \\ -1/2 \\ 1 \end{bmatrix}, \quad \mathbf{r}_4 = \begin{bmatrix} 1 \\ 0 \\ 0 \\ 0 \end{bmatrix}.$$

For practice working with homogenous systems, work out the details. Check that each eigenvector satisfies $A\mathbf{r}_i = \lambda_i \mathbf{r}_i$.

If some of the eigenvalues are multiple zeroes of the characteristic polynomial, for example, $\lambda_1 = \lambda_2 = \lambda$, then we have two identical homogeneous systems $[A - \lambda I]\mathbf{r} = \mathbf{0}$. Each has the same solution vector $\mathbf{r}$ (instead of distinct solution vectors $\mathbf{r}_1, \mathbf{r}_2$ for distinct eigenvalues). We thus see that repeated eigenvalues reduce the number of eigenvectors.

Example 15.4

Let

$$A = \begin{bmatrix} 1 & 2 & 3 \\ 0 & 2 & 0 \\ 0 & 0 & 2 \end{bmatrix}.$$

This matrix has eigenvalues $\lambda_i = 2, 2, 1$. Finding the eigenvector corresponding to $\lambda_1 = \lambda_2 = 2$, we get two identical homogeneous systems

$$\begin{bmatrix} -1 & 2 & 3 \\ 0 & 0 & 0 \\ 0 & 0 & 0 \end{bmatrix} \mathbf{r}_1 = \mathbf{0}.$$

We set $r_{3,1} = r_{2,1} = 1$, and back substitution gives $r_{1,1} = 5$. The homogeneous system corresponding to $\lambda_3 = 1$ is

$$\begin{bmatrix} 0 & 2 & 3 \\ 0 & 1 & 0 \\ 0 & 0 & 1 \end{bmatrix} \mathbf{r}_3 = \mathbf{0}.$$

Thus the two fixed directions for A are

$$\mathbf{r}_1 = \begin{bmatrix} 5 \\ 1 \\ 1 \end{bmatrix} \quad \text{and} \quad \mathbf{r}_3 = \begin{bmatrix} 1 \\ 0 \\ 0 \end{bmatrix}.$$

Check that each eigenvector satisfies $A\mathbf{r}_i = \lambda_i \mathbf{r}_i$.

Example 15.5

Let a rotation matrix be given by

$$A = \begin{bmatrix} c & -s & 0 \\ s & c & 0 \\ 0 & 0 & 1 \end{bmatrix}$$

with $c = \cos \alpha$ and $s = \sin \alpha$. It rotates around the $\mathbf{e}_3$-axis by α degrees. We should thus expect that $\mathbf{e}_3$ is an eigenvector—and indeed, one easily verifies that $A\mathbf{e}_3 = \mathbf{e}_3$. Thus, the eigenvalue corresponding to $\mathbf{e}_3$ is 1.

Symmetric matrices are special again. Not only do they have real eigenvalues, but their eigenvectors are orthogonal. This can be shown in exactly the same way as we did for the 2D case in Section 7.5. Recall that in this case, A is said to be *diagonalizable* because it is possible to transform A to the diagonal matrix $\Lambda = R^{-1}AR$, where the columns of R are A's eigenvectors and Λ is a diagonal matrix of A's eigenvalues.

Example 15.6

The symmetric matrix

$$S = \begin{bmatrix} 3 & 0 & 1 \\ 0 & 3 & 0 \\ 1 & 0 & 3 \end{bmatrix}$$

has eigenvalues $\lambda_1 = 4, \lambda_2 = 3, \lambda_3 = 2$ and corresponding eigenvectors

$$\mathbf{r}_1 = \begin{bmatrix} 1 \\ 0 \\ 1 \end{bmatrix}, \qquad \mathbf{r}_2 = \begin{bmatrix} 0 \\ 1 \\ 0 \end{bmatrix}, \qquad \mathbf{r}_3 = \begin{bmatrix} -1 \\ 0 \\ 1 \end{bmatrix}.$$

The eigenvalues are the diagonal elements of

$$\Lambda = \begin{bmatrix} 4 & 0 & 0 \\ 0 & 3 & 0 \\ 0 & 0 & 2 \end{bmatrix},$$

and the $\mathbf{r}_i$ are normalized for the columns of

$$R = \begin{bmatrix} 1/\sqrt{2} & 0 & -1/\sqrt{2} \\ 0 & 1 & 0 \\ 1/\sqrt{2} & 0 & 1/\sqrt{2} \end{bmatrix};$$

then A is said to be diagonizable since $A = R\Lambda R^{\mathrm{T}}$. This is the eigendecomposition of A.

Projection matrices have eigenvalues that are one or zero. A zero eigenvalue indicates that the determinant of the matrix is zero, thus a projection matrix is singular. When the eigenvalue is one, the eigenvector is projected to itself, and when the eigenvalue is zero, the eigenvector is projected to the zero vector. If $\lambda_1 = \ldots = \lambda_k = 1$, then the column space is made up of the eigenvectors and its dimension is k, and the null space of the matrix is dimension $n - k$.

Example 15.7

Define a 3×3 projection matrix $P = \mathbf{u}\mathbf{u}^T$, where

$$\mathbf{u} = \begin{bmatrix} 1/\sqrt{2} \\ 0 \\ 1/\sqrt{2} \end{bmatrix}, \qquad \text{then} \qquad P = \begin{bmatrix} 1/2 & 0 & 1/2 \\ 0 & 0 & 0 \\ 1/2 & 0 & 1/2 \end{bmatrix}.$$

By design, we know that this matrix is rank one and singular. The characteristic polynomial is $p(\lambda) = \lambda^2(1 - \lambda)$ from which we conclude that $\lambda_1 = 1$ and $\lambda_2 = \lambda_3 = 0$. The eigenvector corresponding to λ_1 is

$$\mathbf{r}_1 = \begin{bmatrix} 1 \\ 0 \\ 1 \end{bmatrix},$$

and it spans the column space of P. The zero eigenvalue leads to the system

$$\begin{bmatrix} 1/2 & 0 & 1/2 \\ 0 & 0 & 0 \\ 0 & 0 & 0 \end{bmatrix} \mathbf{r} = \mathbf{0}.$$

To find one eigenvector associated with this eigenvalue, we can simply assign $r_3 = r_2 = 1$, and back substitution results in $r_1 = -1$.

Alternatively, we can find two vectors that span the two dimensional null space of P,

$$\mathbf{r}_2 = \begin{bmatrix} -1 \\ 0 \\ 1 \end{bmatrix}, \qquad \mathbf{r}_3 = \begin{bmatrix} 0 \\ 1 \\ 0 \end{bmatrix}.$$

They are orthogonal to $\mathbf{r}_1$. All linear combinations of elements of the null space are also in the null space, and thus $\mathbf{r} = 1\mathbf{r}_1 + 1\mathbf{r}_2$. (Normally, the eigenvectors are normalized, but for simplicity they are not here.)

The trace of A is defined as

$$\text{tr}(A) = \lambda_1 + \lambda_2 + \ldots + \lambda_n$$
$$= a_{1,1} + a_{2,2} + \ldots + a_{n,n}.$$

Immediately it is evident that the trace is a quick and easy way to learn a bit about the eigenvalues without computing them directly. For example, if we have a real symmetric matrix, thus real, positive eigenvalues, and the trace is zero, then the eigenvalues must all be zero. For 2×2 matrices, we have

$$\det[A - \lambda I] = \lambda^2 - \lambda(a_{1,1} + a_{2,2}) + (a_{1,1}a_{2,2} - a_{1,2}a_{2,1})$$
$$= \lambda^2 - \lambda\text{tr}(A) + \det A,$$

thus

$$\lambda_{1,2} = \frac{\text{tr}(A) \pm \sqrt{\text{tr}(A)^2 - 4\det A}}{2}. \tag{15.4}$$

Example 15.8

Let

$$A = \begin{bmatrix} 1 & -2 \\ 0 & -2 \end{bmatrix},$$

then simply by observation, we see that $\lambda_1 = -2$ and $\lambda_2 = 1$. Let's compare this to the eigenvalues from (15.4). The trace is the sum of the diagonal elements, $\text{tr}(A) = -1$ and $\det A = -2$. Then

$$\lambda_{1,2} = \frac{-1 \pm 3}{2},$$

resulting in the correct eigenvalues.

In Section 7.6 we introduced 2-dimensional quadratic forms, and they were easy to visualize. This idea applies to n-dimensions as well. The vector $\mathbf{v}$ lives in $\mathbb{R}^n$. For an $n \times n$ symmetric, positive definite matrix C, the quadratic form (7.14) becomes

$$f(\mathbf{v}) = \mathbf{v}^{\mathrm{T}} C \mathbf{v}$$
$$= c_{1,1} v_1^2 + 2 c_{1,2} v_1 v_2 + \ldots + c_{n,n} v_n^2,$$

where each v_i^2 term is paired with diagonal element $c_{i,i}$ and each $v_i v_j$ term is paired with $2 c_{i,j}$ due to the symmetry in C. Just as before, all terms are quadratic. Now, the contour $f(\mathbf{v}) = 1$ is an n-dimensional ellipsoid. The semi-minor axis corresponds to the dominant eigenvector $\mathbf{r}_1$ and its length is $1/\sqrt{\lambda_1}$, and the semi-major axis corresponds to $\mathbf{r}_n$ and its length is $1/\sqrt{\lambda_n}$.

A real matrix is *positive definite* if

$$f(\mathbf{v}) = \mathbf{v}^{\mathrm{T}} A \mathbf{v} > 0 \tag{15.5}$$

for any nonzero vector $\mathbf{v}$ in $\mathbb{R}^n$. This means that the quadratic form is positive everywhere except for $\mathbf{v} = \mathbf{0}$. This is the same condition we encountered in (7.17).

15.2 The Power Method

Let A be a symmetric $n \times n$ matrix.[1] Further, let the eigenvalues of A be ordered such that $|\lambda_1| \geq |\lambda_2| \geq \ldots \geq |\lambda_n|$. Then λ_1 is called the *dominant eigenvalue* of A. To simplify the notation to come, refer to this dominant eigenvalue as λ, and let $\mathbf{r}$ be its corresponding eigenvector. In Section 7.7, we considered repeated applications of a matrix; we restricted ourselves to the 2D case. We encountered an equation of the form

$$A^i \mathbf{r} = \lambda^i \mathbf{r}, \tag{15.6}$$

which clearly holds for matrices of arbitrary size. This equation may be interpreted as follows: if A has an eigenvalue λ, then A^i has an eigenvalue λ^i with the same corresponding eigenvector $\mathbf{r}$.

This property may be used to *find* the dominant eigenvalue and eigenvector. Consider the vector sequence

$$\mathbf{r}^{(i+1)} = A \mathbf{r}^{(i)}; \qquad i = 1, 2, \ldots \tag{15.7}$$

where $\mathbf{r}^{(1)}$ is an arbitrary (nonzero) vector.

---

[1] The method discussed in this section may be extended to nonsymmetric matrices, but since those eigenvalues may be complex, we will avoid them here.

After a sufficiently large i, the $\mathbf{r}^{(i)}$ will begin to line up with $\mathbf{r}$, as illustrated in the two leftmost examples in Figure 15.2. Here is how to utilize that fact for finding λ: for sufficiently large i, we will approximately have

$$\mathbf{r}^{(i+1)} = \lambda \mathbf{r}^{(i)}.$$

This means that all components of $\mathbf{r}^{(i+1)}$ and $\mathbf{r}^{(i)}$ are (approximately) related by

$$\frac{r_j^{(i+1)}}{r_j^{(i)}} = \lambda \quad \text{for } j = 1, \ldots, n. \tag{15.8}$$

In the algorithm to follow, rather than checking each ratio, we will use the ∞-norm to define λ upon each iteration.

Algorithm:

Initialization:
    Estimate dominant eigenvector $\mathbf{r}^{(1)} \neq \mathbf{0}$.
    Find j where $|r_j^{(1)}| = \|\mathbf{r}^{(1)}\|_\infty$ and set $\mathbf{r}^{(1)} = \mathbf{r}^{(1)}/r_j^{(1)}$.
    Set $\lambda^{(1)} = 0$.
    Set tolerance ϵ.
    Set maximum number of iterations m.
For $k = 2, \ldots, m$,
    $\mathbf{y} = A\mathbf{r}^{(k-1)}$,
    $\lambda^{(k)} = y_j$.
    Find j where $|y_j| = \|\mathbf{y}\|_\infty$.
    If $y_j = 0$ then output: "eigenvalue zero; select new $\mathbf{r}^{(1)}$
    and restart"; exit.
    $\mathbf{r}^{(k)} = \mathbf{y}/y_j$.
    If $|\lambda^{(k)} - \lambda^{(k-1)}| < \epsilon$ then output: $\lambda^{(k)}$ and $\mathbf{r}^{(k)}$; exit.
    If $k = m$ output: "maximum iterations exceeded."

Some remarks on using this method:

- If $|\lambda|$ is either "large" or "close" to zero, the $\mathbf{r}^{(k)}$ will either become unbounded or approach zero in length, respectively. This has the potential to cause numerical problems. It is prudent, therefore, to *scale* the $\mathbf{r}^{(k)}$, so in the algorithm above, at each step, the eigenvector is scaled by its element with the largest absolute value—with respect to the ∞-norm.

- Convergence seems impossible if $\mathbf{r}^{(1)}$ is perpendicular to $\mathbf{r}$, the eigenvector of λ. Theoretically, no convergence will kick in, but for once numerical round-off is on our side: after a few iterations, $\mathbf{r}^{(k)}$ will not be perpendicular to $\mathbf{r}$ and we will converge—if slowly!

- Very slow convergence will also be observed if $|\lambda_1| \approx |\lambda_2|$.

- The power method as described here is limited to symmetric matrices with one dominant eigenvalue. It may be generalized to more cases, but for the purpose of this exposition, we decided to outline the principle rather than to cover all details.

Example 15.9

Figure 15.2 illustrates three cases, A_1, A_2, A_3, from left to right. The three matrices and their eigenvalues are as follows:

$$A_1 = \begin{bmatrix} 2 & 1 \\ 1 & 2 \end{bmatrix}, \qquad \lambda_1 = 3, \qquad \lambda_2 = 1,$$

$$A_2 = \begin{bmatrix} 2 & 0.1 \\ 0.1 & 2 \end{bmatrix}, \qquad \lambda_1 = 2.1, \qquad \lambda_2 = 1.9,$$

$$A_3 = \begin{bmatrix} 2 & -0.1 \\ 0.1 & 2 \end{bmatrix}, \qquad \lambda_1 = 2 + 0.1i, \quad \lambda_2 = 2 - 0.1i,$$

and for all of them, we used

$$\mathbf{r}^{(1)} = \begin{bmatrix} 1.5 \\ -0.1 \end{bmatrix}.$$

In all three examples, the vectors $\mathbf{r}^{(i)}$ were scaled relative to the ∞-norm, thus $\mathbf{r}^{(1)}$ is scaled to

$$\mathbf{r}^{(1)} = \begin{bmatrix} 1 \\ -0.066667 \end{bmatrix}.$$

An ϵ tolerance of 5.0×10^{-4} was used for each matrix.

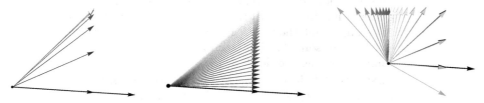

Figure 15.2.
The power method: three examples whose matrices are given in Example 15.9. The longest black vector is the initial (guess) eigenvector. Successive iterations are in lighter shades of gray. Each iteration is scaled with respect to the ∞-norm.

The first matrix, A_1 is symmetric and the dominant eigenvalue is reasonably separated from λ_2, hence the rapid convergence in 11 iterations. The estimate for the dominant eigenvalue: $\lambda = 2.99998$.

The second matrix, A_2, is also symmetric, however λ_1 is close in value to λ_2, hence convergence is slower, needing 41 iterations. The estimate for the dominant eigenvalue: $\lambda = 2.09549$.

The third matrix, a rotation matrix that is not symmetric, has complex eigenvalues, hence no convergence. By following the change in gray scale, we can follow the path of the iterative eigenvector estimate. The wide variation in eigenvectors makes clear the outline of the ∞-norm. (Consult Figure 13.2 for an illustration of unit vectors with respect to the ∞-norm.)

A more realistic numerical example is presented next with the Google eigenvector.

15.3   Application: Google Eigenvector

We now study a linear algebra aspect of search engines. While many search engine techniques are highly proprietary, they all share the basic idea of *ranking* webpages. The concept was first introduced by S. Brin and L. Page in 1998, and forms the basis of the search engine Google. Here we will show how ranking webpages is essentially a straightforward eigenvector problem.

The web (at some frozen point in time) consists of N webpages, most of them pointing to (having links to) other webpages. A page that is pointed to very often would be considered important, whereas a page with none or only very few other pages pointing to it would be considered not important. How can we rank all webpages according to how important they are? In the sequel, we assume all webpages are ordered in some fashion (such as lexicographic) so we can assign a number, such as i, to each page.

First, two definitions: if page i points to page j, then we say this is an *outlink* for page i, whereas if page j points to page i, then this is an *inlink* for page i. A page is not supposed to link to itself. We can represent this connectivity structure of the web by an $N \times N$ *adjacency matrix* C: Each outlink for page i is recorded by setting $c_{j,i} = 1$. If page i does not have an outlink to page j, then $c_{j,i} = 0$.

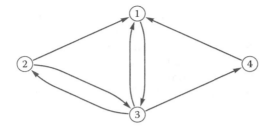

Figure 15.3.
Directed graph: represents the webpage connectivity defined by C in (15.9).

An example matrix is the following:

$$C = \begin{bmatrix} 0 & 1 & 1 & 1 \\ 0 & 0 & 1 & 0 \\ 1 & 1 & 0 & 1 \\ 0 & 0 & 1 & 0 \end{bmatrix}. \tag{15.9}$$

In this example, page 1 has one outlink since $c_{3,1} = 1$ and three inlinks since $c_{1,2} = c_{1,3} = c_{1,4} = 1$. Thus, the ith column describes the outlinks of page i and the ith row describes the inlinks of page i. This connectivity structure is illustrated by the *directed graph* of Figure 15.3.

The ranking r_i of any page i is entirely defined by C. Here are some rules with increasing sophistication:

1. The ranking r_i should grow with the number of page i's inlinks.

2. The ranking r_i should be weighted by the ranking of each of page i's inlinks.

3. Let page i have an inlink from page j. Then the more outlinks page j has, the less it should contribute to r_i.

Let's elaborate on these rules. Rule 1 says that a page that is pointed to very often deserves high ranking. But rule 2 says that if all those inlinks to page i prove to be low-ranked, then their sheer number is mitigated by their low rankings. Conversely, if they are mostly high-ranked, then they should boost page i's ranking. Rule 3 implies that if page j has only one outlink and it points to page i, then page i should be "honored" for such trust from page j. Conversely, if page j points to a large number of pages, page i among them, this does not give page i much pedigree.

Although not realistic, assume for now that each page as at least one outlink and at least one inlink so that the matrix C is structured nicely. Let o_i represent the total number of outlinks of page i. This is simply the sum of all elements of the ith column of C. The more outlinks page i has, the lower its contribution to page j's ranking it will have. Thus we scale every element of column i by $1/o_i$. The resulting matrix D with

$$d_{j,i} = \frac{c_{j,i}}{o_i}$$

is called the *Google matrix*. Note that all columns of D have nonnegative entries and sum to one. Matrices with that property (or with respect to the rows) are called *stochastic*.[2]

In our example above, we have

$$D = \begin{bmatrix} 0 & 1/2 & 1/3 & 1/2 \\ 0 & 0 & 1/3 & 0 \\ 1 & 1/2 & 0 & 1/2 \\ 0 & 0 & 1/3 & 0 \end{bmatrix}.$$

We note that finding r_i involves knowing the ranking of all pages, including r_i. This seems like an ill-posed circular problem, but a little more analysis leads to the following matrix problem. Using the vector $\mathbf{r} = [r_1, \ldots, r_N]^\mathrm{T}$, one can find

$$\mathbf{r} = D\mathbf{r}. \tag{15.10}$$

This states that we are looking for the eigenvector of D corresponding to the eigenvalue 1. But how do we know that D does indeed have an eigenvalue 1? The answer is that all stochastic matrices have that property. One can even show that 1 is D's largest eigenvalue. This vector $\mathbf{r}$ is called a *stationary vector*.

To find $\mathbf{r}$, we simply employ the power method from Section 15.2. This method needs an initial guess for $\mathbf{r}$, and setting all $r_i = 1$, which corresponds to equal ranking for all pages, is not too bad for that. As the iterations converge, the solution is found. The entries of $\mathbf{r}$ are real, since they correspond to a real eigenvalue.

The vector $\mathbf{r}$ now contains the ranking of every page, called *page rank* by Google. If Google retrieves a set of pages all containing a link to a term you are searching for, it presents them to you in decreasing order of the pages' ranking.

---

[2]More precisely, matrices for which the columns sum to one are called *left stochastic*. If the rows sum to one, the matrix is *right stochastic* and if the rows and columns sum to one, the matrix is *doubly stochastic*.

Back to our 4×4 example from above: D has an eigenvalue 1 and the corresponding eigenvector

$$\mathbf{r} = [0.67, 0.33, 1, 0.33]^{\mathrm{T}},$$

which was calculated with the power method algorithm in Section 15.2. Notice that $r_3 = 1$ is the largest component, therefore page 3 has the highest ranking. Even though pages 1 and 3 have the same number of inlinks, the solitary outlink from page 1 to page 3 gives page 3 the edge in the ranking.

In the real world, in 2013, there were approximately 50 billion webpages. This was the world's largest matrix to be used ever. Luckily, it contains mostly zeroes and thus is extremely sparse. Without taking advantage of that, Google (and other search engines) could not function. Figure 15.1 illustrates a portion of a Google matrix for approximately 3 million pages. We gave three simple rules for building D, but in the real world, many more rules are needed. For example, webpages with no inlinks or outlinks must be considered. We would want to modify D to ensure that the ranking $\mathbf{r}$ has only nonnegative components. In order for the power method to converge, other modifications of D are required as well, but that topic falls into numerical analysis.

15.4 Eigenfunctions

An eigenvalue λ of a matrix A is typically thought of as a solution of the matrix equation

$$A\mathbf{r} = \lambda\mathbf{r}.$$

In Section 14.5, we encountered more general spaces than those formed by finite-dimensional vectors: those are spaces formed by *polynomials*. Now, we will even go beyond that: we will explore the space of all real-valued functions. Do eigenvalues and eigenvectors have meaning there? Let's see.

Let f be a function, meaning that $y = f(x)$ assigns the output value y to an input value x, and we assume both x and y are real numbers. We also assume that f is smooth, or differentiable. An example might be $f(x) = \sin(x)$.

The set of all such functions f forms a linear space as observed in Section 14.5.

We can define linear maps L for elements of this function space. For example, setting $Lf = 2f$ is such a map, albeit a bit trivial. A more interesting linear map is that of taking derivatives: $Df = f'$. Thus, to any function f, the map D assigns another function, namely the derivative of f. For example, let $f(x) = \sin(x)$. Then $Df(x) = \cos(x)$.

How can we marry the concept of eigenvalues and linear maps such as D? First of all, D will not have eigen*vectors*, since our linear space consists of functions, not vectors. So we will talk of eigen*functions* instead. A function f is an eigenfunction of D (or any other linear map) if it is mapped to a multiple of itself:

$$Df = \lambda f.$$

The scalar multiple λ is, as usual, referred to as an eigenvalue of f. Note that D may have many eigenfunctions, each corresponding to a different λ.

Since we know that D means taking derivatives, this becomes

$$f' = \lambda f. \tag{15.11}$$

Any function f satisfying (15.11) is thus an eigenfunction of the derivative map D.

Now you have to recall your calculus: the function $f(x) = e^x$ satisfies

$$f'(x) = e^x,$$

which may be written as

$$Df = f = 1 \times f.$$

Hence 1 is an eigenvalue of the derivative map D. More generally, all functions $f(x) = e^{\lambda x}$ satisfy (for $\lambda \neq 0$):

$$f'(x) = \lambda e^{\lambda x},$$

which may be written as

$$Df = \lambda f.$$

Hence all real numbers $\lambda \neq 0$ are eigenvalues of D, and the corresponding eigenfunctions are $e^{\lambda x}$. We see that our map D has infinitely many eigenfunctions!

Let's look at another example: the map is the second derivative, $Lf = f''$. A set of eigenfunctions for this map is $\cos(kx)$ for $k = 1, 2, \ldots$ since

$$\frac{\mathrm{d}^2 \cos(kx)}{\mathrm{d}x^2} = -k\frac{\mathrm{d}\sin(kx)}{\mathrm{d}x} = -k^2 \cos(kx), \tag{15.12}$$

and the eigenvalues are $-k^2$. Can you find another set of eigenfunctions?

This may seem a bit abstract, but eigenfunctions actually have many uses, for example in differential equations and mathematical physics. In engineering mathematics, *orthogonal functions* are key for applications such as data fitting and vibration analysis. Some well-known sets of orthogonal functions arise as the result of the solution to a Sturm-Liouville equation such as

$$y''(x) + \lambda y(x) = 0 \quad \text{such that} \quad y(0) = 0 \quad \text{and} \quad y(\pi) = 0. \quad (15.13)$$

This is a linear second order differential equation with boundary conditions, and it defines a *boundary value problem*. Unknown are the functions $y(x)$ that satisfy this equation. Solving this boundary value problem is out of the scope of this book, so we simply report that $y(x) = \sin(ax)$ for $a = 1, 2, \ldots$. These are eigenfunctions of (15.13) and the corresponding eigenvalues are $\lambda = a^2$.

- eigenvalue
- eigenvector
- characteristic polynomial
- eigenvalues and eigenvectors of a symmetric matrix
- dominant eigenvalue
- eigendecomposition
- trace
- quadratic form
- positive definite matrix
- power method
- max-norm
- adjacency matrix
- directed graph
- stochastic matrix
- stationary vector
- eigenfunction

15.5   Exercises

1. What are the eigenvalues and eigenvectors for

$$A = \begin{bmatrix} 2 & 1 \\ 1 & 2 \end{bmatrix}?$$

2. What are the eigenvalues and eigenvectors for

$$A = \begin{bmatrix} 1 & 1 & 1 \\ 0 & 1 & 0 \\ 1 & 0 & 1 \end{bmatrix}?$$

3. Find the eigenvalues of the matrix

$$\begin{bmatrix} 4 & 2 & -3 & 2 \\ 0 & 3 & 1 & 3 \\ 0 & 0 & 2 & 1 \\ 0 & 0 & 0 & 1 \end{bmatrix}.$$

4. If

$$\mathbf{r} = \begin{bmatrix} 1 \\ 1 \\ 1 \end{bmatrix}$$

is an eigenvector of

$$A = \begin{bmatrix} 0 & 1 & 1 \\ 1 & 1 & 0 \\ 1 & 0 & 1 \end{bmatrix},$$

what is the corresponding eigenvalue?

5. The matrices A and A^T have the same eigenvalues. Why?

6. If A has eigenvalues $4, 2, 0$, what is the rank of A? What is the determinant?

7. Suppose a matrix A has a zero eigenvalue. Will forward elimination change this eigenvalue?

8. Let a rotation matrix be given by

$$R = \begin{bmatrix} \cos\alpha & 0 & -\sin\alpha \\ 0 & 1 & 0 \\ \sin\alpha & 0 & \cos\alpha \end{bmatrix},$$

which rotates around the $\mathbf{e}_2$-axis by $-\alpha$ degrees. What is one eigenvalue and eigenvector?

9. For the matrix

$$A = \begin{bmatrix} 1 & -2 \\ -2 & 1 \end{bmatrix}$$

show that the determinant equals the product of the eigenvalues and the trace equals the sum of the eigenvalues.

10. What is the dominant eigenvalue in Exercise 9?

11. Compute the eigenvalues of A and A^2 where

$$A = \begin{bmatrix} 3 & 0 & 1 \\ 0 & 1 & 0 \\ 0 & 0 & 3 \end{bmatrix}.$$

12. Let A be the matrix

$$A = \begin{bmatrix} 4 & 0 & 1 \\ 0 & 1 & 2 \\ 1 & 2 & 2 \end{bmatrix}.$$

Starting with the vector

$$\mathbf{r}^{(1)} = \begin{bmatrix} 1 & 1 & 1 \end{bmatrix}$$

carry out three steps of the power method with (15.7), and use $\mathbf{r}^{(3)}$ and $\mathbf{r}^{(4)}$ in (15.8) to estimate A's dominant eigenvalue. If you are able to program, then try implementing the power method algorithm.

13. Let A be the matrix

$$A = \begin{bmatrix} -8 & 0 & 8 \\ 0 & 1 & -2 \\ 8 & -2 & 0 \end{bmatrix}.$$

Starting with the vector

$$\mathbf{r}^{(1)} = \begin{bmatrix} 1 & 1 & 1 \end{bmatrix}$$

carry out three steps of the power method with (15.7), and use $\mathbf{r}^{(3)}$ and $\mathbf{r}^{(4)}$ in (15.8) to estimate A's dominant eigenvalue. If you are able to program, then try implementing the power method algorithm.

14. Of the following matrices, which one(s) are stochastic matrices?

$$A = \begin{bmatrix} 0 & 1 & 0 & 0 \\ 2 & 0 & 0 & 1/2 \\ -1 & 0 & 0 & 1/2 \\ 0 & 0 & 1 & 0 \end{bmatrix}, \quad B = \begin{bmatrix} 1 & 1/3 & 1/4 & 0 \\ 0 & 0 & 1/4 & 0 \\ 0 & 1/3 & 1/4 & 1 \\ 0 & 1/3 & 1/4 & 0 \end{bmatrix},$$

$$C = \begin{bmatrix} 1 & 0 & 0 \\ 0 & 0 & 0 \\ 0 & 0 & 1 \end{bmatrix}, \quad D = \begin{bmatrix} 1/2 & 0 & 0 & 1/2 \\ 0 & 1/2 & 1/2 & 0 \\ 1/3 & 1/3 & 1/3 & 0 \\ 1/2 & 0 & 0 & 1/2 \end{bmatrix}.$$

15. The directed graph in Figure 15.4 describes inlinks and outlinks to webpages. What is the corresponding adjacency matrix C and stochastic (Google) matrix D?

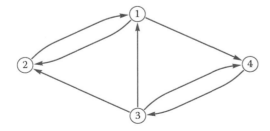

Figure 15.4.

Graph showing the connectivity defined by C.

16. For the adjacency matrix

$$C = \begin{bmatrix} 0 & 0 & 0 & 1 & 0 \\ 1 & 0 & 0 & 0 & 1 \\ 0 & 1 & 0 & 1 & 1 \\ 1 & 0 & 0 & 0 & 1 \\ 1 & 0 & 1 & 0 & 0 \end{bmatrix},$$

draw the corresponding directed graph that describes these inlinks and outlinks to webpages. What is the corresponding stochastic matrix D?

17. The Google matrix in Exercise 16 has dominant eigenvalue 1 and corresponding eigenvector $\mathbf{r} = [1/5, 2/5, 14/15, 2/5, 1]^{\mathrm{T}}$. Which page has the highest ranking? Based on the criteria for page ranking described in Section 15.3, explain why this is so.

18. Find the eigenfunctions and eigenvalues for $Lf(x) = xf'$.

19. For the map $Lf = f''$, a set of eigenfunctions is given in (15.12). Find another set of eigenfunctions.

16

The Singular Value Decomposition

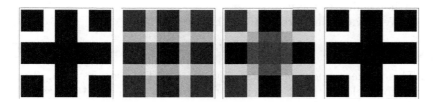

Figure 16.1.
Image compression: a method that uses the SVD. Far left: original image; second from left: highest compression; third from left: moderate compression; far right: method recovers original image. See Section 16.7 for details.

Matrix decomposition is a fundamental tool in linear algebra for understanding the action of a matrix, establishing its suitability to solve a problem, and for solving linear systems more efficiently and effectively. We have encountered an important decomposition already, the eigendecomposition for symmetric matrices (see Section 7.5). The topic of this chapter, the singular value decomposition (SVD), is a tool for more general, even nonsquare matrices. Figure 16.1 demonstrates one application of SVD, image compression.

This chapter allows us to revisit several themes from past chapters: eigenvalues and eigenvectors, the condition number, the least squares solution to an overdetermined system, and more! It provides a good review of some important ideas in linear algebra.

16.1   The Geometry of the 2 × 2 Case

Let A be a 2×2 matrix, nonsingular for now. Let $\mathbf{v}_1$ and $\mathbf{v}_2$ be two unit vectors that are perpendicular to each other, thus they are *orthonormal*. This means that $V = [\mathbf{v}_1 \ \mathbf{v}_2]$ is an *orthogonal* matrix: $V^{-1} = V^{\mathrm{T}}$. In general, A will not map two orthonormal vectors $\mathbf{v}_1, \mathbf{v}_2$ to two orthonormal image vectors $\mathbf{u}_1, \mathbf{u}_2$. However, it is possible for A to map two particular vectors $\mathbf{v}_1$ and $\mathbf{v}_2$ to two *orthogonal* image vectors $\sigma_1 \mathbf{u}_1$ and $\sigma_2 \mathbf{u}_2$. The two vectors $\mathbf{u}_1$ and $\mathbf{u}_2$ are assumed to be of unit length, i.e., they are orthonormal.

We formalize this as

$$AV = U\Sigma \tag{16.1}$$

with an orthogonal matrix $U = [\mathbf{u}_1 \ \mathbf{u}_2]$ and a diagonal matrix

$$\Sigma = \begin{bmatrix} \sigma_1 & 0 \\ 0 & \sigma_2 \end{bmatrix}.$$

If $A\mathbf{v}_i = \sigma \mathbf{u}_i$, then each $\mathbf{u}_i$ is parallel to $A\mathbf{v}_i$, and A preserves the orthogonality of the $\mathbf{v}_i$.

We now conclude from (16.1) that

$$A = U\Sigma V^{\mathrm{T}}. \tag{16.2}$$

This is the *singular value decomposition*, SVD for short, of A. The diagonal elements of Σ are called the *singular values* of A. Let's now find out how to determine U, Σ, V.

In Section 7.6, we established that symmetric positive definite matrices, such as $A^{\mathrm{T}}A$, have real and positive eigenvalues and their eigenvectors are orthogonal. Considering the SVD in (16.2), we can write

$$\begin{aligned} A^{\mathrm{T}}A &= (U\Sigma V^{\mathrm{T}})^{\mathrm{T}}(U\Sigma V^{\mathrm{T}}) \\ &= V\Sigma^{\mathrm{T}}U^{\mathrm{T}}U\Sigma V^{\mathrm{T}} \\ &= V\Sigma^{\mathrm{T}}\Sigma V^{\mathrm{T}} \\ &= V\Lambda' V^{\mathrm{T}}, \end{aligned} \tag{16.3}$$

where

$$\Lambda' = \begin{bmatrix} \lambda_1' & 0 \\ 0 & \lambda_2' \end{bmatrix} = \Sigma^{\mathrm{T}}\Sigma = \begin{bmatrix} \sigma_1^2 & 0 \\ 0 & \sigma_2^2 \end{bmatrix}.$$

Equation (16.3) states the following: The symmetric positive definite matrix $A^{\mathrm{T}}A$ has eigenvalues that are the diagonal entries of Λ' and eigenvectors as columns of V, which are called the *right singular vectors* of A. This is the *eigendecomposition* of $A^{\mathrm{T}}A$.

The symmetric positive definite matrix AA^T also has an eigende-composition,

$$
\begin{aligned}
AA^T &= (U\Sigma V^T)(U\Sigma V^T)^T \\
&= U\Sigma V^T V\Sigma^T U^T \\
&= U\Sigma\Sigma^T U^T \\
&= U\Lambda' U^T,
\end{aligned}
\tag{16.4}
$$

observing that $\Sigma^T\Sigma = \Sigma\Sigma^T$. Equation (16.4) states that the symmetric positive definite matrix AA^T has eigenvalues that are the diagonal entries of Λ' and eigenvectors as the columns of U, and they are called the *left singular vectors* of A.

It should be no surprise that Λ' appears in both (16.3) and (16.4) since a matrix and its transpose share the same (nonzero) eigenvalues.

Now we understand a bit more about the elements of the SVD in (16.2): the singular values, σ_i, of A are the square roots of the eigenvalues of $A^T A$ and AA^T, that is

$$
\sigma_i = \sqrt{\lambda_i'}.
$$

The columns of V are the eigenvectors of $A^T A$ and the columns of U are the eigenvectors of AA^T. Observe from (16.1) that we can compute $\mathbf{u}_i = A\mathbf{v}_i/\|\cdot\|$.

Example 16.1

Let's start with a very simple example:

$$
A = \begin{bmatrix} 3 & 0 \\ 0 & 1 \end{bmatrix},
$$

a symmetric, positive definite matrix that scales in the $\mathbf{e}_1$-direction. Then

$$
AA^T = A^T A = \begin{bmatrix} 9 & 0 \\ 0 & 1 \end{bmatrix},
$$

and we can easily calculate the eigenvalues of $A^T A$ as $\lambda_1' = 9$ and $\lambda_2' = 1$. This means that $\sigma_1 = 3$ and $\sigma_2 = 1$. The eigenvectors of AA^T and $A^T A$ are identical and happen to be the columns of the identity matrix for this simple example,

$$
U = V = \begin{bmatrix} 1 & 0 \\ 0 & 1 \end{bmatrix}.
$$

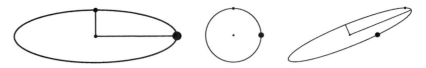

Figure 16.2.
Action of a map: the unit circle is mapped to the action ellipse with semi-major axis length σ_1 and semi-minor axis length σ_2. Left: ellipse from matrix in Example 16.1; middle: circle; right: ellipse from Example 16.2.

Now we can form the SVD of A, $A = U\Sigma V^{\mathrm{T}}$:

$$\begin{bmatrix} 3 & 0 \\ 0 & 1 \end{bmatrix} = \begin{bmatrix} 1 & 0 \\ 0 & 1 \end{bmatrix} \begin{bmatrix} 3 & 0 \\ 0 & 1 \end{bmatrix} \begin{bmatrix} 1 & 0 \\ 0 & 1 \end{bmatrix}.$$

This says that the action of A on a vector $\mathbf{x}$, that is $A\mathbf{x}$, simply amounts to a scaling in the $\mathbf{e}_1$-direction, which we observed to begin with!

Notice that the eigenvalues of A are identical to the singular values. In fact, because this matrix is positive definite, the SVD is identical to the eigendecomposition.

Throughout Chapter 5 we examined the action of a 2×2 matrix using an illustration of the circular Phoenix mapping to an elliptical Phoenix. This *action ellipse* can now be described more precisely: the semi-major axis has length σ_1 and the semi-minor axis has length σ_2. Figure 16.2 illustrates this point for Examples 16.1 and 16.2. In that figure, you see: the semi-axes, the map of $[1\ 0]^{\mathrm{T}}$ (thick point), and the map of $[0\ 1]^{\mathrm{T}}$ (thin point).

Example 16.2

This example is a little more interesting, as the matrix is now a shear,

$$A = \begin{bmatrix} 1 & 2 \\ 0 & 1 \end{bmatrix}.$$

We compute

$$A^{\mathrm{T}}A = \begin{bmatrix} 1 & 2 \\ 2 & 5 \end{bmatrix}, \qquad AA^{\mathrm{T}} = \begin{bmatrix} 5 & 2 \\ 2 & 1 \end{bmatrix},$$

and we observe that these two matrices are no longer identical, but they are both symmetric. As they are 2×2 matrices, we can easily calculate the eigenvalues as $\lambda'_1 = 5.82$ and $\lambda'_2 = 0.17$. (Remember: the nonzero eigenvalues are the same for a matrix and its transpose.) These eigenvalues result in singular values $\sigma_1 = 2.41$ and $\sigma_2 = 0.41$. The eigenvectors of $A^T A$ are the orthonormal column vectors of

$$V = \begin{bmatrix} 0.38 & -0.92 \\ 0.92 & 0.38 \end{bmatrix}$$

and the eigenvectors of AA^T are the orthonormal column vectors of

$$U = \begin{bmatrix} 0.92 & -0.38 \\ 0.38 & 0.92 \end{bmatrix}.$$

Now we can form the SVD of A, $A = U\Sigma V^T$:

$$\begin{bmatrix} 1 & 2 \\ 0 & 1 \end{bmatrix} = \begin{bmatrix} 0.92 & -0.38 \\ 0.38 & 0.92 \end{bmatrix} \begin{bmatrix} 2.41 & 0 \\ 0 & 0.41 \end{bmatrix} \begin{bmatrix} 0.38 & -0.92 \\ 0.92 & 0.38 \end{bmatrix}.$$

Figure 16.3 will help us break down the action of A in terms of the SVD. It is now clear that V and U are rotation or reflection matrices and Σ scales, deforming the circle into an ellipse.

Notice that the eigenvalues of A are $\lambda_1 = \lambda_2 = 1$, making the point that, in general, the singular values are not the eigenvalues!

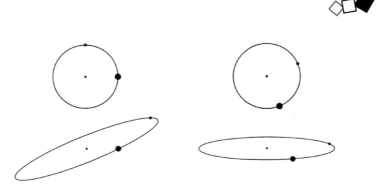

Figure 16.3.
SVD breakdown: shear matrix A from Example 16.2. Clockwise from top left: Initial point set forming a circle with two reference points; $V^T\mathbf{x}$ rotates clockwise 67.5°; $\Sigma V^T\mathbf{x}$ stretches in $\mathbf{e}_1$ and shrinks in $\mathbf{e}_2$; $U\Sigma V^T\mathbf{x}$ rotates counterclockwise 22.5°, illustrating the action of A.

Now we come full circle and look at what we have solved in terms of our original question that was encapsulated by (16.1): What orthonormal vectors $\mathbf{v}_i$ are mapped to orthogonal vectors $\sigma_i \mathbf{u}_i$? The SVD provides a solution to this question by providing V, U, and Σ. Furthermore, note that for this nonsingular case, the columns of V form a basis for the row space of A and the columns of U form a basis for the column space of A.

It should be clear that the SVD is not limited to invertible 2×2 matrices, so let's look at the SVD more generally.

16.2 The General Case

The SVD development of the previous section assumed that A was square and invertible. This had the effect that $A^{\mathrm{T}}A$ had nonzero eigenvalues and well-defined eigenvectors. However, everything still works if A is neither square nor invertible. Just a few more aspects of the decomposition come into play.

For the general case, A will be a rectangular matrix with m rows and n columns mapping $\mathbb{R}^n$ to $\mathbb{R}^m$. As a result of this freedom in the dimensions, the matrices of the SVD (16.2) must be modified. Illustrated in Figure 16.4 is each scenario, $m < n$, $m = n$, or $m > n$. The matrix dimensions are as follows: A is $m \times n$, U is $m \times m$, Σ is $m \times n$, and V^{T} is $n \times n$. The matrix Λ' in (16.3) is $n \times n$ and in (16.4) is $m \times m$, but they still hold the same nonzero eigenvalues because the rank of a matrix cannot exceed $\min\{m, n\}$.

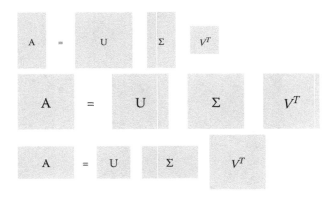

Figure 16.4.

SVD matrix dimensions: an overview of the SVD of an $m \times n$ matrix A. Top: $m > n$; middle: $m = n$; bottom: $m < n$.

Again we ask, what orthonormal vectors $\mathbf{v}_i$ are mapped by A to orthogonal vectors $\sigma_i \mathbf{u}_i$, where the $\mathbf{u}_i$ are orthonormal? This is encapsulated in (16.1). In the general case as well, the matrices U and V form bases, however, as we are considering rectangular and singular matrices, the rank r of A plays a part in the interpretation of the SVD. The following are the main SVD properties, for a detailed exposition, see Strang [16].

- The matrix Σ has nonzero singular values, $\sigma_1, \ldots, \sigma_r$, and all other entries are zero.

- The first r columns of U form an orthonormal basis for the column space of A.

- The last $m - r$ columns of U form an orthonormal basis for the null space of A^{T}.

- The first r columns of V form an orthonormal basis for the row space of A.

- The last $n - r$ columns of V form an orthonormal basis for the null space of A.

Examples will make this clearer.

Example 16.3

Let A be given by

$$A = \begin{bmatrix} 1 & 0 \\ 0 & 2 \\ 0 & 1 \end{bmatrix}.$$

The first step is to form $A^{\mathrm{T}}A$ and AA^{T} and find their eigenvalues and (normalized) eigenvectors, which make up the columns of an orthogonal matrix.

$$A^{\mathrm{T}}A = \begin{bmatrix} 1 & 0 \\ 0 & 5 \end{bmatrix}, \qquad \begin{matrix} \lambda_1' = 5, \\ \lambda_2' = 1, \end{matrix} \qquad V = \begin{bmatrix} 0 & 1 \\ 1 & 0 \end{bmatrix};$$

$$AA^{\mathrm{T}} = \begin{bmatrix} 1 & 0 & 0 \\ 0 & 4 & 2 \\ 0 & 2 & 1 \end{bmatrix}, \qquad \begin{matrix} \lambda_1' = 5, \\ \lambda_2' = 1, \\ \lambda_3' = 0, \end{matrix} \qquad U = \begin{bmatrix} 0 & 1 & 0 \\ 0.89 & 0 & -0.44 \\ 0.44 & 0 & 0.89 \end{bmatrix}.$$

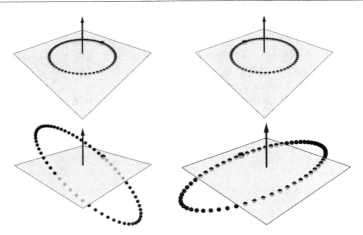

Figure 16.5.
SVD of a 3×2 matrix A: see Example 16.3. Clockwise from top left: Initial point set forming a circle with one reference point; $V^{\mathsf{T}}\mathbf{x}$ reflects; $\Sigma V^{\mathsf{T}}\mathbf{x}$ stretches in $\mathbf{e}_1$; $U\Sigma V^{\mathsf{T}}\mathbf{x}$ rotates counterclockwise $26.5°$, illustrating the action of A.

The rank of A is 2, thus there are two singular values, and

$$\Sigma = \begin{bmatrix} 2.23 & 0 \\ 0 & 1 \\ 0 & 0 \end{bmatrix}.$$

The SVD of $A = U\Sigma V^{\mathsf{T}}$:

$$\begin{bmatrix} 1 & 0 \\ 0 & 2 \\ 0 & 1 \end{bmatrix} = \begin{bmatrix} 0 & 1 & 0 \\ 0.89 & 0 & -0.44 \\ 0.44 & 0 & 0.89 \end{bmatrix} \begin{bmatrix} 2.23 & 0 \\ 0 & 1 \\ 0 & 0 \end{bmatrix} \begin{bmatrix} 0 & 1 \\ 1 & 0 \end{bmatrix}.$$

Figure 16.5 illustrates the elements of the SVD and the action of A. Because $m > n$, $\mathbf{u}_3$ is in the null space of A^{T}, that is $A^{\mathsf{T}}\mathbf{u}_3 = \mathbf{0}$.

Example 16.4

For a matrix of different dimensions, we pick

$$A = \begin{bmatrix} -0.8 & 0 & 0.8 \\ 1 & 1.5 & -0.3. \end{bmatrix}.$$

The first step is to form $A^{\mathsf{T}}A$ and AA^{T} and find their eigenvalues and (normalized) eigenvectors, which are made by the columns of an orthogonal matrix.

$$A^{\mathrm{T}}A = \begin{bmatrix} 1.64 & 1.5 & -0.94 \\ 1.5 & 2.25 & -0.45 \\ -0.94 & -0.45 & 0.73 \end{bmatrix}, \quad \begin{matrix} \lambda'_1 = 3.77, \\ \lambda'_2 = 0.84, \\ \lambda'_3 = 0, \end{matrix} \quad V = \begin{bmatrix} -0.63 & 0.38 & 0.67 \\ -0.71 & -0.62 & -0.31 \\ 0.30 & -0.68 & 0.67 \end{bmatrix};$$

$$AA^{\mathrm{T}} = \begin{bmatrix} 1.28 & -1.04 \\ -1.04 & 3.34 \end{bmatrix}, \quad \begin{matrix} \lambda'_1 = 3.77, \\ \lambda'_2 = 0.84, \end{matrix} \quad U = \begin{bmatrix} 0.39 & -0.92 \\ -0.92 & -0.39 \end{bmatrix}.$$

The matrix A is rank 2, thus there are two singular values, and

$$\Sigma = \begin{bmatrix} 1.94 & 0 & 0 \\ 0 & 0.92 & 0 \end{bmatrix}.$$

The SVD of $A = U\Sigma V^{\mathrm{T}}$:

$$\begin{bmatrix} -0.8 & 0 & 0.8 \\ 1 & 1.5 & -0.3. \end{bmatrix} = \begin{bmatrix} 0.39 & -0.92 \\ -0.92 & -0.39 \end{bmatrix} \begin{bmatrix} 1.94 & 0 & 0 \\ 0 & 0.92 & 0 \end{bmatrix} \begin{bmatrix} -0.63 & -0.71 & 0.3 \\ 0.38 & -0.62 & -0.68 \\ 0.67 & -0.31 & 0.67 \end{bmatrix}.$$

Figure 16.6 illustrates the elements of the SVD and the action of A. Because $m < n$, $\mathbf{v}_3$ is in the null space of A, that is, $A\mathbf{v}_3 = \mathbf{0}$.

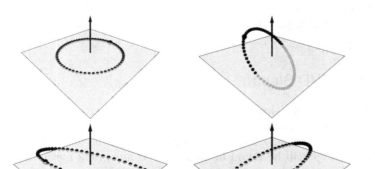

Figure 16.6.
The SVD of a 2 × 3 matrix A: see Example 16.4. Clockwise from top left: Initial point set forming a circle with one reference point; $V^{\mathrm{T}}\mathbf{x}$; $\Sigma V^{\mathrm{T}}\mathbf{x}$; $U\Sigma V^{\mathrm{T}}\mathbf{x}$, illustrating the action of A.

Example 16.5

Let's look at a fairly simple example, a rank deficient matrix,

$$A = \begin{bmatrix} 1 & 0 & 0 \\ 0 & 1 & 0 \\ 0 & 0 & 0 \end{bmatrix},$$

which is a projection into the $[\mathbf{e}_1, \mathbf{e}_2]$-plane. Because A is symmetric and idempotent, $A = A^{\mathrm{T}}A = AA^{\mathrm{T}}$. It is easy to see that $A = U\Sigma V^{\mathrm{T}}$ with

$$\begin{bmatrix} 1 & 0 & 0 \\ 0 & 1 & 0 \\ 0 & 0 & 0 \end{bmatrix} = \begin{bmatrix} 1 & 0 & 0 \\ 0 & 1 & 0 \\ 0 & 0 & 1 \end{bmatrix} \begin{bmatrix} 1 & 0 & 0 \\ 0 & 1 & 0 \\ 0 & 0 & 0 \end{bmatrix} \begin{bmatrix} 1 & 0 & 0 \\ 0 & 1 & 0 \\ 0 & 0 & 1 \end{bmatrix}.$$

Since the rank is two, the first two columns of U and V form an orthonormal basis for the column and row spaces of A, respectively. The $\mathbf{e}_3$ vector is projected to the zero vector by A, and thus this vector spans the null space of A and A^{T}.

Generalizing the 2×2 case, the σ_i are the lengths of the semi-axes of an ellipsoid. As before, the semi-major axis is length σ_1 and the length of the semi-minor axis is equal to the smallest singular value.

The SVD is an important tool for dealing with rank deficient matrices.

16.3   SVD Steps

This section is titled "Steps" rather than "Algorithm" to emphasize that the description here is simply a review of our introduction to the SVD. A robust algorithm that is efficient in terms of computing time and storage will be found in an advanced numerical methods text.

Let's summarize the steps for finding the SVD of a matrix A.

> *Input:* an $m \times n$ matrix A.
> *Output:* U, V, Σ such that $A = U\Sigma V^{\mathrm{T}}$.
>
> Find the eigenvalues $\lambda'_1, \ldots, \lambda'_n$ of $A^{\mathrm{T}}A$.
>   Order the λ'_i so that $\lambda'_1 \geq \lambda'_2 \geq \ldots \geq \lambda'_n$.
>   Suppose $\lambda'_1, \ldots, \lambda'_r > 0$, then the rank of A is r.
> Create an $m \times n$ diagonal matrix Σ with $\sigma_{i,i} = \sqrt{\lambda'_i}, i = 1, \ldots, r$.

Find the corresponding (normalized) eigenvectors $\mathbf{v}_i$ of $A^{\mathrm{T}}A$.
Create an $n \times n$ matrix V with column vectors $\mathbf{v}_i$.
Find the (normalized) eigenvectors $\mathbf{u}_i$ of AA^{T}.
Create an $m \times m$ matrix U with column vectors $\mathbf{u}_i$.

You have now found the singular valued decomposition of A, which is $A = U\Sigma V^{\mathrm{T}}$.

A note on U: instead of finding the columns as the eigenvectors of $A^{\mathrm{T}}A$, one can compute $\mathbf{u}_i, i = 1, r$ as $\mathbf{u}_i = A\mathbf{v}_i / \| \cdot \|$. If $m > n$ then the remaining $\mathbf{u}_i$ are found from the null space of A^{T}.

The only "hard" task in this is finding the λ_i'. But there are several highly efficient algorithms for this task, taking advantage of the fact that $A^{\mathrm{T}}A$ is symmetric. Many of these algorithms will return the corresponding eigenvectors as well.

As we discussed in the context of least squares in Section 13.1, forming $A^{\mathrm{T}}A$ can result in an ill-posed problem because the condition number of this matrix is the square of the condition number of A. Thus, numerical methods will avoid direct computation of this matrix by employing a method such as Householder.

16.4 Singular Values and Volumes

As a practical application, we will use the SVD to compute the determinant of a matrix A. We observe that in (16.2), the matrices U and V, being orthogonal, have determinants equal to ± 1. Thus,

$$| \det A| = \det \Sigma = \sigma_1 \cdot \ldots \cdot \sigma_n. \qquad (16.5)$$

If a 2D triangle has area φ, it will have area $\pm\sigma_1\sigma_2\varphi$ after being transformed by a 2D linear map with singular values σ_1, σ_2. Similarly, if a 3D object has volume φ, it will have volume $\pm\sigma_1\sigma_2\sigma_3\varphi$ after being transformed by a linear map with singular values $\sigma_1, \sigma_2, \sigma_3$.

Of course one can compute determinants without using singular values. Recall from Section 15.1, the characteristic polynomial of A,

$$p(\lambda) = \det[A - \lambda I] = (\lambda_1 - \lambda)(\lambda_2 - \lambda) \cdot \ldots \cdot (\lambda_n - \lambda).$$

Evaluating p at $\lambda = 0$, we have

$$\det A = \lambda_1 \cdot \ldots \cdot \lambda_n, \qquad (16.6)$$

where the λ_i are A's eigenvalues.

16.5   The Pseudoinverse

The inverse of a matrix, introduced in Sections 5.9 and 12.4, is mainly
a theoretical tool for analyzing the solution to a linear system. Additionally, the inverse is limited to square, nonsingular matrices. What
might the inverse of a more general type of matrix be? The answer
is in the form of a *generalized inverse*, or the so-called *pseudoinverse*,
and it is denoted as $A^\dagger$ ("A dagger"). The SVD is a very nice tool
for finding the pseudoinverse, and it is suited for practical use as we
shall see in Section 16.6.

Let's start with a special matrix, an $m \times n$ diagonal matrix Σ with
diagonal elements σ_i. The pseudoinverse, $\Sigma^\dagger$, has diagonal elements
$\sigma_i^\dagger$ given by

$$\sigma_i^\dagger = \left\{ \begin{array}{ll} 1/\sigma_i & \text{if } \sigma_i > 0, \\ 0 & \text{else.} \end{array} \right\},$$

and its dimension is $n \times m$. If the rank of Σ is r, then the product
$\Sigma^\dagger \Sigma$ holds the $r \times r$ identity matrix, and all other elements are zero.

We can use this very simple expression for the pseudoinverse of Σ
to express the pseudoinverse for a general $m \times n$ matrix A using its
SVD,

$$A^\dagger = (U\Sigma V^\mathrm{T})^{-1} = V\Sigma^\dagger U^\mathrm{T}, \tag{16.7}$$

recalling that U and V are orthogonal matrices.

If A is square and invertible, then $A^\dagger = A^{-1}$. Otherwise, $A^\dagger$ still
has some properties of an inverse:

$$A^\dagger A A^\dagger = A^\dagger, \tag{16.8}$$
$$A A^\dagger A = A. \tag{16.9}$$

Example 16.6

Let's find the pseudoinverse of the matrix from Example 16.3,

$$A = \begin{bmatrix} 1 & 0 \\ 0 & 2 \\ 0 & 1 \end{bmatrix}.$$

We find

$$\Sigma^\dagger = \begin{bmatrix} 1/2.23 & 0 & 0 \\ 0 & 1 & 0 \end{bmatrix},$$

then (16.7) results in

$$A^\dagger = \begin{bmatrix} 1 & 0 & 0 \\ 0 & 2/5 & 1/5 \end{bmatrix} = \begin{bmatrix} 0 & 1 \\ 1 & 0 \end{bmatrix} \begin{bmatrix} 1/2.23 & 0 & 0 \\ 0 & 1 & 0 \end{bmatrix} \begin{bmatrix} 0 & 0.89 & 0.44 \\ 1 & 0 & 0 \\ 0 & -0.44 & 0.89 \end{bmatrix}.$$

And we check that (16.8) holds:

$$A^\dagger = \begin{bmatrix} 1 & 0 & 0 \\ 0 & 2/5 & 1/5 \end{bmatrix} \begin{bmatrix} 1 & 0 \\ 0 & 2 \\ 0 & 1 \end{bmatrix} \begin{bmatrix} 1 & 0 & 0 \\ 0 & 2/5 & 1/5 \end{bmatrix},$$

and also (16.9):

$$A = \begin{bmatrix} 1 & 0 \\ 0 & 2 \\ 0 & 1 \end{bmatrix} \begin{bmatrix} 1 & 0 & 0 \\ 0 & 2/5 & 1/5 \end{bmatrix} \begin{bmatrix} 1 & 0 \\ 0 & 2 \\ 0 & 1 \end{bmatrix}.$$

Example 16.7

The matrix from Example 16.1 is square and nonsingular, therefore the pseudoinverse is equal to the inverse. Just by visual inspection:

$$A = \begin{bmatrix} 3 & 0 \\ 0 & 1 \end{bmatrix}, \qquad A^{-1} = \begin{bmatrix} 1/3 & 0 \\ 0 & 1 \end{bmatrix},$$

and the pseudoinverse is

$$A^\dagger = \begin{bmatrix} 1 & 0 \\ 0 & 1 \end{bmatrix} \begin{bmatrix} 1/3 & 0 \\ 0 & 1 \end{bmatrix} \begin{bmatrix} 1 & 0 \\ 0 & 1 \end{bmatrix} = \begin{bmatrix} 1/3 & 0 \\ 0 & 1 \end{bmatrix}.$$

This generalization of the inverse of a matrix is often times called the *Moore-Penrose generalized inverse*. Least squares approximation is a primary application for this pseudoinverse.

16.6   Least Squares

The pseudoinverse allows for a concise approximate solution, specifically the least squares solution, to an overdetermined linear system. (A detailed introduction to least squares may be found in Section 12.7.)

In an overdetermined linear system

$$A\mathbf{x} = \mathbf{b},$$

A is a rectangular matrix with dimension $m \times n$, $m > n$. This means that the linear system has m equations in n unknowns and it is inconsistent because it is unlikely that $\mathbf{b}$ lives in the subspace $\mathcal{V}$ defined by the columns of A. The least squares solution finds the orthogonal projection of $\mathbf{b}$ into $\mathcal{V}$, which we will call $\mathbf{b}'$. Thus the solution to $A\mathbf{x} = \mathbf{b}'$ produces the vector closest to $\mathbf{b}$ that lives in $\mathcal{V}$. This leads us to the *normal equations*

$$A^{\mathrm{T}}A\mathbf{x} = A^{\mathrm{T}}\mathbf{b}, \tag{16.10}$$

whose solution minimizes

$$\|A\mathbf{x} - \mathbf{b}\|. \tag{16.11}$$

In our introduction to condition numbers in Section 13.4, we discussed that this system can be ill-posed because the condition number of $A^{\mathrm{T}}A$ is the square of that of A. To avoid forming $A^{\mathrm{T}}A$, in Section 13.1 we proposed the Householder algorithm as a method to find the least squares solution.

As announced at the onset of this section, the SVD and pseudoinverse provide a numerically stable method for finding the least squares solution to an overdetermined linear system. We find an approximate solution rather easily:

$$\mathbf{x} = A^{\dagger}\mathbf{b}. \tag{16.12}$$

But why is this the least squares solution?

Again, we want to find $\mathbf{x}$ to minimize (16.11). Let's frame the linear system in terms of the SVD of A and take advantage of the orthogonality of U,

$$\begin{aligned} A\mathbf{x} - \mathbf{b} &= U\Sigma V^{\mathrm{T}}\mathbf{x} - \mathbf{b} \\ &= U\Sigma V^{\mathrm{T}}\mathbf{x} - UU^{\mathrm{T}}\mathbf{b} \\ &= U(\Sigma\mathbf{y} - \mathbf{z}). \end{aligned}$$

This new framing of the problem exposes that

$$\|A\mathbf{x} - \mathbf{b}\| = \|\Sigma\mathbf{y} - \mathbf{z}\|,$$

and leaves us with an easier diagonal least squares problem to solve. The steps involved are as follows.

1. Compute the SVD $A = U\Sigma V^{\mathrm{T}}$.

2. Compute the $m \times 1$ vector $\mathbf{z} = U^{\mathrm{T}}\mathbf{b}$.

3. Compute the $n \times 1$ vector $\mathbf{y} = \Sigma^{\dagger}\mathbf{z}$.
   This is the least squares solution to the $m \times n$ problem $\Sigma\mathbf{y} = \mathbf{z}$.
   The least squares solution requires minimizing $\mathbf{v} = \Sigma\mathbf{y} - \mathbf{z}$, which
   has the simple form:

$$
\mathbf{v} = \begin{bmatrix} \sigma_1 y_1 - z_1 \\ \sigma_2 y_2 - z_2 \\ \vdots \\ \sigma_r y_r - z_r \\ -z_{r+1} \\ \vdots \\ -z_m \end{bmatrix},
$$

   where the rank of Σ is r.
   It is easy to see that the $\mathbf{y}$ that will minimize $\mathbf{v}$ is $y_i = z_i/\sigma_i$ for
   $i = 1, \ldots, r$, hence $\mathbf{y} = \Sigma^{\dagger}\mathbf{z}$.

4. Compute the $n \times 1$ solution vector $\mathbf{x} = V\mathbf{y}$.

To summarize, the least squares solution is reduced to a diagonal least
squares problem ($\Sigma\mathbf{y} = \mathbf{z}$), which requires only simple matrix-vector
multiplications. The calculations in reverse order include

$$
\begin{aligned}
\mathbf{x} &= V\mathbf{y} \\
\mathbf{x} &= V(\Sigma^{\dagger}\mathbf{z}) \\
\mathbf{x} &= V\Sigma^{\dagger}(U^{\mathrm{T}}\mathbf{b}).
\end{aligned}
$$

We have rediscovered the pseudoinverse, and we have come back to
(16.12), while verifying that this is indeed the least squares solution.

Example 16.8

Let's revisit the least squares problem that we solved using the normal
equations in Example 12.13 and the Householder method in Exam-

ple 13.3. The 7×2 overdetermined linear system, $A\mathbf{x} = \mathbf{b}$, is

$$\begin{bmatrix} 0 & 1 \\ 10 & 1 \\ 20 & 1 \\ 30 & 1 \\ 40 & 1 \\ 50 & 1 \\ 60 & 1 \end{bmatrix} \mathbf{x} = \begin{bmatrix} 30 \\ 25 \\ 40 \\ 40 \\ 30 \\ 5 \\ 25 \end{bmatrix}.$$

The best fit line coefficients are found in four steps.

1. Compute the SVD, $A = U\Sigma V^{\mathrm{T}}$. (The matrix dimensions are as follows: U is 7×7, Σ is 7×2, V is 2×2.)

$$\Sigma = \begin{bmatrix} 95.42 & 0 \\ 0 & 1.47 \\ 0 & 0 \\ 0 & 0 \\ 0 & 0 \\ 0 & 0 \\ 0 & 0 \end{bmatrix}, \qquad \Sigma^{\dagger} = \begin{bmatrix} 0.01 & 0 & 0 & 0 & 0 & 0 & 0 \\ 0 & 0.68 & 0 & 0 & 0 & 0 & 0 \end{bmatrix}.$$

2. Compute

$$\mathbf{z} = U^{\mathrm{T}}\mathbf{b} = \begin{bmatrix} 54.5 \\ 51.1 \\ 3.2 \\ -15.6 \\ 9.6 \\ 15.2 \\ 10.8 \end{bmatrix}.$$

3. Compute

$$\mathbf{y} = \Sigma^{\dagger}\mathbf{z} = \begin{bmatrix} 0.57 \\ 34.8 \end{bmatrix}.$$

4. Compute

$$\mathbf{x} = V\mathbf{y} = \begin{bmatrix} -0.23 \\ 34.8 \end{bmatrix},$$

resulting in the same best fit line, $x_2 = -0.23x_1 + 34.8$, as we found via the normal equations and the Householder method.

The normal equations (16.10) give a best approximation

$$\mathbf{x} = (A^{\mathrm{T}}A)^{-1}A^{\mathrm{T}}\mathbf{b} \tag{16.13}$$

to the original problem $A\mathbf{x} = \mathbf{b}$. This approximation was developed by considering a new right-hand side vector $\mathbf{b}'$ in the subspace of A, called $\mathcal{V}$. If we substitute the expression for $\mathbf{x}$ in (16.13) into $A\mathbf{x} = \mathbf{b}'$, we have an expression for $\mathbf{b}'$ in relation to $\mathbf{b}$,

$$\begin{aligned} \mathbf{b}' &= A(A^{\mathrm{T}}A)^{-1}A^{\mathrm{T}}\mathbf{b} \\ &= AA^{\dagger}\mathbf{b} \\ &= \mathrm{proj}_{\mathcal{V}}\mathbf{b} \end{aligned}$$

Geometrically, we can see that $AA^{\dagger}$ is a projection because the goal is to project $\mathbf{b}$ into the subspace $\mathcal{V}$. A projection must be idempotent as well, and a property of the pseudoinverse in (16.8) ensures this.

This application of the SVD gave us the opportunity to bring together several important linear algebra topics!

16.7 Application: Image Compression

Suppose the $m \times n$ matrix A has $k = \min(m, n)$ singular values σ_i and as before, $\sigma_1 \geq \sigma_2 \geq \ldots \geq \sigma_k$. Using the SVD, it is possible to write A as a sum of k rank one matrices:

$$A = \sigma_1\mathbf{u}_1\mathbf{v}_1^{\mathrm{T}} + \sigma_2\mathbf{u}_2\mathbf{v}_2^{\mathrm{T}} + \ldots + \sigma_k\mathbf{u}_k\mathbf{v}_k^{\mathrm{T}}. \tag{16.14}$$

This is analogous to what we did in (7.13) for the eigendecomposition.

We can use (16.14) for *image compression*. An image is comprised of a grid of colored pixels.[1] Figure 16.7 (left) is a very simple example; it has only 4×4 pixels. Each grayscale is associated with a number, thus this grid can be thought of as a matrix. The singular values for this matrix are $\sigma_i = 7.1, 3.8, 1.3, 0.3$. Let's refer to the images from left to right as I_0, I_1, I_2, I_3. The original image is I_0. The matrix

$$A_1 = \sigma_1\mathbf{u}_1\mathbf{v}_1^{\mathrm{T}}$$

results in image I_1 and the matrix

$$A_2 = \sigma_1\mathbf{u}_1\mathbf{v}_1^{\mathrm{T}} + \sigma_2\mathbf{u}_2\mathbf{v}_2^{\mathrm{T}}$$

results in image I_2. Notice that the original image is nearly replicated incorporating only half the singular values. This is due to the fact

---

[1] We use grayscales here.

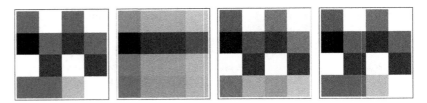

Figure 16.7.

Image compression: a method that uses SVD. The input matrix has singular values $\sigma_i = 7.1, 3.8, 1.3, 0.3$. Far left: original image; from left to right: recovering the image by adding projection terms.

that σ_1 and σ_2 are large in comparison to σ_3 and σ_4. Image I_3 is created from $A_3 = A_2 + \sigma_3 \mathbf{u}_3 \mathbf{v}_3^{\mathrm{T}}$. Image I_4, which is not displayed, is identical to I_0.

The change in an image by adding only those I_i corresponding to small σ_i can be hardly noticeable. Thus omitting images I_k corresponding to small σ_k amounts to compressing the original image; there is no severe quality loss for small σ_k. Furthermore, if some σ_i are zero, compression can clearly be achieved. This is the case in Figure 16.1, an 8×8 matrix with $\sigma_i = 6.2, 1.7, 1.49, 0, \ldots, 0$. The figure illustrates images corresponding to each nonzero σ_i. The last image is identical to the input, making it clear that the five remaining $\sigma_i = 0$ are unimportant to image quality.

16.8 Principal Components Analysis

Figure 16.8 illustrates a 2D *scatter plot*: each circle represents a coordinate pair (point) in the $[\mathbf{e}_1, \mathbf{e}_2]$-system. For example, we might want to determine if gross domestic product (GDP) and the poverty rate (PR) for countries in the World Trade Organization are related. We would then record $[\text{GDP} \;\; \text{PR}]^{\mathrm{T}}$ as coordinate pairs. How might we reveal trends in this data set?

Let our data set be given as $\mathbf{x}_1, \ldots, \mathbf{x}_n$, a set of 2D points such that $\mathbf{x}_1 + \ldots + \mathbf{x}_n = \mathbf{0}$. This condition simply means that the origin is the centroid of the points. The data set in Figure 16.8 has been translated to produce Figure 16.9 (left). Let $\mathbf{d}$ be a unit vector. Then the projection of $\mathbf{x}_i$ onto a line through the origin containing $\mathbf{d}$ is a vector with (signed) length $\mathbf{x}_i \cdot \mathbf{d}$. Let $l(\mathbf{d})$ be the sum of the squares of all these lengths:

$$l(\mathbf{d}) = [\mathbf{x}_1 \cdot \mathbf{d}]^2 + \ldots + [\mathbf{x}_n \cdot \mathbf{d}]^2.$$

Figure 16.8.

Scatter plot: data pairs recorded in Cartesian coordinates.

Imagine rotating $\mathbf{d}$ around the origin. For each position of $\mathbf{d}$, we compute the value $l(\mathbf{d})$. For the longest line in the left part of Figure 16.9, the value of $l(\mathbf{d})$ is large in comparison to $\mathbf{d}$ orthogonal to it, demonstrating that the value of $l(\mathbf{d})$ indicates the variation in the point set from the line generated by $\mathbf{d}$. Higher variation results in larger $l(\mathbf{d})$.

Let's arrange the data $\mathbf{x}_i$ in a matrix

$$X = \begin{bmatrix} \mathbf{x}_1^{\mathrm{T}} \\ \mathbf{x}_2^{\mathrm{T}} \\ \vdots \\ \mathbf{x}_n^{\mathrm{T}} \end{bmatrix}.$$

Then we rewrite $l(\mathbf{d})$ as

$$\begin{aligned} l(\mathbf{d}) &= \|X\mathbf{d}\|^2 \\ &= (X\mathbf{d}) \cdot (X\mathbf{d}) \\ &= \mathbf{d}^{\mathrm{T}} X^{\mathrm{T}} X \mathbf{d}. \end{aligned} \tag{16.15}$$

We further abbreviate $C = X^{\mathrm{T}} X$ and note that C is a symmetric, positive definite 2×2 matrix. Hence (16.15) describes a quadratic form just as we discussed in Section 7.6.

For which $\mathbf{d}$ is $l(\mathbf{d})$ maximal? The answer is simple: for $\mathbf{d}$ being the eigenvector corresponding to C's largest eigenvalue. Similarly, $l(\mathbf{d})$ is minimal for $\mathbf{d}$ being the eigenvector corresponding to C's smallest

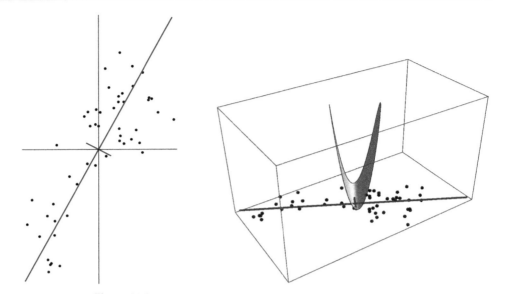

Figure 16.9.
PCA: Analysis of a data set. Left: given data with centroid translated to the origin. Thick lines are coincident with the eigenvectors scaled by their corresponding eigenvalue. Right: points, eigenvector lines, and quadratic form over the unit circle.

eigenvalue. Recall that these eigenvectors form the major and minor axis of the *action ellipse* of C, and the thick lines in Figure 16.9 (left) are precisely these axes. In the right part of Figure 16.9, the quadratic form for the data set is shown along with the data and action ellipse axes. We see that the dominant eigenvector corresponds to highest variance in the data and this is reflected in the quadratic form as well. If $\lambda_1 = \lambda_2$, then there is no preferred direction in the point set and the quadratic form is spherical. We are guaranteed that the eigenvectors will be orthogonal because C is a symmetric matrix.

Looking more closely at C, we see its very simple form,

$$c_{1,1} = x_{1,1}^2 + x_{2,1}^2 + \ldots + x_{n,1}^2,$$
$$c_{1,2} = c_{2,1} = x_{1,1}x_{1,2} + x_{2,1}x_{2,2} + \ldots + x_{n,1}x_{n,2},$$
$$c_{2,2} = x_{1,2}^2 + x_{2,2}^2 + \ldots + x_{n,2}^2.$$

If each element of C is divided by n, then this is called the *covariance matrix*. This matrix is a summary of the variation in each coordinate and between coordinates. Dividing by n will result in scaled eigenvalues; the eigenvectors will not change.

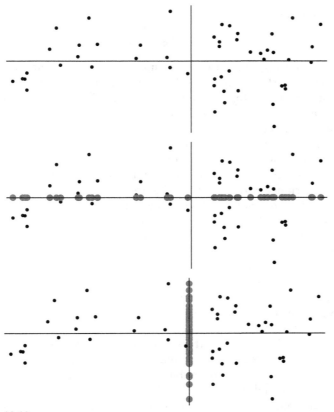

Figure 16.10.
PCA data transformations: three possible data transformations based on PCA analysis. Top: data points transformed to the principal components coordinate system. This set appears in all images. Middle: data compression by keeping dominant eigenvector component. Bottom: data compression by keeping the nondominant eigenvector.

The eigenvectors provide a convenient *local coordinate frame* for the data set. This isn't a new idea: it is exactly the principle of the *eigendecomposition*. This frame is commonly called the *principal axes*. Now we can construct an orthogonal transformation of the data, expressing them in terms of a new, more meaningful coordinate system. Let the matrix $V = [\mathbf{v}_1 \ \mathbf{v}_2]$ hold the normalized eigenvectors as column vectors, where $\mathbf{v}_1$ is the dominant eigenvector. The orthogonal transformation of the data X that aligns $\mathbf{v}_1$ with $\mathbf{e}_1$ and $\mathbf{v}_2$ with $\mathbf{e}_2$ is simply

$$\hat{X} = XV.$$

This results in

$$\hat{\mathbf{x}}_i = \begin{bmatrix} \mathbf{x}_i \cdot \mathbf{v}_1 \\ \mathbf{x}_i \cdot \mathbf{v}_2 \end{bmatrix}.$$

Figure 16.10 (top) illustrates the result of this transformation. Revisit Example 7.6: this is precisely the transformation we used to align the contour ellipse to the coordinate axes. (The transformation is written a bit differently here to accommodate the point set organized as transposed vectors.)

In summary: the data coordinates are now in terms of the trend lines, defined by the eigenvectors of the covariance matrix, and the coordinates directly measure the distance from each trend line. The greatest variance corresponds to the first coordinate in this *principal components coordinate system*. This leads us to the name of this method: principal components analysis (PCA).

So far, PCA has worked with all components of the given data. However, it can also be used for data compression by reducing dimensionality. Instead of constructing V to hold all eigenvectors, we may use only the most significant, so suppose $V = \mathbf{v}_1$. This transformation produces the middle image shown in Figure 16.10. If instead, $V = \mathbf{v}_2$, then the result is the bottom image, and for clarity, we chose to display these points on the $\mathbf{e}_2$-axis, but this is arbitrary. Comparing these results: there is greater spread of the data in the middle image, which corresponds to a trend line with higher variance.

Here we focused on 2D data, but the real power of PCA comes with higher dimensional data for which it is very difficult to visualize and understand relationships between dimensions. PCA makes it possible to identify insignificant dimensions and eliminate them.

- singular value decomposition (SVD)
- singular values
- right singular vector
- left singular vector
- SVD matrix dimensions
- SVD column, row, and null spaces
- SVD steps
- volume in terms of singular values
- eigendecomposition
- matrix decomposition
- action ellipse axes length
- pseudoinverse
- generalized inverse
- least squares solution via the pseudoinverse
- quadratic form
- contour ellipse
- principal components analysis (PCA)
- covariance matrix

16.9 Exercises

1. Find the SVD for

$$A = \begin{bmatrix} 1 & 0 \\ 0 & 4 \end{bmatrix}.$$

2. What is the eigendecomposition of matrix A in Exercise 1.

3. For what type of matrix are the eigenvalues the same as the singular values?

4. The action of a 2×2 linear map can be described by the mapping of the unit circle to an ellipse. Figure 4.3 illustrates such an ellipse. What are the lengths of the semi-axes? What are the singular values of the corresponding matrix,

$$A = \begin{bmatrix} 1/2 & 0 \\ 0 & 2 \end{bmatrix}?$$

5. Find the SVD for the matrix

$$A = \begin{bmatrix} 0 & -2 \\ 1 & 0 \\ 0 & 0 \end{bmatrix}.$$

6. Let

$$A = \begin{bmatrix} -1 & 0 & 1 \\ 0 & 1 & 0 \\ 1 & 1 & -2 \end{bmatrix},$$

and $C = A^{\mathrm{T}}A$. Is one of the eigenvalues of C negative?

7. For the matrix

$$A = \begin{bmatrix} -2 & 0 & 0 \\ 0 & 1 & 0 \\ 0 & 0 & 1 \end{bmatrix},$$

show that both (16.5) and (16.6) yield the same result for the absolute value of the determinant of A.

8. For the matrix

$$A = \begin{bmatrix} 2 & 0 \\ 2 & 0.5 \end{bmatrix},$$

show that both (16.5) and (16.6) yield the same result for the determinant of A.

9. For the matrix

$$A = \begin{bmatrix} 1 & 0 & 1 \\ 0 & 1 & 0 \\ 0 & 1 & 0 \end{bmatrix},$$

show that both (16.5) and (16.6) yield the same result for the absolute value of the determinant of A.

10. What is the pseudoinverse of the matrix from Exercise 5?

11. What is the pseudoinverse of the matrix

$$\begin{bmatrix} 0 & 1 \\ 3 & 0 \\ 0 & 0 \end{bmatrix}?$$

12. What is the pseudoinverse of the matrix

$$A = \begin{bmatrix} 2 \\ 0 \\ 0 \end{bmatrix}?$$

Note that this matrix is actually a vector.

13. What is the pseudoinverse of

$$A = \begin{bmatrix} 3 & 0 & 0 \\ 0 & 2 & 0 \\ 0 & 0 & 1 \end{bmatrix}?$$

14. What is the least squares solution to the linear system $A\mathbf{v} = \mathbf{b}$ given by:

$$\begin{bmatrix} 1 & 0 \\ 0 & 2 \\ 0 & 1 \end{bmatrix} \begin{bmatrix} v_1 \\ v_2 \end{bmatrix} = \begin{bmatrix} 0 \\ 1 \\ 0 \end{bmatrix}.$$

Use the pseudoinverse and the enumerated steps in Section 16.6. The SVD of A may be found in Example 16.3.

15. What is the least squares solution to the linear system $A\mathbf{v} = \mathbf{b}$ given by:

$$\begin{bmatrix} 4 & 0 \\ 0 & 1 \\ 0 & 0 \end{bmatrix} \begin{bmatrix} v_1 \\ v_2 \end{bmatrix} = \begin{bmatrix} 1 \\ 1 \\ 1 \end{bmatrix}.$$

Use the pseudoinverse and the enumerated steps in Section 16.6.

16. What is the least squares solution to the linear system $A\mathbf{v} = \mathbf{b}$ given by:

$$\begin{bmatrix} 3 & 0 & 0 \\ 0 & 1 & 2 \\ 0 & 1 & 0 \end{bmatrix} \begin{bmatrix} v_1 \\ v_2 \\ v_3 \end{bmatrix} = \begin{bmatrix} 3 \\ 3 \\ 1 \end{bmatrix}.$$

Use the pseudoinverse and the enumerated steps in Section 16.6.

17. For the following data set X, apply PCA using all eigenvectors. Give the covariance matrix and the final components $\hat{X}$ in the principal components coordinate system.

$$X = \begin{bmatrix} -2 & 2 \\ -1 & -1 \\ 0 & 0 \\ 1 & 1 \\ 2 & 2 \\ -1 & 0 \\ 1 & 0 \end{bmatrix}$$

Make a sketch of the data; this will help with finding the solution.

18. For the data set in Exercise 17, apply PCA using the dominant eigenvector only.

Breaking It Up: Triangles

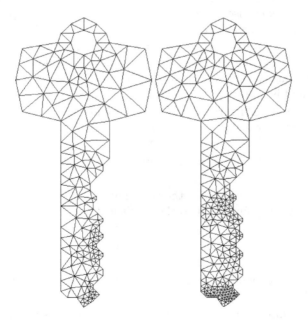

Figure 17.1.
2D finite element method: refinement of a triangulation based on stress and strain calculations. (Source: J. Shewchuk, http://www.cs.cmu.edu/~quake/triangle.html.)

Triangles are as old as geometry. They were of interest to the ancient Greeks, and in fact the roots of trigonometry can be found in

367

their study. Triangles also became an indispensable tool in computer graphics and advanced disciplines such as *finite element analysis*. In graphics, objects are broken down into triangular facets for display purposes; in the finite element method (FEM), 2D shapes are broken down into triangles in order to facilitate complicated algorithms. Figure 17.1 illustrates a refinement procedure based on stress and strain calculations. For both applications, reducing the geometry to linear or piecewise planar makes computations more tractable.

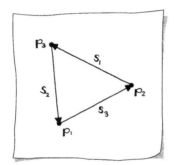

Sketch 17.1.
Vertices and edges of a triangle.

17.1 Barycentric Coordinates

A *triangle* T is given by three points, its *vertices*, $\mathbf{p}_1$, $\mathbf{p}_2$, and $\mathbf{p}_3$. The vertices may live in 2D or 3D. Three points define a plane, thus a triangle is a 2D element. We use the convention of labeling the $\mathbf{p}_i$ in a counterclockwise sense. The edge, or side, opposite point $\mathbf{p}_i$ is labeled $\mathbf{s}_i$. (See Sketch 17.1.)

When we study the properties of this triangle, it is more convenient to work in terms of a local coordinate system that is closely tied to the triangle. This type of coordinate system was invented by F. Moebius in 1827 and is known as *barycentric coordinates*.

Let $\mathbf{p}$ be an arbitrary point inside T. Our aim is to write it as a combination of the vertices $\mathbf{p}_i$ in a form such as

$$\mathbf{p} = u\mathbf{p}_1 + v\mathbf{p}_2 + w\mathbf{p}_3. \tag{17.1}$$

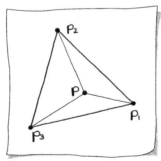

Sketch 17.2.
Barycentric coordinates.

We know one thing already: the right-hand side of this equation is a combination of points, and so the coefficients must sum to one:

$$u + v + w = 1. \tag{17.2}$$

Otherwise, we would not have a barycentric combination! (See Sketch 17.2.)

We can simply write (17.1) and (17.2) as a linear system

$$\begin{bmatrix} \mathbf{p}_1 & \mathbf{p}_2 & \mathbf{p}_3 \end{bmatrix} \begin{bmatrix} u \\ v \\ w \end{bmatrix} = \begin{bmatrix} p_1 \\ p_2 \\ 1 \end{bmatrix}$$

Using Cramer's rule, the solution to this 3×3 linear system is formed as ratios of determinants or areas, thus

$$u = \frac{\text{area}(\mathbf{p}, \mathbf{p}_2, \mathbf{p}_3)}{\text{area}(\mathbf{p}_1, \mathbf{p}_2, \mathbf{p}_3)}, \qquad (17.3)$$

$$v = \frac{\text{area}(\mathbf{p}, \mathbf{p}_3, \mathbf{p}_1)}{\text{area}(\mathbf{p}_1, \mathbf{p}_2, \mathbf{p}_3)}, \qquad (17.4)$$

$$w = \frac{\text{area}(\mathbf{p}, \mathbf{p}_1, \mathbf{p}_2)}{\text{area}(\mathbf{p}_1, \mathbf{p}_2, \mathbf{p}_3)}. \qquad (17.5)$$

Recall that for linear interpolation in Section 2.5, barycentric coordinates on a line segment were defined in terms of ratios of lengths. Here, for a triangle, we have the analogous ratios of areas. To review determinants and areas, see Sections 8.2, 8.5, and 9.8.

Let's see why this works. First, we observe that (u, v, w) do indeed sum to one. Next, let $\mathbf{p} = \mathbf{p}_2$. Now (17.3)–(17.5) tell us that $v = 1$ and $u = w = 0$, just as expected. One more check: if $\mathbf{p}$ is on the edge $\mathbf{s}_1$, say, then $u = 0$, again as expected. Try to show for yourself that the remaining vertices and edges work the same way.

We call (u, v, w) *barycentric coordinates* and denote them by boldface: $\mathbf{u} = (u, v, w)$. Although they are not independent of each other (e.g., we may set $w = 1 - u - v$), they behave much like "normal" coordinates: if $\mathbf{p}$ is given, then we can find $\mathbf{u}$ from (17.3)–(17.5). If $\mathbf{u}$ is given, then we can find $\mathbf{p}$ from (17.1).

The three vertices of the triangle have barycentric coordinates

$$\mathbf{p}_1 \cong (1, 0, 0),$$
$$\mathbf{p}_2 \cong (0, 1, 0),$$
$$\mathbf{p}_3 \cong (0, 0, 1).$$

The $\cong$ symbol will be used to indicate the barycentric coordinates of a point. These and several other examples are shown in Sketch 17.3.

As you see, even points outside of T can be given barycentric coordinates! This works since the areas involved in (17.3)–(17.5) are *signed*. So points inside T have positive barycentric coordinates, and those outside have mixed signs.[1]

This observation is the basis for one of the most frequent uses of barycentric coordinates: the *triangle inclusion test*. If a triangle T and a point $\mathbf{p}$ are given, how do we determine if $\mathbf{p}$ is inside T

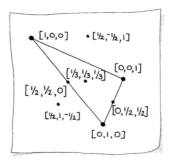

Sketch 17.3.

Examples of barycentric coordinates.

---

[1]This assumes the triangle to be oriented counterclockwise. If it is oriented clockwise, then the points inside have all negative barycentric coordinates, and the outside ones still have mixed signs.

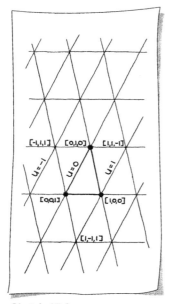

Sketch 17.4.

Barycentric coordinates coordinate lines.

or not? We simply compute $\mathbf{p}$'s barycentric coordinates and check their signs. If they are all of the same sign, inside—else, outside. Theoretically, one or two of the barycentric coordinates could be zero, indicating that $\mathbf{p}$ is on one of the edges. In "real" situations, you are not likely to encounter values that are *exactly* equal to zero; be sure not to test for a barycentric coordinate to be *equal* to zero! Instead, use a tolerance ϵ, and flag a point as being on an edge if one of its barycentric coordinates is less than ϵ in absolute value. A good value for ϵ? Obviously, this is application dependent, but something like $1.0E-6$ should work for most cases.

Finally, Sketch 17.4 shows how we may think of the whole plane as being covered by a grid of coordinate lines. Note that the plane is divided into seven regions by the (extended) edges of T.

Example 17.1

Let's work with a simple example that is easy to sketch. Suppose the three triangle vertices are given by

$$\mathbf{p}_1 = \begin{bmatrix} 0 \\ 0 \end{bmatrix}, \quad \mathbf{p}_2 = \begin{bmatrix} 1 \\ 0 \end{bmatrix}, \quad \mathbf{p}_3 = \begin{bmatrix} 0 \\ 1 \end{bmatrix}.$$

The points $\mathbf{q}$, $\mathbf{r}$, $\mathbf{s}$ with barycentric coordinates

$$\mathbf{q} \cong \left(0, \frac{1}{2}, \frac{1}{2}\right), \quad \mathbf{r} \cong (-1, 1, 1), \quad \mathbf{s} \cong \left(\frac{1}{3}, \frac{1}{3}, \frac{1}{3}\right)$$

have the following coordinates in the plane:

$$\mathbf{q} = 0 \times \mathbf{p}_1 + \frac{1}{2} \times \mathbf{p}_2 + \frac{1}{2} \times \mathbf{p}_3 = \begin{bmatrix} 1/2 \\ 1/2 \end{bmatrix},$$

$$\mathbf{r} = -1 \times \mathbf{p}_1 + 1 \times \mathbf{p}_2 + 1 \times \mathbf{p}_3 = \begin{bmatrix} 1 \\ 1 \end{bmatrix},$$

$$\mathbf{s} = \frac{1}{3} \times \mathbf{p}_1 + \frac{1}{3} \times \mathbf{p}_2 + \frac{1}{3} \times \mathbf{p}_3 = \begin{bmatrix} 1/3 \\ 1/3 \end{bmatrix}.$$

17.2   Affine Invariance

In this short section, we will discuss the statement: *barycentric coordinates are affinely invariant.*

Let $\hat{T}$ be an affine image of T, having vertices $\hat{\mathbf{p}}_1, \hat{\mathbf{p}}_2, \hat{\mathbf{p}}_3$. Let $\mathbf{p}$ be a point with barycentric coordinates $\mathbf{u}$ relative to T. We may apply the affine map to $\mathbf{p}$ also, and then we ask: What are the barycentric coordinates of $\hat{\mathbf{p}}$ with respect to $\hat{T}$?

While at first sight this looks like a daunting task, simple geometry yields the answer quickly. Note that in (17.3)–(17.5), we employ *ratios of areas*. These are, as introduced in Section 6.2, unchanged by affine maps. So while the individual areas in (17.3)–(17.5) do change, their quotients do not. Thus, $\hat{\mathbf{p}}$ also has barycentric coordinates $\mathbf{u}$ with respect to $\hat{T}$.

This fact, namely that affine maps do not change barycentric coordinates, is what is meant by the statement at the beginning of this section. (See Sketch 17.5 for an illustration.)

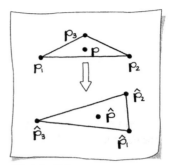

Sketch 17.5.
Affine invariance of barycentric coordinates.

Example 17.2

Let's revisit the simple triangle in Example 17.1 and look at the affine invariance of barycentric coordinates. Suppose we apply a 90° rotation,

$$R = \begin{bmatrix} 0 & -1 \\ 1 & 0 \end{bmatrix}$$

to the triangle vertices, resulting in $\hat{\mathbf{p}}_i = R\mathbf{p}_i$. Apply this rotation to $\mathbf{s}$ from Example 17.1,

$$\hat{\mathbf{s}} = R\mathbf{s} = \begin{bmatrix} -1/3 \\ 1/3 \end{bmatrix}.$$

Due to the affine invariance of barycentric coordinates, we could have found the coordinates of $\hat{\mathbf{s}}$ as

$$\hat{\mathbf{s}} = \frac{1}{3} \times \hat{\mathbf{p}}_1 + \frac{1}{3} \times \hat{\mathbf{p}}_2 + \frac{1}{3} \times \hat{\mathbf{p}}_3 = \begin{bmatrix} -1/3 \\ 1/3 \end{bmatrix}.$$

17.3   Some Special Points

In classical geometry, many special points relative to a triangle have been discovered, but for our purposes, just three will do: the centroid, the incenter, and the circumcenter. They are used for a multitude of geometric computations.

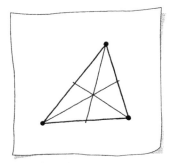

Sketch 17.6.
The centroid.

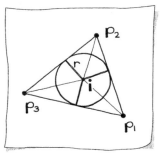

Sketch 17.7.
The incenter.

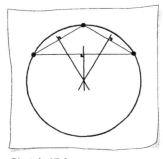

Sketch 17.8.
The circumcenter.

The *centroid* **c** of a triangle is given by the intersection of the three medians. (A median is the connection of an edge midpoint to the opposite vertex.) Its barycentric coordinates (see Sketch 17.6) are given by

$$\mathbf{c} \cong \left(\frac{1}{3}, \frac{1}{3}, \frac{1}{3} \right). \qquad (17.6)$$

We verify this by writing

$$\left(\frac{1}{3}, \frac{1}{3}, \frac{1}{3} \right) = \frac{1}{3}(0,1,0) + \frac{2}{3}\left(\frac{1}{2}, 0, \frac{1}{2} \right),$$

thus asserting that $\left(\frac{1}{3}, \frac{1}{3}, \frac{1}{3} \right)$ lies on the median associated with $\mathbf{p}_2$. In the same way, we show that it is also on the remaining two medians. We also observe that a triangle and its centroid are related in an affinely invariant way.

The *incenter* **i** of a triangle is the intersection of the three angle bisectors (see Sketch 17.7). There is a circle, called the *incircle*, that has **i** as its center and touches all three triangle edges. Let s_i be the length of the triangle edge opposite vertex $\mathbf{p}_i$. Let r be the radius of the incircle—there is a formula for it, but we won't need it here.

If the barycentric coordinates of **i** are (i_1, i_2, i_3), then we see that

$$i_1 = \frac{\text{area}(\mathbf{i}, \mathbf{p}_2, \mathbf{p}_3)}{\text{area}(\mathbf{p}_1, \mathbf{p}_2, \mathbf{p}_3)}.$$

This may be rewritten as

$$i_1 = \frac{rs_1}{rs_1 + rs_2 + rs_3},$$

using the "1/2 base times height" rule for triangle areas.

Simplifying, we obtain

$$i_1 = \frac{s_1}{c}, \qquad i_2 = \frac{s_2}{c}, \qquad i_3 = \frac{s_3}{c},$$

where $c = s_1 + s_2 + s_3$ is the circumference of T. A triangle is not affinely related to its incenter—affine maps change the barycentric coordinates of **i**.

The *circumcenter* **cc** of a triangle is the center of the circle through its vertices. It is obtained as the intersection of the edge bisectors. (See Sketch 17.8.) Notice that the circumcenter might not be inside the triangle. This circle is called the *circumcircle* and we will refer to its radius as R.

The barycentric coordinates (cc_1, cc_2, cc_3) of the circumcenter are

$$cc_1 = \frac{d_1(d_2 + d_3)}{D}, \qquad cc_2 = \frac{d_2(d_1 + d_3)}{D}, \qquad cc_3 = \frac{d_3(d_1 + d_2)}{D},$$

where

$$d_1 = (\mathbf{p}_2 - \mathbf{p}_1) \cdot (\mathbf{p}_3 - \mathbf{p}_1),$$
$$d_2 = (\mathbf{p}_1 - \mathbf{p}_2) \cdot (\mathbf{p}_3 - \mathbf{p}_2),$$
$$d_3 = (\mathbf{p}_1 - \mathbf{p}_3) \cdot (\mathbf{p}_2 - \mathbf{p}_3),$$
$$D = 2(d_1 d_2 + d_2 d_3 + d_3 d_1).$$

Furthermore,

$$R = \frac{1}{2}\sqrt{\frac{(d_1 + d_2)(d_2 + d_3)(d_3 + d_1)}{D/2}}.$$

These formulas are due to [8]. Confirming our observation in Sketch 17.8 that the circumcircle might be outside of the triangle, note that some of the cc_i may be negative. If T has an angle close to $180°$, then the corresponding cc_i will be *very* negative, leading to serious numerical problems. As a result, the circumcenter will be far away from the vertices, and thus not be of practical use. As with the incenter, affine maps of the triangle change the barycentric coordinates of the circumcenter.

Example 17.3

Yet again, let's visit the simple triangle in Example 17.1. Be sure to make a sketch to check the results of this example. Let's compute the incenter. The lengths of the edges of the triangle are $s_1 = \sqrt{2}$, $s_2 = 1$, and $s_3 = 1$. The circumference of the triangle is $c = 2 + \sqrt{2}$. The barycentric coordinates of the incenter are then

$$\mathbf{i} \cong \left(\frac{\sqrt{2}}{2 + \sqrt{2}}, \frac{1}{2 + \sqrt{2}}, \frac{1}{2 + \sqrt{2}} \right).$$

(These barycentric coordinates are approximately $(0.41, 0.29, 0.29)$.) The coordinates of the incenter are

$$\mathbf{i} = 0.41 \times \mathbf{p}_1 + 0.29 \times \mathbf{p}_2 + 0.29 \times \mathbf{p}_3 = \begin{bmatrix} 0.29 \\ 0.29 \end{bmatrix}.$$

The circumcircle's circumcenter is easily calculated, too. First compute $d_1 = 0$, $d_2 = 1$, $d_3 = 1$, and $D = 2$. Then the barycentric

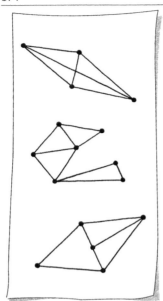

Sketch 17.9.

Examples of illegal
triangulations.

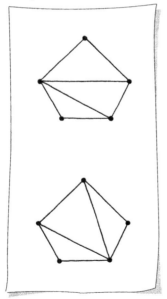

Sketch 17.10.

Nonuniqueness of
triangulations.

coordinates of the circumcenter are $\mathbf{c} \cong (0, 1/2, 1/2)$. This is the midpoint of the "diagonal" edge of the triangle.

Now the radius of the circumcircle is easily computed with the equation above, $R = \sqrt{2}/2$.

More interesting constructions based on triangles may be found in Coxeter [3].

17.4   2D Triangulations

The study of *one* triangle is the realm of classical geometry; in modern applications, one often encounters millions of triangles. Typically, they are connected in some well-defined way; the most basic one being the 2D *triangulation*. Triangulations have been used in surveying for centuries; more modern applications rely on satellite data, which are collected in triangulations called *TINs* (*triangular irregular networks*).

Here is the formal definition of a 2D triangulation. A triangulation of a set of 2D points $\{\mathbf{p}_i\}_{i=1}^{N}$ is a connected set of triangles meeting the following criteria:

1. The vertices of the triangles consist of the given points.

2. The interiors of any two triangles do not intersect.

3. If two triangles are not disjoint, then they share a vertex or have coinciding edges.

4. The union of all triangles equals the convex hull of the $\mathbf{p}_i$.

These rules sound abstract, but some examples will shed light on them. Figure 17.2 shows a triangulation that satisfies the 2D triangulation definition. Evident from this example: the number of triangles surrounding a vertex, or *valence*, varies from vertex to vertex. These triangles make up the *star* of a vertex. In contrast, Sketch 17.9 shows three illegal triangulations, violating the above rules. The top example involves overlapping triangles. In the middle example, the boundary of the triangulation is not the convex hull of the point set. (A lot more on convex hulls may be found in [4]; also see Section 18.3.) The bottom example violates condition 3.

If we are given a point set, is there a unique triangulation? Certainly not, as Sketch 17.10 shows. Among the many possible triangulations, there is one that is most commonly agreed to be the

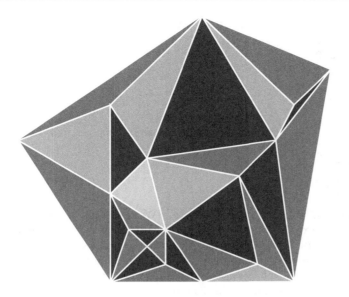

Figure 17.2.
Triangulation: a valid triangulation of the convex hull.

"best." This is the *Delaunay triangulation*. Describing the details of this method is beyond the scope of this text, however a wealth of information can be found on the Web.

17.5   A Data Structure

What is the best data structure for storing a triangulation? The factors that determine the best structure include storage requirements and accessibility. Let's build the "best" structure based on the point set and triangulation illustrated in Sketch 17.11.

In order to minimize storage, it is an accepted practice to store each point only once. Since these are floating point values, they take up the most space. Thus, a basic triangulation structure would be a listing of the point set followed by the triangulation information. This constitutes pointers into the point set, indicating which points are joined to form a triangle. Store the triangles in a counterclockwise orientation! This is the data structure for the triangulation in Sketch 17.11:

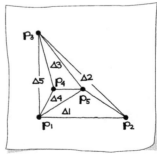

Sketch 17.11.
A sample triangulation.

```
5                   (number of points)
0.0  0.0            (point #1)
1.0  0.0
0.0  1.0
0.25 0.3
0.5  0.3
5                   (number of triangles)
1 2 5               (first triangle - connects points #1,2,5)
2 3 5
4 5 3
1 5 4
1 4 3
```

We can improve this structure. We will encounter applications that require knowledge of the connectivity of the triangulation, as described in Section 17.6. To facilitate this, it is not uncommon to also see the *neighbor information* of the triangulation stored. This means that for each triangle, the indices of the triangles surrounding it are stored. For example, in Sketch 17.11, triangle 1 defined by points $1, 2, 5$ is surrounded by triangles $2, 4, -1$. The neighboring triangles are listed corresponding to the point across from the shared edge. Triangle -1 indicates that there is not a neighboring triangle across this edge. Immediately, we see that this gives us a fast method for determining the boundary of the triangulation. Listing the neighbor information after each triangle, the final data structure is as follows.

```
5                   (number of points)
0.0  0.0            (point #1)
1.0  0.0
0.0  1.0
0.25 0.3
0.5  0.3
5                   (number of triangles)
1 2 5   2 4 -1      (first triangle and neighbors)
2 3 5   3 1 -1
4 5 3   2 5  4
1 5 4   3 5  1
1 4 3   3 -1 4
```

This is but one of many possible data structures for a triangulation. Based on the needs of particular applications, researchers have developed a variety of structures to optimize searches. One such structure that has proved to be popular is called the *winged-edge data structure* [14].

17.6 Application: Point Location

Given a triangulation of points $\mathbf{p}_i$, assume we are given a point $\mathbf{p}$ that has not been used in building the triangulation. Question: In which triangle is $\mathbf{p}$, if any? The easiest way is to compute $\mathbf{p}$'s barycentric coordinates with respect to all triangles; if all of them are positive with respect to some triangle, then that is the desired one; else, $\mathbf{p}$ is in none of the triangles.

While simple, this algorithm is expensive. In the worst case, every triangle has to be considered; on average, half of all triangles have to be considered. A much more efficient algorithm may be based upon the following observation. Suppose $\mathbf{p}$ is not in a particular triangle T. Then at least one of its barycentric coordinates with respect to T must be negative; let's assume it is u. We then know that $\mathbf{p}$ has no chance of being inside T's two neighbors along edges $\mathbf{s}_2$ or $\mathbf{s}_3$ (see Sketch 17.12).

So a likely candidate to check is the neighbor along $\mathbf{s}_1$—recall that we have stored the neighboring information in a data structure. In this way—always searching in the direction of the currently most negative barycentric coordinate—we create a path from a starting triangle to the one that actually contains $\mathbf{p}$.

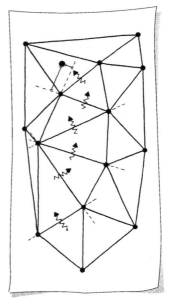

Sketch 17.12.
Point location triangle search.

Point Location Algorithm

Input: Triangulation and neighbor information, plus one point $\mathbf{p}$.

Output: Triangle that $\mathbf{p}$ is in.

Step 0: Set the "current triangle" to be the first triangle in the triangulation.

Step 1: Perform the *triangle inclusion test* (see Section 17.1) for $\mathbf{p}$ and the current triangle. If all barycentric coordinates are positive, output the current triangle. If the barycentric coordinates are mixed in sign, then determine the barycentric coordinate of $\mathbf{p}$ with respect to the current triangle that has the most negative value. Set the current triangle to be the corresponding neighbor and repeat Step 1.

Notes:

- Try improving the speed of this algorithm by not completing the division for determining the barycentric coordinates in (17.3)–(17.5). This division does not change the sign. Keep in mind the test for which triangle to move to changes.

- Suppose the algorithm is to be executed for more than one point. Consider using the triangle that was output from the previous run as input, rather than always using the first triangle. Many times a data set has some *coherence*, and the output for the next run might be the same triangle, or one very near to the triangle from the previous run.

17.7   3D Triangulations

In computer applications, one often encounters millions of triangles, connected in some well-defined way, describing a geometric object. In particular, shading algorithms require this type of structure.

The rules for 3D triangulations are the same as for 2D. Additionally, the data structure is the same, except that now each point has three instead of two coordinates.

Figure 17.3 shows a 3D surface that is composed of triangles. Another example is provided in Figure 8.4. Shading requires a 3D unit vector, called a *normal*, to be associated with each triangle or vertex. A normal is perpendicular to an object's surface at a particular point. This normal is used to calculate how light is reflected, and in turn the illumination of the object. (See [14] for details on such illumi-

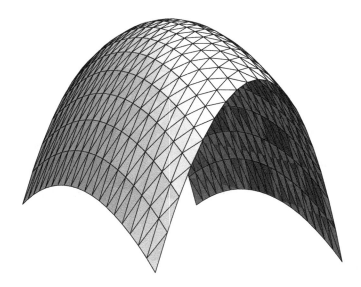

Figure 17.3.
3D triangulated surface: a wireframe and shaded renderings superimposed.

nation methods.) We investigated just how to calculate normals in Section 8.6.

- barycentric coordinates
- triangle inclusion test
- affine invariance of barycentric coordinates
- centroid, barycenter
- incenter
- circumcenter
- 2D triangulation criteria
- star

- valence
- triangulation data structure
- point location algorithm
- 3D triangulation criteria
- 3D triangulation data structure
- normal

17.8 Exercises

Let a triangle T_1 be given by the vertices

$$\mathbf{p}_1 = \begin{bmatrix} 1 \\ 1 \end{bmatrix}, \quad \mathbf{p}_2 = \begin{bmatrix} 2 \\ 2 \end{bmatrix}, \quad \mathbf{p}_3 = \begin{bmatrix} -1 \\ 2 \end{bmatrix}.$$

Let a triangle T_2 be given by the vertices

$$\mathbf{q}_1 = \begin{bmatrix} 0 \\ 0 \end{bmatrix}, \quad \mathbf{q}_2 = \begin{bmatrix} 0 \\ -1 \end{bmatrix}, \quad \mathbf{q}_3 = \begin{bmatrix} -1 \\ 0 \end{bmatrix}.$$

1. Using T_1:

   (a) What are the barycentric coordinates of $\mathbf{p} = \begin{bmatrix} 0 \\ 1.5 \end{bmatrix}$?

   (b) What are the barycentric coordinates of $\mathbf{p} = \begin{bmatrix} 0 \\ 0 \end{bmatrix}$?

   (c) Find the triangle's incenter.

   (d) Find the triangle's circumcenter.

   (e) Find the centroid of the triangle.

2. Using T_2:

   (a) What are the barycentric coordinates of $\mathbf{p} = \begin{bmatrix} 0 \\ 1.5 \end{bmatrix}$?

   (b) What are the barycentric coordinates of $\mathbf{p} = \begin{bmatrix} 0 \\ 0 \end{bmatrix}$?

   (c) Find the triangle's incenter.

(d) Find the triangle's circumcenter.

(e) Find the centroid of the triangle.

3. What are the areas of T_1 and T_2?

4. Let an affine map be given by

$$\mathbf{x}' = \begin{bmatrix} 1 & 2 \\ -1 & 2 \end{bmatrix} \mathbf{x} + \begin{bmatrix} -1 \\ 0 \end{bmatrix}.$$

What are the areas of mapped triangles T_1' and T_2'? Compare the ratios

$$\frac{T_1}{T_2} \quad \text{and} \quad \frac{T_1'}{T_2'}.$$

5. What is the unit normal to the triangle T_1?

6. What is the unit normal to the triangle with vertices

$$\begin{bmatrix} 1 \\ 0 \\ 0 \end{bmatrix}, \quad \begin{bmatrix} 1 \\ 1 \\ 0 \end{bmatrix}, \quad \begin{bmatrix} 0 \\ 0 \\ 1 \end{bmatrix}?$$

18

Putting Lines Together: Polylines and Polygons

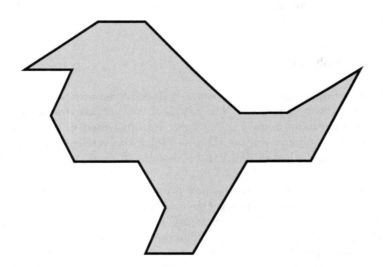

Figure 18.1.
Polygon: straight line segments forming a bird shape.

Figure 18.1 shows a *polygon*. It is the outline of a shape, drawn with straight line segments. Since such shapes are all a printer or plotter can draw, just about every computer-generated drawing consists of polygons. If we add an "eye" to the bird-shaped polygon from

Figure 18.2.

Mixing maps: a pattern is created by composing rotations and translations.

Figure 18.1, and if we apply a sequence of rotations and translations to it, then we arrive at Figure 18.2. It turns out copies of our special bird polygon can cover the whole plane! This technique is also present in the Escher illustration in Figure 6.8.

18.1 Polylines

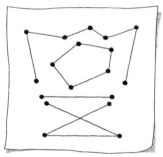

Sketch 18.1.

2D polyline examples.

Straight line segments, called *edges*, connecting an ordered set of *vertices* constitute a *polyline*. The first and last vertices are not necessarily connected. Some 2D examples are illustrated in Sketch 18.1, however, a polyline can be 3D, too. Since the vertices are ordered, the edges are oriented and can be thought of as vectors. Let's call them *edge vectors*.

Polylines are a primary output primitive, and thus they are included in graphics standards. One example of a graphics standard is the *GKS (Graphical Kernel System)*; this is a specification of what belongs in a graphics package. The development of PostScript was based on GKS. Polylines have many uses in computer graphics and modeling. Whether in 2D or 3D, they are typically used to outline a shape. The power of polylines to reveal a shape is illustrated in Figure 18.3 in the display of a 3D surface. The surface is *evaluated* in an organized fashion so that the points can be logically connected as polylines, giving the observer a feeling of the "flow" of the surface. In modeling, a polyline is often used to approximate a complex curve or data, which in turn makes analysis easier and less costly. An example of this is illustrated in Figure 12.3.

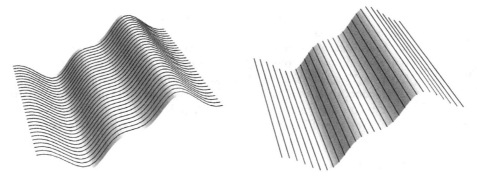

Figure 18.3.
Polylines: the display of a 3D surface. Two different directions for the polyline sets give different impressions of the surface shape.

18.2 Polygons

When the first and last vertices of a polyline are connected, it is called a *polygon*. Normally, a polygon is thought to enclose an area. For this reason, unless a remark is made, we will consider planar polygons only. Just as with polylines, polygons constitute an ordered set of vertices and we will continue to use the term edge vectors. Thus, a polygon with n edges is given by an ordered set of 2D points

$$\mathbf{p}_1, \mathbf{p}_2, \ldots, \mathbf{p}_n$$

and has edge vectors $\mathbf{v}_i = \mathbf{p}_{i+1} - \mathbf{p}_i; i = 1, \ldots, n$. Note that the edge vectors sum to the *zero vector*.

If you look at the edge vectors carefully, you'll discover that $\mathbf{v}_n = \mathbf{p}_{n+1} - \mathbf{p}_n$, but there is no vertex $\mathbf{p}_{n+1}$! This apparent problem is resolved by defining $\mathbf{p}_{n+1} = \mathbf{p}_1$, a convention called *cyclic numbering*. We'll use this convention throughout, and will not mention it every time. We also add one more topological characterization of polygons: the number of vertices equals the number of edges.

Since a polygon is closed, it divides the plane into two parts: a finite part, the polygon's *interior*, and an infinite part, the polygon's *exterior*.

As you traverse a polygon, you follow the path determined by the vertices and edge vectors. Between vertices, you'll move along straight lines (the edges), but at the vertices, you'll have to perform a rotation before resuming another straight line path. The angle α_i by which you rotate at vertex $\mathbf{p}_i$ is called the *turning angle* or *exterior angle* at $\mathbf{p}_i$. The *interior angle* is then given by $\pi - \alpha_i$ (see Sketch 18.2).

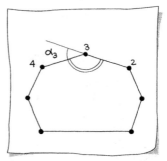

Sketch 18.2.
Interior and exterior angles.

Sketch 18.3.

A minmax box is a polygon.

Polygons are used a lot! For instance, in Chapter 1 we discussed the *extents* of geometry in a 2D coordinate system. Another name for these extents is a *minmax box* (see Sketch 18.3). It is a special polygon, namely a rectangle. Another type of polygon studied in Chapter 17 is the triangle. This type of polygon is often used to define a *polygonal mesh* of a 3D model. The triangles may then be filled with color to produce a shaded image. A polygonal mesh from triangles is illustrated in Figure 17.3, and one from rectangles is illustrated in Figure 8.3.

18.3 Convexity

Polygons are commonly classified by their shape. There are many ways of doing this. One important classification is as *convex* or *nonconvex*. The latter is also referred to as *concave*. Sketch 18.4 gives an example of each. How do you describe the shape of a convex polygon? As in Sketch 18.5, stick nails into the paper at the vertices. Now take a rubberband and stretch it around the nails, then let go. If the rubberband shape follows the outline of the polygon, it is convex. The points on the rubberband define the *convex hull*. Another definition: take any two points in the polygon (including on the edges) and connect them with a straight line. If the line never leaves the polygon, then it is convex. This must work for *all* possible pairs of points!

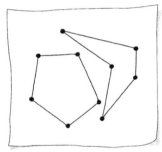

Sketch 18.4.

Convex (left) and nonconvex (right) polygons.

The issue of convexity is important, because algorithms that involve polygons can be simplified if the polygons are known to be convex. This is true for algorithms for the problem of *polygon clipping*. This problem starts with two polygons, and the goal is to find the intersection of the polygon areas. Some examples are illustrated in Sketch 18.6. The intersection area is defined in terms of one or more polygons. If both polygons are convex, then the result will be just one convex polygon. However, if even one polygon is not convex then the result might be two or more, possibly disjoint or nonconvex, polygons. Thus, nonconvex polygons need more record keeping in order to properly define the intersection area(s). Not all algorithms are designed for nonconvex polygons. See [10] for a detailed description of clipping algorithms.

An n-sided convex polygon has a sum of interior angles I equal to

$$I = (n - 2)\pi. \tag{18.1}$$

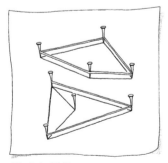

Sketch 18.5.

Rubberband test for convexity of a polygon.

To see this, take one polygon vertex and form triangles with the other vertices, as illustrated in Sketch 18.7. This forms $n - 2$ triangles. The sum of interior angles of a triangle is known to be π. Thus, we get the above result.

The sum of the exterior angles of a convex polygon is easily found with this result. Each interior and exterior angle sums to π. Suppose the ith interior angle is α_i radians, then the exterior angle is $\pi - \alpha_i$ radians. Sum over all angles, and the exterior angle sum E is

$$E = n\pi - (n-2)\pi = 2\pi. \tag{18.2}$$

To test if an n-sided polygon is convex, we'll use the *barycenter* of the vertices $\mathbf{p}_1, \ldots, \mathbf{p}_n$. The barycenter $\mathbf{b}$ is a special barycentric combination (see Sections 17.1 and 17.3):

$$\mathbf{b} = \frac{1}{n}(\mathbf{p}_1 + \ldots + \mathbf{p}_n).$$

It is the center of gravity of the vertices. We need to construct the implicit line equation for each edge vector in a consistent manner. If the polygon is convex, then the point $\mathbf{b}$ will be on the "same" side of every line. The implicit equation will result in all positive or all negative values. (Sketch 3.3 illustrates this concept.) This will not work for some unusual, nonsimple, polygons. (See Section 18.5 and the Exercises.)

Another test for convexity is to check if there is a *reentrant angle*. This is an interior angle that is greater than π.

18.4 Types of Polygons

There are a variety of special polygons. First, we introduce two terms to help describe these polygons:

- *equilateral* means that all sides are of equal length, and

- *equiangular* means that all *interior angles* at the vertices are equal.

In the following illustrations, edges with the same number of tick marks are of equal length and angles with the same number of arc markings are equal.

A very special polygon is the *regular polygon*: it is equilateral and equiangular. Examples are illustrated in Sketch 18.8. A regular polygon is also referred to as an *n-gon*, indicating it has n edges. We list the names of the "classical" n-gons:

- a 3-gon is an equilateral triangle,

- a 4-gon is a square,

- a 5-gon is a regular pentagon,

- a 6-gon is a regular hexagon, and

- an 8-gon is a regular octagon.

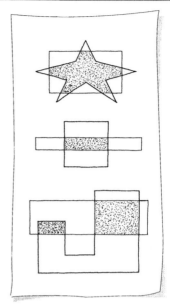

Sketch 18.6.
Polygon clipping.

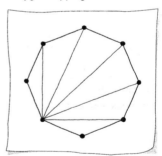

Sketch 18.7.
Sum of interior angles using triangles.

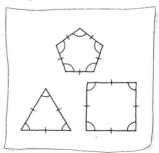

Sketch 18.8.
Regular polygons.

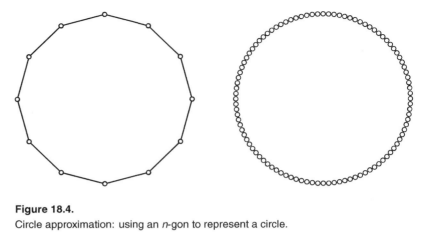

Figure 18.4.
Circle approximation: using an *n*-gon to represent a circle.

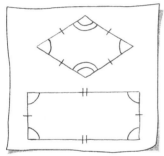

Sketch 18.9.
Rhombus and rectangle.

An *n*-gon is commonly used to approximate a circle in computer graphics, as illustrated in Figure 18.4.

A *rhombus* is equilateral but not equiangular, whereas a *rectangle* is equiangular but not equilateral. These are illustrated in Sketch 18.9.

18.5   Unusual Polygons

Most often, applications deal with *simple* polygons, as opposed to *nonsimple* polygons. A nonsimple polygon, as illustrated in Sketch 18.10, is characterized by edges intersecting other than at the vertices. Topology is the reason nonsimple polygons can cause havoc in some algorithms. For convex and nonconvex simple polygons, as you traverse along the boundary of the polygon, the interior remains on one side. This is not the case for nonsimple polygons. At the mid-edge intersections, the interior switches sides. In more concrete terms, recall how the implicit line equation could be used to determine if a point is on the line, and more generally, which side it is on. Suppose you have developed an algorithm that associates the + side of the line with being inside the polygon. This rule will work fine if the polygon is simple, but not otherwise.

Sometimes nonsimple polygons can arise due to an error. The polygon clipping algorithms, as discussed in Section 18.3, involve sorting vertices to form the final polygon. If this sorting goes haywire, then you could end up with a nonsimple polygon rather than a simple one as desired.

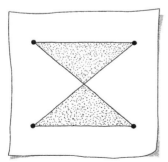

Sketch 18.10.
Nonsimple polygon.

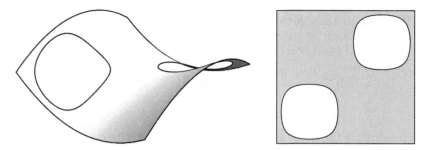

Figure 18.5.
Trimmed surface: an application of polygons with holes. Left: trimmed surface. Right: rectangular parametric domain with polygonal holes.

In applications, it is not uncommon to encounter polygons with holes. Such a polygon is illustrated in Sketch 18.11. As you see, this is actually more than one polygon. An example of this, illustrated in Figure 18.5, is a special CAD/CAM (computer-aided manufacturing) surface called a *trimmed surface*. The polygons define parts of the material to be cut or punched out. This allows other parts to fit to this one.

For trimmed surfaces and other CAD/CAM applications, a certain convention is accepted in order to make more sense out of this multi-polygon geometry. The polygons must be oriented a special way. The *visible region*, or the region that is not cut out, is to the "left." As a result, the outer boundary is oriented counterclockwise and the inner boundaries are oriented clockwise. More on the visible region in Section 18.9.

Sketch 18.11.
Polygon with holes.

18.6 Turning Angles and Winding Numbers

The *turning angle* of a polygon or polyline is essentially another name for the *exterior angle*, which is illustrated in Sketch 18.12. Notice that this sketch illustrates the turning angles for a convex and a nonconvex polygon. Here the difference between a turning angle and an exterior angle is illustrated. The turning angle has an orientation as well as an angle measure. All turning angles for a convex polygon have the same orientation, which is not the case for a nonconvex polygon. This fact will allow us to easily differentiate the two types of polygons.

Here is an application of the turning angle. Suppose for now that a given polygon is 2D and lives in the $[\mathbf{e}_1, \mathbf{e}_2]$-plane. Its n vertices are labeled

$$\mathbf{p}_1, \mathbf{p}_2, \cdots \mathbf{p}_n.$$

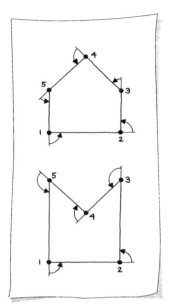

Sketch 18.12.
Turning angles.

We want to know if the polygon is convex. We only need to look at the orientation of the turning angles, not the actual angles. First, let's embed the 2D vectors in 3D by adding a zero third coordinate, for example:

$$\mathbf{p}_1 = \begin{bmatrix} p_{1,1} \\ p_{2,1} \\ 0 \end{bmatrix}.$$

Recall that the cross product of two vectors in the $\mathbf{e}_1, \mathbf{e}_2$-plane will produce a vector that points "in" or "out," that is in the $+\mathbf{e}_3$ or $-\mathbf{e}_3$ direction. Therefore, by taking the cross product of successive edge vectors

$$\mathbf{u}_i = (\mathbf{p}_{i+1} - \mathbf{p}_i) \wedge (\mathbf{p}_{i+2} - \mathbf{p}_{i+1}), \qquad (18.3)$$

we'll encounter $\mathbf{u}_i$ of the form

$$\begin{bmatrix} 0 \\ 0 \\ u_{3,i} \end{bmatrix}.$$

If the sign of the $u_{3,i}$ value is the *same* for all angles, then the polygon is convex. A mathematical way to describe this is by using the scalar triple product (see Section 8.5). The turning angle orientation is determined by the scalar

$$u_{3,i} = \mathbf{e}_3 \cdot ((\mathbf{p}_{i+1} - \mathbf{p}_i) \wedge (\mathbf{p}_{i+2} - \mathbf{p}_{i+1})).$$

Notice that the sign is dependent upon the traversal direction of the polygon, but only a change of sign is important. The determinant of the 2D vectors would have worked just as well, but the 3D approach is more useful for what follows.

If the polygon lies in an arbitrary plane, having a normal $\mathbf{n}$, then the above convex/concave test is changed only a bit. The cross product in (18.3) produces a vector $\mathbf{u}_i$ that has direction $\pm\mathbf{n}$. Now we need the dot product, $\mathbf{n} \cdot \mathbf{u}_i$ to extract a signed scalar value.

If we actually computed the turning angle at each vertex, we could form an accumulated value called the *total turning angle*. Recall from (18.2) that the total turning angle for a convex polygon is 2π. For a polygon that is not known to be convex, assign a sign using the scalar triple product as above, to each angle measurement. The sum E will then be used to compute the *winding number* of the polygon. The winding number W is

$$W = \frac{E}{2\pi}.$$

Thus, for a convex polygon, the winding number is one. Sketch 18.13 illustrates a few examples. A non-self-intersecting polygon is essentially one loop. A polygon can have more than one loop, with different orientations: clockwise versus counterclockwise. The winding number gets decremented for each clockwise loop and incremented for each counterclockwise loop, or vice versa depending on how you assign signs to your angles.

18.7   Area

A simple method for calculating the area of a 2D polygon is to use the *signed area* of a triangle as in Section 4.9. First, *triangulate* the polygon. For example, choose one vertex of the polygon and form all triangles from it and successive pairs of vertices, as is illustrated in Sketch 18.14. The sum of the signed areas of the triangles results in the area of the polygon. For this method to work, we must form the triangles with a consistent orientation. For example, in Sketch 18.14, triangles $(\mathbf{p}_1, \mathbf{p}_2, \mathbf{p}_3)$, $(\mathbf{p}_1, \mathbf{p}_3, \mathbf{p}_4)$, and $(\mathbf{p}_1, \mathbf{p}_4, \mathbf{p}_5)$ are all counterclockwise or right-handed, and therefore have positive area. More precisely, if we form $\mathbf{v}_i = \mathbf{p}_i - \mathbf{p}_1$, then the area of the polygon in Sketch 18.14 is

$$A = \frac{1}{2}(\det[\mathbf{v}_2, \mathbf{v}_3] + \det[\mathbf{v}_3, \mathbf{v}_4] + \det[\mathbf{v}_4, \mathbf{v}_5]).$$

In general, if a polygon has n vertices, then this area calculation becomes

$$A = \frac{1}{2}(\det[\mathbf{v}_2, \mathbf{v}_3] + \ldots + \det[\mathbf{v}_{n-1}, \mathbf{v}_n]). \tag{18.4}$$

The use of signed area makes this idea work for nonconvex polygons as in Sketch 18.15. As illustrated in the sketch, the negative areas cancel duplicate and extraneous areas.

Equation (18.4) takes an interesting form if its terms are expanded. We observe that the determinants that represent edges of triangles within the polygon cancel. So this leaves us with

$$A = \frac{1}{2}(\det[\mathbf{p}_1, \mathbf{p}_2] + \ldots + \det[\mathbf{p}_{n-1}, \mathbf{p}_n] + \det[\mathbf{p}_n, \mathbf{p}_1]). \tag{18.5}$$

Equation (18.5) seems to have lost all geometric meaning because it involves the determinant of point pairs, but we can recapture geometric meaning if we consider each point to be $\mathbf{p}_i - \mathbf{o}$.

Is (18.4) or (18.5) the preferred form? The amount of computation for each equation is similar; however, there is one drawback of (18.5).

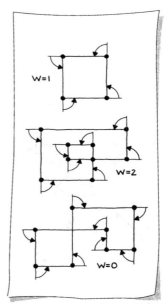

Sketch 18.13.
Winding numbers.

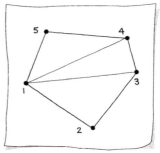

Sketch 18.14.
Area of a convex polygon.

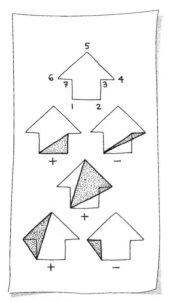

Sketch 18.15.

Area of a nonconvex polygon.

If the polygon is far from the origin then numerical problems can occur because the vectors $\mathbf{p}_i$ and $\mathbf{p}_{i+1}$ will be close to parallel. The form in (18.4) essentially builds a local frame in which to compute the area. For debugging and making sense of intermediate computations, (18.4) is easier to work with. This is a nice example of how reducing an equation to its "simplest" form is not always "optimal"!

An interesting observation is that (18.5) may be written as a *generalized determinant*. The coordinates of the vertices are

$$\mathbf{p}_i = \begin{bmatrix} p_{1,i} \\ p_{2,i} \end{bmatrix}.$$

The area is computed as follows,

$$A = \frac{1}{2} \begin{vmatrix} p_{1,1} & p_{1,2} & \cdots & p_{1,n} & p_{1,1} \\ p_{2,1} & p_{2,2} & \cdots & p_{2,n} & p_{2,1} \end{vmatrix},$$

which is computed by adding the products of all "downward" diagonals, and subtracting the products of all "upward" diagonals.

Example 18.1

Let

$$\mathbf{p}_1 = \begin{bmatrix} 0 \\ 0 \end{bmatrix}, \quad \mathbf{p}_2 = \begin{bmatrix} 1 \\ 0 \end{bmatrix}, \quad \mathbf{p}_3 = \begin{bmatrix} 1 \\ 1 \end{bmatrix}, \quad \mathbf{p}_4 = \begin{bmatrix} 0 \\ 1 \end{bmatrix}.$$

We have

$$A = \frac{1}{2} \begin{vmatrix} 0 & 1 & 1 & 0 & 0 \\ 0 & 0 & 1 & 1 & 0 \end{vmatrix} = \frac{1}{2}[0 + 1 + 1 + 0 - 0 - 0 - 0 - 0] = 1.$$

Since our polygon was a square, this is as expected.

But now take

$$\mathbf{p}_1 = \begin{bmatrix} 0 \\ 0 \end{bmatrix}, \quad \mathbf{p}_2 = \begin{bmatrix} 1 \\ 1 \end{bmatrix}, \quad \mathbf{p}_3 = \begin{bmatrix} 0 \\ 1 \end{bmatrix}, \quad \mathbf{p}_4 = \begin{bmatrix} 1 \\ 0 \end{bmatrix}.$$

This is a nonsimple polygon! Its area computes to

$$A = \frac{1}{2} \begin{vmatrix} 0 & 1 & 0 & 1 & 0 \\ 0 & 1 & 1 & 1 & 0 \end{vmatrix} = \frac{1}{2}[0 + 1 + 0 + 0 - 0 - 0 - 1 - 0] = 0.$$

Draw a sketch and convince yourself this is correct.

Planar polygons are sometimes specified by 3D points. We can adjust the area formula (18.4) accordingly. Recall that the cross product of two vectors results in a vector whose length equals the area of the parallelogram spanned by the two vectors. So we will replace each determinant by a cross product

$$\mathbf{u}_i = \mathbf{v}_i \wedge \mathbf{v}_{i+1} \qquad \text{for} \quad i = 2, n - 1 \qquad (18.6)$$

and as before, each $\mathbf{v}_i = \mathbf{p}_i - \mathbf{p}_1$. Suppose the (unit) normal to the polygon is $\mathbf{n}$. Notice that $\mathbf{n}$ and all $\mathbf{u}_i$ share the same direction, therefore $\|\mathbf{u}\| = \mathbf{u} \cdot \mathbf{n}$. Now we can rewrite (18.5) for 3D points,

$$A = \frac{1}{2}\mathbf{n} \cdot (\mathbf{u}_2 + \ldots + \mathbf{u}_{n-1}), \qquad (18.7)$$

with the $\mathbf{u}_i$ defined in (18.6). Notice that (18.7) is a sum of scalar triple products, which were introduced in Section 8.5.

Example 18.2

Take the four coplanar 3D points

$$\mathbf{p}_1 = \begin{bmatrix} 0 \\ 2 \\ 0 \end{bmatrix}, \quad \mathbf{p}_2 = \begin{bmatrix} 2 \\ 0 \\ 0 \end{bmatrix}, \quad \mathbf{p}_3 = \begin{bmatrix} 2 \\ 0 \\ 3 \end{bmatrix}, \quad \mathbf{p}_4 = \begin{bmatrix} 0 \\ 2 \\ 3 \end{bmatrix}.$$

Compute the area with (18.7), and note that the normal

$$\mathbf{n} = \begin{bmatrix} -1/\sqrt{2} \\ -1/\sqrt{2} \\ 0 \end{bmatrix}.$$

First compute

$$\mathbf{v}_2 = \begin{bmatrix} 2 \\ -2 \\ 0 \end{bmatrix}, \quad \mathbf{v}_3 = \begin{bmatrix} 2 \\ -2 \\ 3 \end{bmatrix}, \quad \mathbf{v}_4 = \begin{bmatrix} 0 \\ 0 \\ 3 \end{bmatrix},$$

and then compute the cross products:

$$\mathbf{u}_2 = \mathbf{v}_2 \wedge \mathbf{v}_3 = \begin{bmatrix} -6 \\ -6 \\ 0 \end{bmatrix} \quad \text{and} \quad \mathbf{u}_3 = \mathbf{v}_3 \wedge \mathbf{v}_4 = \begin{bmatrix} -6 \\ -6 \\ 0 \end{bmatrix}.$$

Then the area is

$$A = \frac{1}{2}\mathbf{n} \cdot (\mathbf{u}_2 + \mathbf{u}_3) = 6\sqrt{2}.$$

(The equality $\sqrt{2}/2 = 1/\sqrt{2}$ was used to eliminate the denominator.) You may also realize that our simple example polygon is just a rectangle, and so you have another way to check the area.

In Section 8.6, we looked at calculating normals to a polygonal mesh specifically for computer graphics lighting models. The results of this section are useful for this task as well. By removing the dot product with the normal, (18.7) provides us with a method of computing a good average normal to a nonplanar polygon:

$$\mathbf{n} = \frac{(\mathbf{u}_2 + \mathbf{u}_3 + \ldots + \mathbf{u}_{n-2})}{\|\mathbf{u}_2 + \mathbf{u}_3 + \ldots + \mathbf{u}_{n-2}\|}.$$

This normal estimation method is a weighted average based on the areas of the triangles. To eliminate this weighting, normalize each $\mathbf{u}_i$ before summing them.

Example 18.3

What is an estimate normal to the nonplanar polygon

$$\mathbf{p}_1 = \begin{bmatrix} 0 \\ 0 \\ 1 \end{bmatrix}, \quad \mathbf{p}_2 = \begin{bmatrix} 1 \\ 0 \\ 0 \end{bmatrix}, \quad \mathbf{p}_3 = \begin{bmatrix} 1 \\ 1 \\ 1 \end{bmatrix}, \quad \mathbf{p}_4 = \begin{bmatrix} 0 \\ 1 \\ 0 \end{bmatrix}?$$

Calculate

$$\mathbf{u}_2 = \begin{bmatrix} 1 \\ -1 \\ 1 \end{bmatrix}, \qquad \mathbf{u}_3 = \begin{bmatrix} -1 \\ 1 \\ 1 \end{bmatrix},$$

and the normal is

$$\mathbf{n} = \begin{bmatrix} 0 \\ 0 \\ 1 \end{bmatrix}.$$

18.8 Application: Planarity Test

Suppose someone sends you a CAD file that contains a polygon. For your application, the polygon must be 2D; however, it is oriented arbitrarily in 3D. How do you verify that the data points are *coplanar*?

There are many ways to solve this problem, although some solutions have clear advantages over the others. Some considerations when comparing algorithms include:

- numerical stability,

- speed,

- ability to define a meaningful tolerance,

- size of data set, and

- maintainability of the algorithm.

The order of importance is arguable.

Let's look at three possible methods to solve this planarity test and then compare them.

- **Volume test:** Choose the first polygon vertex as a base point. Form vectors to the next three vertices. Use the scalar triple product to calculate the volume spanned by these three vectors. If it is less than a given tolerance, then the four points are coplanar. Continue for all other sets.

- **Plane test:** Construct the plane through the first three vertices. Check if all of the other vertices lie in this plane, within a given tolerance.

- **Average normal test:** Find the centroid $\mathbf{c}$ of all points. Compute all normals $\mathbf{n}_i = [\mathbf{p}_i - \mathbf{c}] \wedge [\mathbf{p}_{i+1} - \mathbf{c}]$. Check if all angles formed by two subsequent normals are below a given angle tolerance.

If we compare these three methods, we see that they employ different kinds of tolerances: for volumes, distances, and angles. Which of these is preferable must depend on the application at hand. Clearly the plane test is the fastest of the three; yet it has a problem if the first three vertices are close to being collinear.

18.9   Application: Inside or Outside?

Another important concept for 2D polygons is the *inside/outside test* or *visibility test*. The problem is this: Given a polygon in the $[\mathbf{e}_1, \mathbf{e}_2]$-plane and a point $\mathbf{p}$, determine if the point lies inside the polygon.

One obvious application for this is polygon fill for raster device software, e.g., as with PostScript. Each pixel must be checked to see if it is in the polygon and should be colored. The inside/outside test is also encountered in CAD with *trimmed surfaces*, which were introduced in Section 18.5. With both applications it is not uncommon to have a polygon with one or more holes, as illustrated in Sketch 18.11. In the PostScript fill application, nonsimple polygons are not unusual either.[1]

We will present two similar algorithms, producing different results in some special cases. However we choose to solve this problem, we will want to incorporate a *trivial reject* test. This simply means that if a point is "obviously" not in the polygon, then we output that result immediately, i.e., with a minimal amount of calculation. In this problem, trivial reject refers to constructing a *minmax box* around the polygon. If a point lies outside of this minmax box then it may be trivially rejected. As you see, this involves simple comparison of $\mathbf{e}_1$- and $\mathbf{e}_2$-coordinates.

18.9.1   Even-Odd Rule

From a point $\mathbf{p}$, construct a line in parametric form with vector $\mathbf{r}$ in any direction. The parametric line is

$$\mathbf{l}(t) = \mathbf{p} + t\mathbf{r}.$$

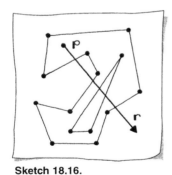

Sketch 18.16.
Even-odd rule.

This is illustrated in Sketch 18.16. Count the number of intersections this line has with the polygon edges for $t \geq 0$ only. This is why the vector is sometimes referred to as a *ray*. The number of intersections will be odd if $\mathbf{p}$ is inside and even if $\mathbf{p}$ is outside. Figure 18.6 illustrates the results of this rule with the polygon fill application.

It can happen that $\mathbf{l}(t)$ coincides with an edge of the polygon or passes through a vertex. Either a more elaborate counting scheme must be developed, or you can choose a different $\mathbf{r}$. As a rule, it is better to not choose $\mathbf{r}$ parallel to the $\mathbf{e}_1$- or $\mathbf{e}_2$-axis because the polygons often have edges parallel to these axes.

---

[1]As an example, take Figure 3.4. The lines inside the bounding rectangle are the edges of a nonsimple polygon.

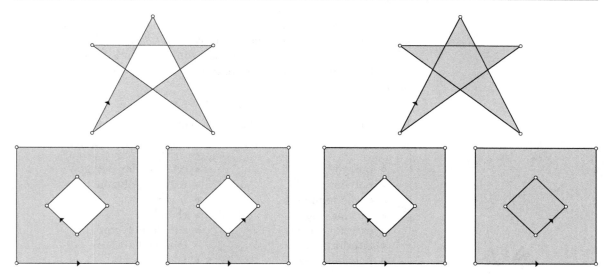

Figure 18.6.
Even-odd rule: applied to polygon fill.

Figure 18.7.
Nonzero winding rule: applied to polygon fill.

18.9.2 Nonzero Winding Number

In Section 18.6 the winding number was introduced. Here is another use for it. This method proceeds similarly to the even-odd rule. Construct a parametric line at a point p and intersect the polygon edges. Again, consider only those intersections for $t \geq 0$. The counting method depends on the orientation of the polygon edges. Start with a winding number of zero. Following Sketch 18.17, if a polygon edge is oriented "right to left" then add one to the winding number. If a polygon edge is oriented "left to right" then subtract one from the winding number. If the final result is zero then the point is outside the polygon. Figure 18.7 illustrates the results of this rule with the same polygons used in the even-odd rule. As with the previous rule, if you encounter edges head-on, then choose a different ray.

The differences in the algorithms are interesting. The PostScript language uses the nonzero winding number rule as the default. The authors (of the PostScript language) feel that this produces better results for the polygon fill application, but the even-odd rule is available with a special command. PostScript must deal with the most general (and crazy) polygons. In the trimmed surface application, the polygons must be simple and polygons cannot intersect; therefore, either algorithm is suitable.

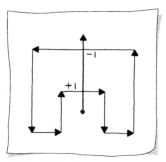

Sketch 18.17.
Nonzero winding number rule.

If you happen to know that you are dealing only with *convex polygons*, another inside/outside test is available. Check which side of the edges the point $\mathbf{p}$ is on. If it is on the same side for all edges, then $\mathbf{p}$ is inside the polygon. All you have to do is to compute all determinants of the form

$$\left| (\mathbf{p}_i - \mathbf{p}) \quad (\mathbf{p}_{i+1} - \mathbf{p}) \right|.$$

If they are all of the same sign, $\mathbf{p}$ is inside the polygon.

• polygon	• equiangular polygon
• polyline	• regular polygon
• cyclic numbering	• n-gon
• turning angle	• rhombus
• exterior angle	• simple polygon
• interior angle	• trimmed surface
• polygonal mesh	• visible region
• convex	• total turning angle
• concave	• winding number
• polygon clipping	• polygon area
• sum of interior angles	• planarity test
• sum of exterior angles	• trivial reject
• reentrant angle	• inside/outside test
• equilateral polygon	• scalar triple product

18.10 Exercises

1. What is the sum of the interior angles of a six-sided polygon? What is the sum of the exterior angles?

2. What type of polygon is equiangular and equilateral?

3. Which polygon is equilateral but not equiangular?

4. Develop an algorithm that determines whether or not a polygon is simple.

5. Calculate the winding number of the polygon with the following vertices:

$$\mathbf{p}_1 = \begin{bmatrix} 0 \\ 0 \end{bmatrix}, \qquad \mathbf{p}_2 = \begin{bmatrix} -2 \\ 0 \end{bmatrix}, \qquad \mathbf{p}_3 = \begin{bmatrix} -2 \\ 2 \end{bmatrix},$$

$$\mathbf{p}_4 = \begin{bmatrix} 0 \\ 2 \end{bmatrix}, \qquad \mathbf{p}_5 = \begin{bmatrix} 2 \\ 2 \end{bmatrix}, \qquad \mathbf{p}_6 = \begin{bmatrix} 2 \\ -2 \end{bmatrix},$$

$$\mathbf{p}_7 = \begin{bmatrix} 3 \\ -2 \end{bmatrix}, \qquad \mathbf{p}_8 = \begin{bmatrix} 3 \\ -1 \end{bmatrix}, \qquad \mathbf{p}_9 = \begin{bmatrix} 0 \\ -1 \end{bmatrix}.$$

6. Compute the area of the polygon with the following vertices:

$$\mathbf{p}_1 = \begin{bmatrix} -1 \\ 0 \end{bmatrix}, \qquad \mathbf{p}_2 = \begin{bmatrix} 0 \\ 1 \end{bmatrix}, \qquad \mathbf{p}_3 = \begin{bmatrix} 1 \\ 0 \end{bmatrix},$$

$$\mathbf{p}_4 = \begin{bmatrix} 1 \\ 2 \end{bmatrix}, \qquad \mathbf{p}_5 = \begin{bmatrix} -1 \\ 2 \end{bmatrix}.$$

Use both methods from Section 18.7.

7. Give an example of a nonsimple polygon that will pass the test for convexity, which uses the barycenter from Section 18.3.

8. Find an estimate normal to the nonplanar polygon

$$\mathbf{p}_1 = \begin{bmatrix} 0 \\ 0 \\ 1 \end{bmatrix}, \qquad \mathbf{p}_2 = \begin{bmatrix} 1 \\ 0 \\ 0 \end{bmatrix}, \qquad \mathbf{p}_3 = \begin{bmatrix} 1 \\ 1 \\ 1 \end{bmatrix}, \qquad \mathbf{p}_4 = \begin{bmatrix} 0 \\ 1 \\ 0 \end{bmatrix}.$$

9. The following points are the vertices of a polygon that should lie in a plane:

$$\mathbf{p}_1 = \begin{bmatrix} 2 \\ 0 \\ -2 \end{bmatrix}, \qquad \mathbf{p}_2 = \begin{bmatrix} 3 \\ 2 \\ -4 \end{bmatrix}, \qquad \mathbf{p}_3 = \begin{bmatrix} 2 \\ 4 \\ -4 \end{bmatrix},$$

$$\mathbf{p}_4 = \begin{bmatrix} 0 \\ 3 \\ -1 \end{bmatrix}, \qquad \mathbf{p}_5 = \begin{bmatrix} 0 \\ 2 \\ -1 \end{bmatrix}, \qquad \mathbf{p}_6 = \begin{bmatrix} 0 \\ 0 \\ 0 \end{bmatrix}.$$

However, one point lies outside this plane. Which point is the *outlier*?[2] Which planarity test is the most suited to this problem?

---

[2]This term is frequently used to refer to noisy, inaccurate data from a laser scanner.

19

Conics

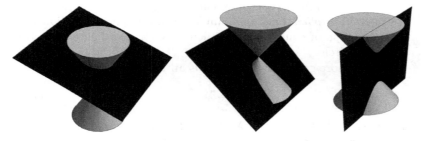

Figure 19.1.
Conic sections: three types of curves formed by the intersection of a plane and a cone.
From left to right: ellipse, parabola, and hyperbola.

Take a flashlight and shine it straight onto a wall. You will see a circle. Tilt the light, and the circle will turn into an ellipse. Tilt further, and the ellipse will become more and more elongated, and will become a parabola eventually. Tilt a little more, and you will have a hyperbola—actually one branch of it. The beam of your flashlight is a *cone*, and the image it generates on the wall is the intersection of that cone with a *plane* (i.e., the wall). Thus, we have the name *conic section* for curves that are the intersections of cones and planes. Figure 19.1 illustrates this idea.

The three types, ellipses, parabolas, and hyperbolas, arise in many situations and are the subject of this chapter. The basic tools for handling them are nothing but the matrix theory developed earlier.

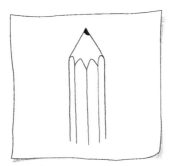

Sketch 19.1.
A pencil with hyperbolic arcs.

Before we delve into the theory of conic sections, we list some "real-life" occurrences.

- The paths of the planets around the sun are ellipses.

- If you sharpen a pencil, you generate hyperbolas (see Sketch 19.1).

- If you water your lawn, the water leaving the hose traces a parabolic arc.

19.1 The General Conic

We know that all points $\mathbf{x}$ satisfying

$$x_1^2 + x_2^2 = r^2 \tag{19.1}$$

are on a *circle* of radius r, centered at the origin. This type of equation is called an *implicit equation*. Similar to the implicit equation for a line, this type of equation is satisfied only for coordinate pairs that lie on the circle.

A little more generality will give us an *ellipse*:

$$\lambda_1 x_1^2 + \lambda_2 x_2^2 = c. \tag{19.2}$$

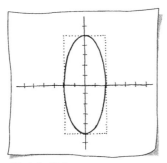

Sketch 19.2.
An ellipse with $\lambda_1 = 1/4$, $\lambda_2 = 1/25$, and c = 1.

The positive factors λ_1 and λ_2 denote how much the ellipse deviates from a circle. For example, if $\lambda_1 > \lambda_2$, the ellipse is more elongated in the x_2-direction. See Sketch 19.2 for the example

$$\frac{1}{4}x_1^2 + \frac{1}{25}x_2^2 = 1.$$

An ellipse in the form of (19.2) is said to be in *standard position* because its *minor and major axes* are coincident with the coordinate axes and the *center* is at the origin. The ellipse is symmetric about the major and minor axes. The semi-major and semi-minor axes are one-half the respective major and minor axes. In standard position, the ellipse lives in the rectangle with

$$x_1 \text{ extents } [-\sqrt{c/\lambda_1}, \sqrt{c/\lambda_1}] \text{ and } x_2 \text{ extents } [-\sqrt{c/\lambda_2}, \sqrt{c/\lambda_2}].$$

We will now rewrite (19.2) in matrix form:

$$\begin{bmatrix} x_1 & x_2 \end{bmatrix} \begin{bmatrix} \lambda_1 & 0 \\ 0 & \lambda_2 \end{bmatrix} \begin{bmatrix} x_1 \\ x_2 \end{bmatrix} - c = 0. \tag{19.3}$$

You will see the wisdom of this in a short while. This equation allows for significant compaction:

$$\mathbf{x}^T D \mathbf{x} - c = 0. \tag{19.4}$$

Example 19.1

Let's start with the ellipse $2x_1^2 + 4x_2^2 - 1 = 0$. In matrix form, corresponding to (19.3), we have

$$\begin{bmatrix} x_1 & x_2 \end{bmatrix} \begin{bmatrix} 2 & 0 \\ 0 & 4 \end{bmatrix} \begin{bmatrix} x_1 \\ x_2 \end{bmatrix} - 1 = 0. \tag{19.5}$$

This ellipse in standard position is shown in Figure 19.2 (left). The major axis is on the $\mathbf{e}_1$-axis with extents $[-1/\sqrt{2}, 1/\sqrt{2}]$ and the minor axis is on the $\mathbf{e}_2$-axis with extents $[-1/2, 1/2]$.

Suppose we encounter an ellipse with center at the origin, but with minor and major axes not aligned with the coordinate axes, as in Figure 19.2 (middle). What is the equation of such an ellipse? Points $\hat{\mathbf{x}}$ on this ellipse are mapped from an ellipse in standard position via a rotation, $\hat{\mathbf{x}} = R\mathbf{x}$. Using the fact that a rotation matrix is orthogonal, we replace $\mathbf{x}$ by $R^T \hat{\mathbf{x}}$ in (19.4), and the rotated conic takes the form

$$[R^T \hat{\mathbf{x}}]^T D [R^T \hat{\mathbf{x}}] - c = 0,$$

which becomes

$$\hat{\mathbf{x}}^T R D R^T \hat{\mathbf{x}} - c = 0. \tag{19.6}$$

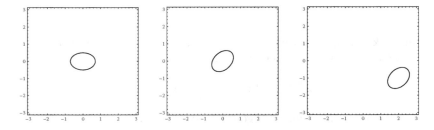

Figure 19.2.

Conic section: an ellipse in three positions. From left to right: standard position as given in (19.5), with a $45°$ rotation, and with a $45°$ rotation and translation of $[2, -1]^T$.

An abbreviation of

$$A = RDR^{\mathrm{T}} \tag{19.7}$$

shortens (19.6) to

$$\hat{\mathbf{x}}^{\mathrm{T}} A \hat{\mathbf{x}} - c = 0. \tag{19.8}$$

There are a couple of things about (19.8) that should look familiar. Note that A is a symmetric matrix. While studying the geometry of symmetric matrices in Section 7.5, we discovered that (19.7) is the eigendecompostion of A. The diagonal matrix D was called Λ there. One slight difference: convention is that the diagonal elements of Λ satisfy $\lambda_{1,1} \geq \lambda_{2,2}$. The matrix D does not; that would result in all ellipses in standard position having major axis on the $\mathbf{e}_2$-axis, as is the case with the example in Sketch 19.2. The curve defined in (19.8) is a contour of a quadratic form as described in Section 7.6. In fact, the figures of conics in this chapter were created as contours of quadratic forms. See Figure 7.8.

Now suppose we encounter an ellipse as in Figure 19.2 (right), rotated and translated out of standard position. What is the equation of this ellipse? Points $\hat{\mathbf{x}}$ on this ellipse are mapped from an ellipse in standard position via a rotation and then a translation, $\hat{\mathbf{x}} = R\mathbf{x} + \mathbf{v}$. Again using the fact that a rotation matrix is orthogonal, we replace $\mathbf{x}$ by $R^{\mathrm{T}}(\hat{\mathbf{x}} - \mathbf{v})$ in (19.4), and the rotated conic takes the form

$$[R^{\mathrm{T}}(\hat{\mathbf{x}} - \mathbf{v})]^T D [R^{\mathrm{T}}(\hat{\mathbf{x}} - \mathbf{v})] - c = 0,$$

or

$$[\hat{\mathbf{x}}^{\mathrm{T}} - \mathbf{v}^{\mathrm{T}}] R D R^{\mathrm{T}} [\hat{\mathbf{x}} - \mathbf{v}] - c = 0.$$

Again, shorten this with the definition of A in (19.7),

$$[\hat{\mathbf{x}}^{\mathrm{T}} - \mathbf{v}^{\mathrm{T}}] A [\hat{\mathbf{x}} - \mathbf{v}] - c = 0.$$

Since $\hat{\mathbf{x}}$ is simply a variable, we may drop the "hat" notation. The symmetry of A results in equality of $\mathbf{x}^{\mathrm{T}} A \mathbf{v}$ and $\mathbf{v}^{\mathrm{T}} A \mathbf{x}$, and we obtain

$$\mathbf{x}^{\mathrm{T}} A \mathbf{x} - 2 \mathbf{x}^{\mathrm{T}} A \mathbf{v} + \mathbf{v}^{\mathrm{T}} A \mathbf{v} - c = 0. \tag{19.9}$$

This denotes an ellipse in general position and it may be slightly abbreviated as

$$\mathbf{x}^{\mathrm{T}} A \mathbf{x} - 2 \mathbf{x}^{\mathrm{T}} \mathbf{b} + d = 0, \tag{19.10}$$

with $\mathbf{b} = A\mathbf{v}$ and $d = \mathbf{v}^{\mathrm{T}} A \mathbf{v} - c$.

If we relabel d and the elements of A and $\mathbf{b}$ from (19.10), the equation takes the form

$$\begin{bmatrix} x_1 & x_2 \end{bmatrix} \begin{bmatrix} c_1 & \frac{1}{2}c_3 \\ \frac{1}{2}c_3 & c_2 \end{bmatrix} \begin{bmatrix} x_1 \\ x_2 \end{bmatrix} - 2 \begin{bmatrix} x_1 & x_2 \end{bmatrix} \begin{bmatrix} -\frac{1}{2}c_4 \\ -\frac{1}{2}c_5 \end{bmatrix} + c_6 = 0. \quad (19.11)$$

Expanding this, we arrive at a familiar equation of a conic,

$$c_1 x_1^2 + c_2 x_2^2 + c_3 x_1 x_2 + c_4 x_1 + c_5 x_2 + c_6 = 0. \quad (19.12)$$

In fact, many texts simply start out by using (19.12) as the initial definition of a conic.

Example 19.2

Let's continue with the ellipse from Example 19.1. We have $\mathbf{x}^T D \mathbf{x} - 1 = 0$, where

$$D = \begin{bmatrix} 2 & 0 \\ 0 & 4 \end{bmatrix}.$$

This ellipse is illustrated in Figure 19.2 (left). Now we rotate by $45°$, using the rotation matrix

$$R = \begin{bmatrix} s & -s \\ s & s \end{bmatrix}$$

with $s = \sin 45° = \cos 45° = 1/\sqrt{2}$. The matrix $A = RDR^T$ becomes

$$A = \begin{bmatrix} 3 & -1 \\ -1 & 3 \end{bmatrix}.$$

This ellipse, $\mathbf{x}^T A \mathbf{x} - 1 = 0$, is illustrated in Figure 19.2 (middle). We could also write it as $3x_1^2 - 2x_1 x_2 + 3x_2^2 - 1 = 0$.

If we now translate by a vector

$$\mathbf{v} = \begin{bmatrix} 2 \\ -1 \end{bmatrix},$$

then (19.10) gives the recipe for adding translation terms, and the ellipse is now

$$\mathbf{x}^T \begin{bmatrix} 3 & -1 \\ -1 & 3 \end{bmatrix} \mathbf{x} - 2\mathbf{x}^T \begin{bmatrix} 7 \\ -5 \end{bmatrix} + 18 = 0.$$

Expanding this equation,

$$3x_1^2 - 2x_1x_2 + 3x_2^2 - 14x_1 + 10x_2 + 18 = 0.$$

This ellipse is illustrated in Figure 19.2 (right).

This was a lot of work just to find the general form of an ellipse! However, as we shall see, a lot more has been achieved here; the form (19.10) does not just represent ellipses, but *any* conic. To see that, let's examine the two remaining conic types: hyperbolas and parabolas.

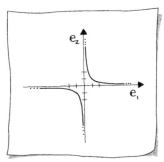

Sketch 19.3.
A hyperbola.

Example 19.3

Sketch 19.3 illustrates the conic

$$x_1x_2 - 1 = 0$$

or the more familiar form,

$$x_2 = \frac{1}{x_1},$$

which is a hyperbola. This may be written in matrix form as

$$\begin{bmatrix} x_1 & x_2 \end{bmatrix} \begin{bmatrix} 0 & \frac{1}{2} \\ \frac{1}{2} & 0 \end{bmatrix} \begin{bmatrix} x_1 \\ x_2 \end{bmatrix} - 1 = 0.$$

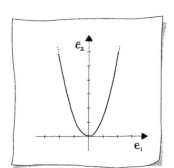

Sketch 19.4.
A parabola.

Example 19.4

The parabola

$$x_1^2 - x_2 = 0$$

or

$$x_2 = x_1^2,$$

is illustrated in Sketch 19.4. In matrix form, it is

$$\begin{bmatrix} x_1 & x_2 \end{bmatrix} \begin{bmatrix} -1 & 0 \\ 0 & 0 \end{bmatrix} \begin{bmatrix} x_1 \\ x_2 \end{bmatrix} + \begin{bmatrix} x_1 & x_2 \end{bmatrix} \begin{bmatrix} 0 \\ 1 \end{bmatrix} = 0.$$

19.2   Analyzing Conics

If you are given the equation of a conic as in (19.12), how can you tell which of the three basic types it is? The determinant of the 2×2 matrix A in (19.11) reveals the type:

- If $\det A > 0$, then the conic is an ellipse.

- If $\det A = 0$, then the conic is a parabola.

- If $\det A < 0$, then the conic is an hyperbola.

If A is the zero matrix and either c_4 or c_5 are nonzero, then the conic is degenerate and simply consists of a straight line.

Since A is a symmetric matrix, it has an eigendecomposition, $A = RDR^{\mathrm{T}}$. The eigenvalues of A, which are the diagonal elements of D, also characterize the conic type.

- Two nonzero entries of the same sign: ellipse.

- One nonzero entry: parabola.

- Two nonzero entries with opposite signs: hyperbola.

Of course these conditions are summarized by the determinant of D. Rotations do not change areas (determinants), thus checking A suffices and it is not necessary to find D.

Example 19.5

Let's check the type for the examples of the last section.

Example 19.2: we encountered the ellipse in this example in two forms, in standard position and rotated,

$$\begin{vmatrix} 2 & 0 \\ 0 & 4 \end{vmatrix} = \begin{vmatrix} 3 & -1 \\ -1 & 3 \end{vmatrix} = 8.$$

The determinant is positive, confirming that we have an ellipse.

Example 19.3:

$$\begin{vmatrix} 0 & \frac{1}{2} \\ \frac{1}{2} & 0 \end{vmatrix} = -\frac{1}{4},$$

confirming that we have a hyperbola. The characteristic equation for this matrix is $(\lambda + 1/2)(\lambda - 1/2) = 0$, thus the eigenvalues have opposite sign.

Example 19.4:

$$\begin{vmatrix} 1 & 0 \\ 0 & 0 \end{vmatrix} = 0,$$

confirming that we have a parabola. The characteristic equation for this matrix is $\lambda(\lambda + 1) = 0$, thus one eigenvalue is zero.

We have derived the general conic and folded this into a tool to determine its type. What might not be obvious: we found that affine maps, $M\mathbf{x} + \mathbf{v}$, where M is invertible, take a particular type of (non-degenerate) conic to another one of the same type. The conic type is determined by the sign of the determinant of A and it is unchanged by affine maps.

19.3   General Conic to Standard Position

If we are given a general conic equation as in (19.12), how do we find its equation in standard position? For an ellipse, this means that the center is located at the origin and the major and minor axes correspond with the coordinate axes. This leaves a degree of freedom: the major axis can coincide with the $\mathbf{e}_1$- or $\mathbf{e}_2$-axis.

Let's work with the ellipse from the previous sections,

$$3x_1^2 - 2x_1 x_2 + 3x_2^2 - 14x_1 + 10x_2 + 18 = 0.$$

It is illustrated in Figure 19.2 (right). Upon converting this equation to the form in (19.11), we have

$$\mathbf{x}^{\mathrm{T}} \begin{bmatrix} 3 & -1 \\ -1 & 3 \end{bmatrix} \mathbf{x} - 2\mathbf{x}^{\mathrm{T}} \begin{bmatrix} 7 \\ -5 \end{bmatrix} + 18 = 0.$$

Breaking down this equation into the elements of (19.10), we have

$$A = \begin{bmatrix} 3 & -1 \\ -1 & 3 \end{bmatrix} \quad \text{and} \quad \mathbf{b} = \begin{bmatrix} 7 \\ -5 \end{bmatrix}.$$

This means that the translation $\mathbf{v}$ may be found by solving the 2×2 linear system

$$A\mathbf{v} = \mathbf{b},$$

which in this case is

$$\mathbf{v} = \begin{bmatrix} 2 \\ -1 \end{bmatrix}.$$

This linear system may be solved if A has full rank. This is equivalent to A having two nonzero eigenvalues, and so the given conic is either an ellipse or a hyperbola.

Calculating $c = \mathbf{v}^T A \mathbf{v} - d$ from (19.10) and removing the translation terms, the ellipse with center at the origin is

$$\mathbf{x}^T \begin{bmatrix} 3 & -1 \\ -1 & 3 \end{bmatrix} \mathbf{x} - 1 = 0.$$

The eigenvalues of this 2×2 matrix are the solutions of the characteristic equation

$$\lambda^2 - 6\lambda + 8 = 0,$$

and thus are $\lambda_1 = 4$ and $\lambda_2 = 2$, resulting in

$$D = \begin{bmatrix} 4 & 0 \\ 0 & 2 \end{bmatrix}.$$

The convention established for building decompositions (e.g., eigen or SVD) is to order the eigenvalues in decreasing value. Either this D or the matrix in (19.5) suffice to define this ellipse in standard position. This one,

$$\mathbf{x}^T \begin{bmatrix} 4 & 0 \\ 0 & 2 \end{bmatrix} \mathbf{x} - 1 = 0,$$

will result in the major axis aligned with the $\mathbf{e}_2$-axis.

If we want to find the rotation that resulted in the general conic, then we find the eigenvectors of A. The homogeneous linear systems introduce degrees of freedom in selection of the eigenvectors. Some choices correspond to reflection and rotation combinations. (We discussed this in Section 7.5.) However, any choice of R will define orthogonal semi-major and semi-minor axis directions. Continuing with our example, R corresponds to a $-45°$ rotation,

$$R = \frac{1}{\sqrt{2}} \begin{bmatrix} 1 & 1 \\ -1 & 1 \end{bmatrix}.$$

Example 19.6

Given the conic section

$$x_1^2 + 2x_2^2 + 8x_1 x_2 - 4x_1 - 16x_2 + 3 = 0,$$

let's find its type and its equation in standard position.

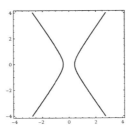 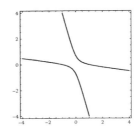 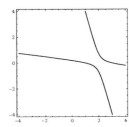

Figure 19.3.

Conic section: a hyperbola in three positions. From left to right: standard position, with rotation, with rotation and translation.

First let's write this conic in matrix form,

$$\mathbf{x}^{\mathrm{T}} \begin{bmatrix} 1 & 4 \\ 4 & 2 \end{bmatrix} \mathbf{x} - 2\mathbf{x}^{\mathrm{T}} \begin{bmatrix} 2 \\ 8 \end{bmatrix} + 3 = 0.$$

Since the determinant of the matrix is negative, we know this is a hyperbola. It is illustrated in Figure 19.3 (right).

We recover the translation $\mathbf{v}$ by solving the linear system

$$\begin{bmatrix} 1 & 4 \\ 4 & 2 \end{bmatrix} \mathbf{v} = \begin{bmatrix} 2 \\ 8 \end{bmatrix},$$

resulting in

$$\mathbf{v} = \begin{bmatrix} 2 \\ 0 \end{bmatrix}.$$

Calculate $c = \mathbf{v}^{\mathrm{T}} A \mathbf{v} - 3 = 1$, then the conic without the translation is

$$\mathbf{x}^{\mathrm{T}} \begin{bmatrix} 1 & 4 \\ 4 & 2 \end{bmatrix} \mathbf{x} - 1 = 0.$$

This hyperbola is illustrated in Figure 19.3 (middle).

The characteristic equation of the matrix is $\lambda^2 - 3\lambda - 14 = 0$; its roots are $\lambda_1 = 5.53$ and $\lambda_2 = -2.53$. The hyperbola in standard position is

$$\mathbf{x}^{\mathrm{T}} \begin{bmatrix} 5.53 & 0 \\ 0 & -2.53 \end{bmatrix} \mathbf{x} - 1 = 0.$$

This hyperbola is illustrated in Figure 19.3 (left).

- conic section
- implicit equation
- circle
- quadratic form
- ellipse
- minor axis
- semi-minor axis
- major axis
- semi-major axis
- center
- standard position
- conic type
- hyperbola
- parabola
- straight line
- eigenvalues
- eigenvectors
- eigendecomposition
- affine invariance

19.4   Exercises

1. What is the matrix form of a circle with radius r in standard position?

2. What is the equation of an ellipse that is centered at the origin and has $\mathbf{e}_1$-axis extents of $[-5, 5]$ and $\mathbf{e}_2$-axis extents of $[-2, 2]$?

3. What are the $\mathbf{e}_1$- and $\mathbf{e}_2$-axis extents of the ellipse

$$16x_1^2 + 4x_2^2 - 4 = 0?$$

4. What is the implicit equation of the ellipse

$$10x_1^2 + 2x_2^2 - 4 = 0$$

when it is rotated $90°$?

5. What is the implicit equation of the hyperbola

$$10x_1^2 - 2x_2^2 - 4 = 0$$

when it is rotated $45°$?

6. What is the implicit equation of the conic

$$10x_1^2 - 2x_2^2 - 4 = 0$$

rotated by $45°$ and translated by $[2 \ \ 2]^{\mathrm{T}}$?

7. Expand the matrix form

$$\mathbf{x}^{\mathrm{T}} \begin{bmatrix} -5 & -1 \\ -1 & -6 \end{bmatrix} \mathbf{x} - 2\mathbf{x}^{\mathrm{T}} \begin{bmatrix} -1 \\ 0 \end{bmatrix} + 3 = 0$$

of the equation of a conic.

8. Expand the matrix form

$$\mathbf{x}^{\mathrm{T}} \begin{bmatrix} 1 & 3 \\ 3 & 3 \end{bmatrix} \mathbf{x} - 2\mathbf{x}^{\mathrm{T}} \begin{bmatrix} 3 \\ 3 \end{bmatrix} + 9 = 0$$

of the equation of a conic.

9. What is the eigendecomposition of

$$A = \begin{bmatrix} 4 & 2 \\ 2 & 4 \end{bmatrix}?$$

10. What is the eigendecomposition of

$$A = \begin{bmatrix} 5 & 0 \\ 0 & 5 \end{bmatrix}?$$

11. Let $x_1^2 - 2x_1x_2 - 4 = 0$ be the equation of a conic section. What type is it?

12. Let $x_1^2 + 2x_1x_2 - 4 = 0$ be the equation of a conic section. What type is it?

13. Let $2x_1^2 + x_2 - 5 = 0$ be the equation of a conic section. What type is it?

14. Let a conic be given by

$$3x_1^2 + 2x_1x_2 + 3x_2^2 + 10x_1 - 2x_2 + 10 = 0.$$

Write it in matrix form. What type of conic is it? What is the rotation and translation that took it out of standard position? Write it in matrix form in standard position.

15. What affine map takes the circle

$$(x_1 - 3)^2 + (x_2 + 1)^2 - 4 = 0$$

to the ellipse

$$2x_1^2 + 4x_2^2 - 1 = 0?$$

16. How many intersections does a straight line have with a conic? Given a conic in the form (19.2) and a parametric form of a line $\mathbf{l}(t)$, what are the t-values of the intersection points? Explain any singularities.

17. If the shear

$$\begin{bmatrix} 1 & 1 \\ 1/2 & 0 \end{bmatrix}$$

is applied to the conic of Example 19.1, what is the type of the resulting conic?

20

Curves

Figure 20.1.
Car design: curves are used to design cars such as the Ford Synergy 2010 concept car. (Source http://www.ford.com.)

Earlier in this book, we mentioned that all letters that you see here were designed by a font designer, and then put into a font library. The font designer's main tool is a cubic curve, also called a cubic Bézier curve. Such curves are handy for font design, but they were initially invented for car design. This happened in France in the early 1960s at Rénault and Citroën in Paris. These techniques are still in use today, as illustrated in Figure 20.1. We will briefly outline this kind of curve, and also apply previous linear algebra and geometric concepts to the study of curves in general. This type of work is called *geometric modeling* or *computer-aided geometric design*; see an introductory text such as [7]. Please keep in mind: this chapter just scratches the surface of the modeling field!

20.1   Parametric Curves

You will recall that one way to write a straight line is the *parametric form*:

$$\mathbf{x}(t) = (1-t)\mathbf{a} + t\mathbf{b}.$$

If we interpret t as time, then this says at time $t = 0$ a moving point is at $\mathbf{a}$. It moves toward $\mathbf{b}$, and reaches it at time $t = 1$. You might have observed that the coefficients $(1-t)$ and t are linear, or degree 1, polynomials, which explains another name for this: *linear interpolation*. So we have a simple example of a *parametric curve*: a curve that can be written as

$$\mathbf{x}(t) = \begin{bmatrix} f(t) \\ g(t) \end{bmatrix},$$

where $f(t)$ and $g(t)$ are functions of the parameter t. For the linear interpolant above, $f(t) = (1-t)a_1 + tb_1$ and $g(t) = (1-t)a_2 + tb_2$. In general, f and g can be any functions, e.g., polynomial, trigonometric, or exponential. However, in this chapter we will be looking at polynomial f and g.

Let us be a bit more ambitious now and study motion along *curves*, i.e., paths that do not have to be straight. The simplest example is that of driving a car along a road. At time $t = 0$, you start, you follow the road, and at time $t = 1$, you have arrived somewhere. It does not really matter what kind of units we use to measure time; the $t = 0$ and $t = 1$ may just be viewed as a normalization of an arbitrary time interval.

We will now attack the problem of modeling curves, and we will choose a particularly simple way of doing this, namely *cubic Bézier curves*. We start with four points in 2D or 3D, $\mathbf{b}_0, \mathbf{b}_1, \mathbf{b}_2,$ and $\mathbf{b}_3$, called *Bézier (control) points*. Connect them with straight lines as shown in Sketch 20.1. The resulting polygon is called a *Bézier (control) polygon*.[1]

The four control points, $\mathbf{b}_0, \mathbf{b}_1, \mathbf{b}_2, \mathbf{b}_3$, define a cubic curve, and some examples are illustrated in Figure 20.2. To create these plots, we *evaluate* the cubic curve at many t-parameters that range between

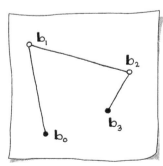

Sketch 20.1.
A Bézier polygon.

---

[1]Bézier polygons are not assumed to be closed as were the polygons of Section 18.2.

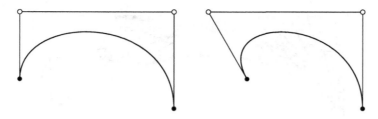

Figure 20.2.
Bézier curves: two examples that differ in the location of one control point, $\mathbf{b}_0$ only.

zero and one, $t \in [0, 1]$. If we evaluate at 50 points, then we would find points on the curve associated with

$$t = 0, 1/50, 2/50, \ldots, 49/50, 1,$$

and then these points are connected by straight line segments to make the curve look smooth. The points are so close together that you cannot detect the line segments. In other words, we plot a *discrete approximation* of the curve.

Here is how you generate one point on a cubic Bézier curve. Pick a parameter value t between 0 and 1. Find the corresponding point on each polygon leg by linear interpolation:

$$\mathbf{b}_0^1(t) = (1 - t)\mathbf{b}_0 + t\mathbf{b}_1,$$
$$\mathbf{b}_1^1(t) = (1 - t)\mathbf{b}_1 + t\mathbf{b}_2,$$
$$\mathbf{b}_2^1(t) = (1 - t)\mathbf{b}_2 + t\mathbf{b}_3.$$

These three points form a polygon themselves. Now repeat the linear interpolation process, and you get two points:

$$\mathbf{b}_0^2 = (1 - t)\mathbf{b}_0^1(t) + t\mathbf{b}_1^1(t),$$
$$\mathbf{b}_1^2 = (1 - t)\mathbf{b}_1^1(t) + t\mathbf{b}_2^1(t).$$

Repeat one more time,

$$\mathbf{b}_0^3(t) = (1 - t)\mathbf{b}_0^2(t) + t\mathbf{b}_1^2(t), \tag{20.1}$$

and you have a point on the Bézier curve defined by $\mathbf{b}_0, \mathbf{b}_1, \mathbf{b}_2, \mathbf{b}_3$ at the parameter value t. The recursive process of applying linear interpolation is called the *de Casteljau algorithm*, and the steps above are shown in Sketch 20.2. Figure 20.3 illustrates all de Casteljau steps

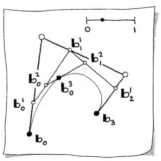

Sketch 20.2.
The de Casteljau algorithm.

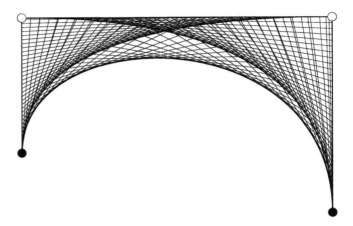

Figure 20.3.

Bézier curves: all intermediate Bézier points generated by the de Casteljau algorithm for 33 evaluations.

for 33 evaluations. The points $\mathbf{b}_i^j$ are often called *intermediate Bézier points*, and the following schematic is helpful in keeping track of how each point is generated.

$$
\begin{array}{ccccc}
\mathbf{b}_0 & & & \\
\mathbf{b}_1 & \mathbf{b}_0^1 & & \\
\mathbf{b}_2 & \mathbf{b}_1^1 & \mathbf{b}_0^2 & \\
\mathbf{b}_3 & \mathbf{b}_2^1 & \mathbf{b}_1^2 & \mathbf{b}_0^3 \\
\text{stage}: & 1 & 2 & 3
\end{array}
$$

Except for the (input) Bézier polygon, each point in the schematic is a function of t.

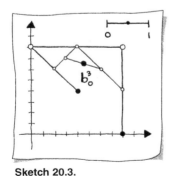

Sketch 20.3.

Evaluation via the de Casteljau algorithm at t = 1/2.

Example 20.1

A numerical counterpart to Sketch 20.3 follows. Let the polygon be given by

$$
\mathbf{b}_0 = \begin{bmatrix} 4 \\ 4 \end{bmatrix}, \quad \mathbf{b}_1 = \begin{bmatrix} 0 \\ 8 \end{bmatrix}, \quad \mathbf{b}_2 = \begin{bmatrix} 8 \\ 8 \end{bmatrix}, \quad \mathbf{b}_3 = \begin{bmatrix} 8 \\ 0 \end{bmatrix}.
$$

For simplicity, let $t = 1/2$. Linear interpolation is then nothing but finding midpoints, and we have the following intermediate Bézier

points,

$$\mathbf{b}_0^1 = \frac{1}{2}\mathbf{b}_0 + \frac{1}{2}\mathbf{b}_1 = \begin{bmatrix} 2 \\ 6 \end{bmatrix},$$

$$\mathbf{b}_1^1 = \frac{1}{2}\mathbf{b}_1 + \frac{1}{2}\mathbf{b}_2 = \begin{bmatrix} 4 \\ 8 \end{bmatrix},$$

$$\mathbf{b}_2^1 = \frac{1}{2}\mathbf{b}_2 + \frac{1}{2}\mathbf{b}_3 = \begin{bmatrix} 8 \\ 4 \end{bmatrix}.$$

Next,

$$\mathbf{b}_0^2 = \frac{1}{2}\mathbf{b}_0^1 + \frac{1}{2}\mathbf{b}_1^1 = \begin{bmatrix} 3 \\ 7 \end{bmatrix},$$

$$\mathbf{b}_1^2 = \frac{1}{2}\mathbf{b}_1^1 + \frac{1}{2}\mathbf{b}_2^1 = \begin{bmatrix} 6 \\ 6 \end{bmatrix},$$

and finally

$$\mathbf{b}_0^3 = \frac{1}{2}\mathbf{b}_0^2 + \frac{1}{2}\mathbf{b}_1^2 = \begin{bmatrix} \frac{9}{2} \\ \frac{13}{2} \end{bmatrix}.$$

This is the point on the curve corresponding to $t = 1/2$.

The equation for a cubic Bézier curve is found by expanding (20.1):

$$\begin{aligned}
\mathbf{b}_0^3 &= (1-t)\mathbf{b}_0^2 + t\mathbf{b}_1^2 \\
&= (1-t)\left[(1-t)\mathbf{b}_0^1 + t\mathbf{b}_1^1\right] + t\left[(1-t)\mathbf{b}_1^1 + t\mathbf{b}_2^1\right] \\
&= (1-t)\left[(1-t)\left[(1-t)\mathbf{b}_0 + t\mathbf{b}_1\right] + t\left[(1-t)\mathbf{b}_1 + t\mathbf{b}_2\right]\right] \\
&\quad + t\left[(1-t)\left[(1-t)\mathbf{b}_1 + t\mathbf{b}_2\right] + t\left[(1-t)\mathbf{b}_2 + t\mathbf{b}_3\right]\right].
\end{aligned}$$

After collecting terms with the same $\mathbf{b}_i$, this becomes

$$\mathbf{b}_0^3(t) = (1-t)^3\mathbf{b}_0 + 3(1-t)^2 t\mathbf{b}_1 + 3(1-t)t^2\mathbf{b}_2 + t^3\mathbf{b}_3. \quad (20.2)$$

This is the general form of a cubic Bézier curve. As t traces out values between 0 and 1, the point $\mathbf{b}_0^3(t)$ traces out a curve.

The polynomials in (20.2) are called the *Bernstein basis functions*:

$$B_0^3(t) = (1-t)^3,$$

$$B_1^3(t) = 3(1-t)^2 t,$$

$$B_2^3(t) = 3(1-t)t^2,$$

$$B_3^3(t) = t^3,$$

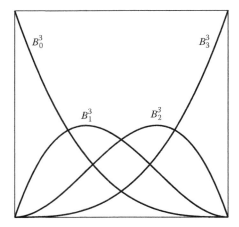

Figure 20.4.
Bernstein polynomials: a plot of the four cubic polynomials for $t \in [0, 1]$.

and they are illustrated in Figure 20.4. Now $\mathbf{b}_0^3(t)$ can be written as

$$\mathbf{b}_0^3(t) = B_0^3(t)\mathbf{b}_0 + B_1^3(t)\mathbf{b}_1 + B_2^3(t)\mathbf{b}_2 + B_3^3(t)\mathbf{b}_3. \qquad (20.3)$$

As we see, the Bernstein basis functions are cubic, or degree 3, polynomials. This set of polynomials is a bit different than the cubic *monomials*, $1, t, t^2, t^3$ that we are used to from calculus. However, either set allow us to write all cubic polynomials. We say that the Bernstein (or monomial) polynomials form a basis for all cubic polynomials, hence the name basis function. See Section 14.1 for more on bases. We'll look at the relationship between these sets of polynomials in Section 20.3.

The original curve was given by the control polygon

$$\mathbf{b}_0, \mathbf{b}_1, \mathbf{b}_2, \mathbf{b}_3.$$

Inspection of Sketch 20.3 suggests that a subset of the intermediate Bézier points form two cubic polygons that also mimic the curve's shape. The curve segment from $\mathbf{b}_0$ to $\mathbf{b}_0^3(t)$ has the polygon

$$\mathbf{b}_0, \mathbf{b}_0^1, \mathbf{b}_0^2, \mathbf{b}_0^3,$$

and the other from $\mathbf{b}_0^3(t)$ to $\mathbf{b}_3$ has the polygon

$$\mathbf{b}_0^3, \mathbf{b}_1^2, \mathbf{b}_2^1, \mathbf{b}_3.$$

This process of generating two Bézier curves from one, is called *sub-division*.

From now on, we will also use the shorter $\mathbf{b}(t)$ instead of $\mathbf{b}_0^3(t)$.

20.2 Properties of Bézier Curves

From inspection of the examples, but also from (20.2), we see that the curve passes through the first and last control points:

$$\mathbf{b}(0) = \mathbf{b}_0 \quad \text{and} \quad \mathbf{b}(1) = \mathbf{b}_3. \qquad (20.4)$$

Another way to say this: the curve *interpolates* to $\mathbf{b}_0$ and $\mathbf{b}_3$.

If we map the control polygon using an affine map, then the curve undergoes the same transformation, as shown in Figure 20.5. This is called *affine invariance*, and it is due to the fact that the cubic coefficients of the control points, the Bernstein polynomials, in (20.3) sum to one. This can be seen as follows:

$$(1-t)^3 + 3(1-t)^2 t + 3(1-t)t^2 + t^3 = [(1-t)+t]^3 = 1.$$

Thus, every point on the curve is a *barycentric combination* of the control points. Such relationships are not changed under affine maps, as per Section 6.2.

The curve, for $t \in [0,1]$, lies in the convex hull of the control polygon—a fact called the *convex hull property*. This can be seen

Figure 20.5.
Affine invariance: as the control polygon rotates, so does the curve.

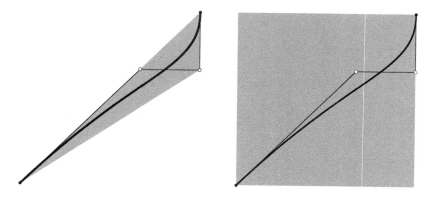

Figure 20.6.

Convex hull property: a Bézier curve lies in the convex hull of its control polygon. Left: shaded area fills in the convex hull of the control polygon. Right: shaded area fills in the minmax box of the control polygon. The convex hull lies inside the minmax box.

by observing in Figure 20.4 that the Bernstein polynomials in (20.2) are nonnegative for $t \in [0,1]$. It follows that every point on the curve is a *convex combination* of the control points, and hence is inside their *convex hull*. For a definition, see Section 17.4; for an illustration, see Figure 20.6 (left). If we evaluate the curve for t-values outside of $[0,1]$, this is called *extrapolation*. We can no longer predict the shape of the curve, so this procedure is normally not recommended.

Clearly, the control polygon is inside its minmax box.[2] Because of the convex hull property, we also know that the curve is inside this box—a property that has numerous applications. See Figure 20.6 (right) for an illustration.

20.3 The Matrix Form

As a preparation for what is to follow, let us rewrite (20.2) using the formalism of dot products. It then looks like this:

$$\mathbf{b}(t) = \begin{bmatrix} \mathbf{b}_0 & \mathbf{b}_1 & \mathbf{b}_2 & \mathbf{b}_3 \end{bmatrix} \begin{bmatrix} (1-t)^3 \\ 3(1-t)^2 t \\ 3(1-t)t^2 \\ t^3 \end{bmatrix}. \qquad (20.5)$$

---

[2]Recall that the minmax box of a polygon is the smallest rectangle with edges parallel to the coordinate axes that contains the polygon.

Instead of the Bernstein polynomials as above, most people think of polynomials as combinations of the *monomials*; they are $1, t, t^2, t^3$ for the cubic case. We can relate the Bernstein and monomial forms by rewriting our expression (20.2) as

$$\mathbf{b}(t) = \mathbf{b}_0 + 3t(\mathbf{b}_1 - \mathbf{b}_0) + 3t^2(\mathbf{b}_2 - 2\mathbf{b}_1 + \mathbf{b}_0) + t^3(\mathbf{b}_3 - 3\mathbf{b}_2 + 3\mathbf{b}_1 - \mathbf{b}_0).$$
$$(20.6)$$

This allows a concise formulation using matrices:

$$\mathbf{b}(t) = \begin{bmatrix} \mathbf{b}_0 & \mathbf{b}_1 & \mathbf{b}_2 & \mathbf{b}_3 \end{bmatrix} \begin{bmatrix} 1 & -3 & 3 & -1 \\ 0 & 3 & -6 & 3 \\ 0 & 0 & 3 & -3 \\ 0 & 0 & 0 & 1 \end{bmatrix} \begin{bmatrix} 1 \\ t \\ t^2 \\ t^3 \end{bmatrix}. \qquad (20.7)$$

This is the matrix form of a Bézier curve.

Equation (20.6) or (20.7) shows how to write a Bézier curve in monomial form. A curve in monomial form looks like this:

$$\mathbf{b}(t) = \mathbf{a}_0 + \mathbf{a}_1 t + \mathbf{a}_2 t^2 + \mathbf{a}_3 t^3.$$

Geometrically, the four control "points" for the curve in monomial form are now a mix of points and vectors: $\mathbf{a}_0 = \mathbf{b}_0$ is a point, but $\mathbf{a}_1, \mathbf{a}_2, \mathbf{a}_3$ are vectors. Using the dot product form, this becomes

$$\mathbf{b}(t) = \begin{bmatrix} \mathbf{a}_0 & \mathbf{a}_1 & \mathbf{a}_2 & \mathbf{a}_3 \end{bmatrix} \begin{bmatrix} 1 \\ t \\ t^2 \\ t^3 \end{bmatrix}.$$

Thus, the monomial coefficients $\mathbf{a}_i$ are defined as

$$\begin{bmatrix} \mathbf{a}_0 & \mathbf{a}_1 & \mathbf{a}_2 & \mathbf{a}_3 \end{bmatrix} = \begin{bmatrix} \mathbf{b}_0 & \mathbf{b}_1 & \mathbf{b}_2 & \mathbf{b}_3 \end{bmatrix} \begin{bmatrix} 1 & -3 & 3 & -1 \\ 0 & 3 & -6 & 3 \\ 0 & 0 & 3 & -3 \\ 0 & 0 & 0 & 1 \end{bmatrix}. \qquad (20.8)$$

How about the inverse process: If we are given a curve in monomial form, how can we write it as a Bézier curve? Simply rearrange (20.8) to solve for the $\mathbf{b}_i$:

$$\begin{bmatrix} \mathbf{b}_0 & \mathbf{b}_1 & \mathbf{b}_2 & \mathbf{b}_3 \end{bmatrix} = \begin{bmatrix} \mathbf{a}_0 & \mathbf{a}_1 & \mathbf{a}_2 & \mathbf{a}_3 \end{bmatrix} \begin{bmatrix} 1 & -3 & 3 & -1 \\ 0 & 3 & -6 & 3 \\ 0 & 0 & 3 & -3 \\ 0 & 0 & 0 & 1 \end{bmatrix}^{-1}.$$

A matrix inversion is all that is needed here. Notice that the square matrix in (20.8) is nonsingular, therefore we can conclude that any cubic curve can be written in either the Bézier or the monomial form.

20.4 Derivatives

Equation (20.2) consists of two (in 2D) or three (in 3D) cubic equations in t. We can take the derivative in each of the components:

$$
\frac{\mathrm{d}\mathbf{b}(t)}{\mathrm{d}t} = -3(1-t)^2\mathbf{b}_0 - 6(1-t)t\mathbf{b}_1 + 3(1-t)^2\mathbf{b}_1
$$
$$
- 3t^2\mathbf{b}_2 + 6(1-t)t\mathbf{b}_2 + 3t^2\mathbf{b}_3.
$$

Rearranging, and using the abbreviation $\frac{\mathrm{d}\mathbf{b}(t)}{\mathrm{d}t} = \dot{\mathbf{b}}(t)$, we have

$$
\dot{\mathbf{b}}(t) = 3(1-t)^2[\mathbf{b}_1 - \mathbf{b}_0] + 6(1-t)t[\mathbf{b}_2 - \mathbf{b}_1] + 3t^2[\mathbf{b}_3 - \mathbf{b}_2]. \quad (20.9)
$$

As expected, the derivative of a degree three curve is one of degree two.[3]

One very nice feature of the de Casteljau algorithm is that the intermediate Bézier points generated by it allow us to express the derivative very simply. A simpler expression than (20.9) for the derivative is

$$
\dot{\mathbf{b}}(t) = 3\left[\mathbf{b}_1^2(t) - \mathbf{b}_0^2(t)\right]. \quad (20.10)
$$

Getting the derivative for free makes the de Casteljau algorithm more efficient than evaluating (20.2) and (20.9) directly to get a point and a derivative.

At the endpoints, the derivative formula is very simple. For $t = 0$, we obtain

$$
\dot{\mathbf{b}}(0) = 3[\mathbf{b}_1 - \mathbf{b}_0],
$$

and, similarly, for $t = 1$,

$$
\dot{\mathbf{b}}(1) = 3[\mathbf{b}_3 - \mathbf{b}_2].
$$

In words, the control polygon is tangent to the curve at the curve's endpoints. This is not a surprising statement if you check Figure 20.2.

---

[3]Note that the derivative curve does not have control *points* anymore, but rather *control vectors!*

Example 20.2

Let us compute the derivative of the curve from Example 20.1 for $t = 1/2$. First, let's evaluate the direct equation (20.9). We obtain

$$\dot{\mathbf{b}}\left(\frac{1}{2}\right) = 3 \cdot \frac{1}{4}\left[\begin{bmatrix} 0 \\ 8 \end{bmatrix} - \begin{bmatrix} 4 \\ 4 \end{bmatrix}\right] + 6 \cdot \frac{1}{4}\left[\begin{bmatrix} 8 \\ 8 \end{bmatrix} - \begin{bmatrix} 0 \\ 8 \end{bmatrix}\right] + 3 \cdot \frac{1}{4}\left[\begin{bmatrix} 8 \\ 0 \end{bmatrix} - \begin{bmatrix} 8 \\ 8 \end{bmatrix}\right],$$

which yields

$$\dot{\mathbf{b}}\left(\frac{1}{2}\right) = \begin{bmatrix} 9 \\ -3 \end{bmatrix}.$$

If instead, we used (20.10), and thus the intermediate control points calculated in Example 20.1, we get

$$\begin{aligned}
\dot{\mathbf{b}}\left(\frac{1}{2}\right) &= 3\left[\mathbf{b}_1^2\left(\frac{1}{2}\right) - \mathbf{b}_0^2\left(\frac{1}{2}\right)\right] \\
&= 3\left[\begin{bmatrix} 6 \\ 6 \end{bmatrix} - \begin{bmatrix} 3 \\ 7 \end{bmatrix}\right] \\
&= \begin{bmatrix} 9 \\ -3 \end{bmatrix},
\end{aligned}$$

which is the same answer but with less work! See Sketch 20.4 for an illustration.

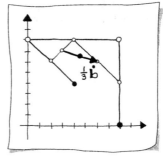

Sketch 20.4.
A derivative vector.

Note that the derivative of a curve is a *vector*. It is tangent to the curve—apparent from our example, but nothing we want to prove here. A convenient way to think about the derivative is by interpreting it as a *velocity vector*. If you interpret the parameter t as time, and you think of traversing the curve such that at time t you have reached $\mathbf{b}(t)$, then the derivative measures your velocity. The larger the magnitude of the tangent vector, the faster you move.

If we rotate the control polygon, the curve will follow, and so will all of its derivative vectors. In calculus, a "horizontal tangent" has a special meaning; it indicates an extreme value of a function. Not here: the very notion of an extreme value is meaningless for parametric curves since the term "horizontal tangent" depends on the curve's orientation and is not a property of the curve itself.

We may take the derivative of (20.9) with respect to t. We then have the *second derivative*. It is given by

$$\ddot{\mathbf{b}}(t) = -6(1-t)[\mathbf{b}_1 - \mathbf{b}_0] - 6t[\mathbf{b}_2 - \mathbf{b}_1] + 6(1-t)[\mathbf{b}_2 - \mathbf{b}_1] + 6t[\mathbf{b}_3 - \mathbf{b}_2]$$

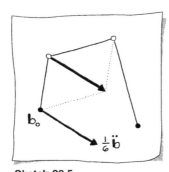

Sketch 20.5.

A second derivative vector.

and may be rearranged to

$$\ddot{\mathbf{b}}(t) = 6(1-t)[\mathbf{b}_2 - 2\mathbf{b}_1 + \mathbf{b}_0] + 6t[\mathbf{b}_3 - 2\mathbf{b}_2 + \mathbf{b}_1]. \qquad (20.11)$$

The de Casteljau algorithm supplies a simple way to write the second derivative, too:

$$\ddot{\mathbf{b}}(t) = 6\left[\mathbf{b}_2^1(t) - 2\mathbf{b}_1^1(t) + \mathbf{b}_0^1(t)\right]. \qquad (20.12)$$

Loosely speaking, we may interpret the second derivative $\ddot{\mathbf{b}}(t)$ as acceleration when traversing the curve.

The second derivative at $\mathbf{b}_0$ (see Sketch 20.5) is particularly simple—it is given by

$$\ddot{\mathbf{b}}(0) = 6[\mathbf{b}_2 - 2\mathbf{b}_1 + \mathbf{b}_0].$$

Notice that this is a scaling of the difference of the vectors $\mathbf{b}_1 - \mathbf{b}_0$ and $\mathbf{b}_2 - \mathbf{b}_1$. Recalling the *parallelogram rule*, the second derivative at $\mathbf{b}_0$ is easy to sketch. A similar equation holds for the second derivative at $t = 1$; now the points involved are $\mathbf{b}_1, \mathbf{b}_2, \mathbf{b}_3$.

20.5 Composite Curves

A Bézier curve is a handsome tool, but one such curve would rarely suffice for describing much of any shape! For "real" shapes, we have to be able to line up many cubic Bézier curves. In order to define a smooth overall curve, these pieces must join smoothly.

This is easily achieved. Let $\mathbf{b}_0, \mathbf{b}_1, \mathbf{b}_2, \mathbf{b}_3$ and $\mathbf{c}_0, \mathbf{c}_1, \mathbf{c}_2, \mathbf{c}_3$ be the control polygons of two Bézier curves with a common point $\mathbf{b}_3 = \mathbf{c}_0$ (see Sketch 20.6). If the two curves are to have the same tangent vector direction at $\mathbf{b}_3 = \mathbf{c}_0$, then all that is required is

$$\mathbf{c}_1 - \mathbf{c}_0 = c[\mathbf{b}_3 - \mathbf{b}_2] \qquad (20.13)$$

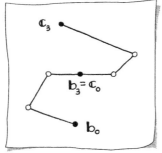

Sketch 20.6.

Smoothly joining Bézier curves.

for some positive real number c, meaning that the three points $\mathbf{b}_2, \mathbf{b}_3 = \mathbf{c}_0, \mathbf{c}_1$ are collinear.

If we use this rule to piece curve segments together, we can design many 2D and 3D shapes. Figure 20.7 gives an example.

20.6 The Geometry of Planar Curves

The geometry of planar curves is centered around one concept: their *curvature*. It is easily understood if you imagine driving a car along a road. For simplicity, let's assume you are driving with constant

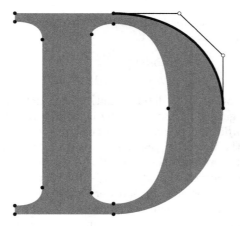

Figure 20.7.
Composite Bézier curves: the letter D as a collection of cubic Bézier curves. Only one Bézier polygon of many is shown.

speed. If the road does not curve, i.e., it is straight, you will not have to turn your steering wheel. When the road does curve, you will have to turn the steering wheel, and more so if the road curves rapidly. The curviness of the road (our model of a curve) is thus proportional to the turning of the steering wheel.

Returning to the more abstract concept of a curve, let us sample its tangents at various points (see Sketch 20.7). Where the curve bends sharply, i.e., where its curvature is high, successive tangents differ from each other significantly. In areas where the curve is relatively flat, or where its curvature is low, successive tangents are almost identical. Curvature may thus be defined as *rate of change of tangents*. (In terms of our car example, the rate of change of tangents is proportional to the turning of the steering wheel.)

Since the tangent is determined by the curve's first derivative, its rate of change should be determined by the second derivative. This is indeed so, but the actual formula for curvature is a bit more complex than can be derived in the context of this book. We denote the curvature of the curve at $\mathbf{b}(t)$ by κ; it is given by

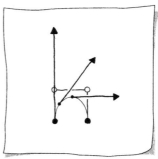

Sketch 20.7.
Tangents on a curve.

$$\kappa(t) = \frac{\|\dot{\mathbf{b}} \wedge \ddot{\mathbf{b}}\|}{\|\dot{\mathbf{b}}\|^3}. \tag{20.14}$$

This formula holds for both 2D and 3D curves. In the 2D case, it may be rewritten as

$$\kappa(t) = \frac{\left| \dot{\mathbf{b}} \quad \ddot{\mathbf{b}} \right|}{\left\| \dot{\mathbf{b}} \right\|^3} \tag{20.15}$$

with the use of a 2×2 determinant. Since determinants may be positive or negative, curvature in 2D is *signed*. A point where $\kappa = 0$ is called an *inflection point*: the 2D curvature changes sign here. In Figure 20.8, the inflection point is marked. In calculus, you learned that a curve has an inflection point if the second derivative vanishes. For parametric curves, the situation is different. An inflection point occurs when the first and second derivative vectors are parallel, or linearly dependent. This can lead to the curious effect of a cubic with *two* inflection points. It is illustrated in Figure 20.9.

Figure 20.8.
Inflection point: an inflection point, a point where the curvature changes sign, is marked on the curve.

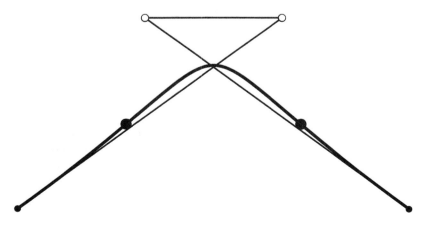

Figure 20.9.
Inflection point: a cubic with two inflection points.

Figure 20.10.
Curve motions: a letter is moved along a curve.

20.7   Moving along a Curve

Take a look at Figure 20.10. You will see the letter B sliding along a curve. If the curve is given in Bézier form, how can that effect be achieved? The answer can be seen in Sketch 20.8. If you want to position an object, such as the letter B, at a point on a curve, all you need to know is the point and the curve's tangent there. If $\dot{\mathbf{b}}$ is the tangent, then simply define $\mathbf{n}$ to be a vector perpendicular to it.[4] Using the local coordinate system with origin $\mathbf{b}(t)$ and $[\dot{\mathbf{b}}, \mathbf{n}]$-axes, you can position any object as in Section 4.1.

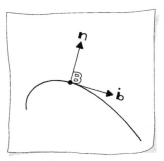

Sketch 20.8.
Sliding along a curve.

The same story is far trickier in 3D! If you had a point on the curve and its tangent, the exact location of your object would not be fixed; it could still rotate around the tangent. Yet there is a unique way to position objects along a 3D curve. At every point on the curve, we may define a *local coordinate system* as follows.

Let the point on the curve be $\mathbf{b}(t)$; we now want to set up a local coordinate system defined by three vectors $\mathbf{f}_1, \mathbf{f}_2, \mathbf{f}_3$. Following the 2D example, we set $\mathbf{f}_1$ to be in the tangent direction; $\mathbf{f}_1 = \dot{\mathbf{b}}(t)$. If the curve does not have an inflection point at t, then $\dot{\mathbf{b}}(t)$ and $\ddot{\mathbf{b}}(t)$ will not be collinear. This means that they span a plane, and that plane's normal is given by $\dot{\mathbf{b}}(t) \wedge \ddot{\mathbf{b}}(t)$. See Sketch 20.9 for some visual information. We make the plane's normal one of our local coordinate axes, namely $\mathbf{f}_3$. The plane, by the way, has a name: it is called the *osculating plane* at $\mathbf{x}(t)$. Since we have two coordinate axes, namely $\mathbf{f}_1$ and $\mathbf{f}_3$, it is not hard to come up with the remaining axis, we just set $\mathbf{f}_2 = \mathbf{f}_1 \wedge \mathbf{f}_3$. Thus, for every point on the curve (as long as it is not an inflection point), there exists an orthogonal coordinate system. It is customary to use coordinate axes of unit length, and then we have

$$\mathbf{f}_1 = \frac{\dot{\mathbf{b}}(t)}{\|\dot{\mathbf{b}}(t)\|}, \tag{20.16}$$

---

[4]If $\dot{\mathbf{b}} = \begin{bmatrix} \dot{b}_1 \\ \dot{b}_2 \end{bmatrix}$ then $\mathbf{n} = \begin{bmatrix} -\dot{b}_2 \\ \dot{b}_1 \end{bmatrix}$.

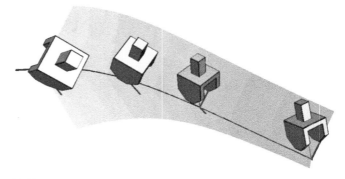

Figure 20.11.
Curve motions: a robot arm is moved along a curve. (Courtesy of M. Wagner, Arizona State University.)

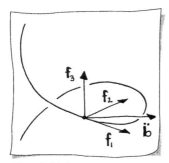

Sketch 20.9.
A Frenet frame.

$$\mathbf{f}_3 = \frac{\dot{\mathbf{b}}(t) \wedge \ddot{\mathbf{b}}(t)}{\|\dot{\mathbf{b}}(t) \wedge \ddot{\mathbf{b}}(t)\|}, \tag{20.17}$$

$$\mathbf{f}_2 = \mathbf{f}_1 \wedge \mathbf{f}_3. \tag{20.18}$$

This system with local origin $\mathbf{b}(t)$ and normalized axes $\mathbf{f}_1, \mathbf{f}_2, \mathbf{f}_3$ is called the *Frenet frame* of the curve at $\mathbf{b}(t)$. Equipped with the tool of Frenet frames, we may now position objects along a 3D curve! See Figure 20.11. Since we are working in 3D, we use the cross product to form the orthogonal frame; however, we could have equally as well used the Gram-Schmidt process from Section 11.8.

Let us now work out exactly how to carry out our object-positioning plan. The object is given in some local coordinate system with axes $\mathbf{u}_1, \mathbf{u}_2, \mathbf{u}_3$. Any point of the object has coordinates

$$\mathbf{u} = \begin{bmatrix} u_1 \\ u_2 \\ u_3 \end{bmatrix}.$$

It is mapped to

$$\mathbf{x}(t, \mathbf{u}) = \mathbf{b}(t) + u_1\mathbf{f}_1 + u_2\mathbf{f}_2 + u_3\mathbf{f}_3.$$

A typical application is *robot motion*. Robots are used extensively in automotive assembly lines; one job is to grab a part and move it to its destination inside the car body. This movement happens along well-defined curves. While the car part is being moved, it has to be oriented into its correct position—exactly the process described in this section!

- linear interpolation
- parametric curve
- de Casteljau algorithm
- cubic Bézier curve
- subdivision
- affine invariance
- convex hull property
- Bernstein polynomials
- basis function
- barycentric combination
- matrix form
- cubic monomial curve
- Bernstein and monomial conversion
- nonsingular
- first derivative
- second derivative
- parallelogram rule
- composite Bézier curves
- curvature
- inflection point
- Frenet frame
- osculating plane

20.8 Exercises

Let a cubic Bézier curve $\mathbf{d}(t)$ be given by the control polygon

$$\mathbf{d}_0 = \begin{bmatrix} 0 \\ 0 \end{bmatrix}, \quad \mathbf{d}_1 = \begin{bmatrix} 6 \\ 3 \end{bmatrix}, \quad \mathbf{d}_2 = \begin{bmatrix} 3 \\ 6 \end{bmatrix}, \quad \mathbf{d}_3 = \begin{bmatrix} 2 \\ 4 \end{bmatrix}.$$

Let a cubic Bézier curve $\mathbf{b}(t)$ be given by the control polygon

$$\mathbf{b}_0 = \begin{bmatrix} 0 \\ 0 \\ 0 \end{bmatrix}, \quad \mathbf{b}_1 = \begin{bmatrix} 4 \\ 0 \\ 0 \end{bmatrix}, \quad \mathbf{b}_2 = \begin{bmatrix} 4 \\ 4 \\ 0 \end{bmatrix}, \quad \mathbf{b}_3 = \begin{bmatrix} 4 \\ 4 \\ 4 \end{bmatrix}.$$

1. Sketch $\mathbf{d}(t)$ manually.

2. Using the de Casteljau algorithm, evaluate $\mathbf{d}(t)$ for $t = 1/2$.

3. Evaluate the first and second derivative of $\mathbf{d}(t)$ for $t = 0$. Add these vectors to the sketch from Exercise 1.

4. For $\mathbf{d}(t)$, what is the control polygon for the curve defined from $t = 0$ to $t = 1/2$ and the curve defined over $t = 1/2$ to $t = 1$?

5. Rewrite $\mathbf{d}(t)$ in monomial form.

6. Find $\mathbf{d}(t)$'s minmax box.

7. Find $\mathbf{d}(t)$'s curvature at $t = 0$ and $t = 1/2$.

8. Find $\mathbf{d}(t)$'s Frenet frame for $t = 1/2$.

9. Using the de Casteljau algorithm, evaluate $\mathbf{d}(t)$ for $t = 2$. This is extrapolation.

10. Attach another curve $c(t)$ at $d(t)$, creating a composite curve with tangent continuity. What are the constraints on $c(t)$'s polygon?

11. Sketch $b(t)$ manually.

12. Using the de Casteljau algorithm, evaluate $b(t)$ for $t = 1/4$.

13. Evaluate the first and second derivative of $b(t)$ for $t = 1/4$. Add these vectors to the sketch from Exercise 11.

14. For $b(t)$, what is the control polygon for the curve defined from $t = 0$ to $t = 1/4$ and the curve defined over $t = 1/4$ to $t = 1$?

15. Rewrite $b(t)$ in monomial form.

16. Find $b(t)$'s minmax box.

17. Find $b(t)$'s curvature at $t = 0$ and $t = 1/2$.

18. Find $b(t)$'s Frenet frame for $t = 1$.

19. Using the de Casteljau algorithm, evaluate $b(t)$ for $t = 2$. This is extrapolation.

20. Attach another curve $c(t)$ at b_0, creating a composite curve with tangent continuity. What are the constraints on $c(t)$'s polygon?

A

Glossary

In this Glossary, we give brief definitions of the major concepts in the book. We avoid equations here, so that we give a slightly different perspective compared to what you find in the text. For more information on any item, reference the Index for full descriptions in the text. Each of these items is in one or more WYSK sections. However, more terms may be found in WYSK.

Action ellipse   The image of the unit circle under a linear map.

Adjacency matrix   An $n \times n$ matrix that has ones or zeroes as elements, representing the connectivity between n nodes in a directed graph.

Affine map   A map that leaves linear relationships between points unchanged. For instance, midpoints are mapped to midpoints. An affine map is described by a linear map and a translation.

Affine space   A set of points with the property that any barycentric combination of two points is again in the space.

Aspect ratio   The ratio of width to height of a rectangle.

Barycentric combination   A weighted average of points where the sum of the weights equals one.

Barycentric coordinates   When a point is expressed as a barycentric combination of other points, the coefficients in that combination are called barycentric coordinates.

Basis   For a linear space of dimension n, any set of n linearly independent vectors is a basis, meaning that every vector in the space may be uniquely expressed as a linear combination of these n basis vectors.

Basis transformation   The linear map taking one set of n basis vectors to another set of n basis vectors.

Bernstein polynomial   A polynomial basis function. The set of degree n Bernstein polynomials are used in the degree n Bézier curve representation.

Best approximation   The best approximation in a given subspace is the one that minimizes distance. Orthogonal projections produce the best approximation.

Bézier curve   A polynomial curve representation, which is based on the Bernstein basis and control points.

Centroid   The center of mass, or the average of a set of points, with all weights being equal and summing to one.

Characteristic equation   Every eigenvalue problem may be stated as finding the zeroes of a polynomial, giving the characteristic equation.

Coefficient matrix   The matrix in a linear system that holds the coefficients of the variables in a set of linear equations.

Collinear   A set of points is called collinear if they all lie on the same straight line.

Column space   The columns of a matrix form a set of vectors. Their span is the column space.

Condition number   A function measuring how sensitive a map is to changes in its input. If small changes in the input cause large changes in the output, then the condition number is large.

Conic section   The intersection curve of a double cone with a plane. A nondegenerate conic section is either an ellipse, a parabola, or a hyperbola.

Consistent linear system   A linear system with one or many solutions.

Contour   A level curve of a bivariate function.

Convex A point set is convex if the straight line segment through any of its points is completely contained inside the set. Example: all points on and inside of a sphere form a convex set; all points on and inside of an hourglass do not.

Convex combination A linear combination of points or vectors with nonnegative coefficients that sum to one.

Coordinates A vector in an n-dimensional linear space may be uniquely written as a linear combination of a set of basis vectors. The coefficients in that combination are the vector's coordinates with respect to that basis.

Coplanar A set of points is coplanar if all points lie on the same plane.

Covariance matrix An n-dimensional matrix that summarizes the variation in each coordinate and between coordinates of given n-dimensional data.

Cramer's rule A method for solving a linear system explicitly using ratios of determinants.

Cross product The cross product of two linearly independent 3D vectors results in a third vector that is perpendicular to them.

Cubic curve A degree three polynomial curve.

Curvature The rate of change of the tangent.

Curve The locus of a moving point.

de Casteljau algorithm A recursive process of linear interpolation for evaluating a Bézier curve.

Decomposition Expression of an element of a linear space, such as a vector or a matrix, in terms of other, often more fundamental, elements of the linear space. Examples: orthogonal decomposition, LU decomposition, eigendecomposition, or SVD.

Determinant A linear map takes a geometric object to another geometric object. The ratio of their volumes is the map's determinant.

Diagonalizable matrix A matrix A is diagonalizable if there exists a matrix R such that $R^{-1}AR$ is a diagonal matrix. The diagonal matrix holds the eigenvalues of A and the columns of R hold the eigenvectors of A.

Dimension The number of linearly independent vectors needed to span a linear space.

Directed graph A set of nodes and edges with associated direction.

Domain A linear map maps from one space to another. The "from" space is the domain.

Dominant eigenvalue The eigenvalue of a matrix with the largest absolute value.

Dominant eigenvector The eigenvector of a matrix corresponding to the dominant eigenvalue.

Dot product Assigns a value to the product of two vectors that is equal to the product of the magnitudes of the vectors and the cosine of the angle between the vectors. Also called the scalar product.

Dual space Consider all linear maps from a linear space into the 1D linear space of scalars. All these maps form a linear space themselves, the dual space of the original space.

Dyadic matrix Symmetric matrix of rank one, obtained from multiplying a column vector by its transpose.

Eigendecomposition The decomposition of a matrix A into $R\Lambda R^{\mathrm{T}}$, where the columns of R hold the eigenvectors of A and Λ is a diagonal matrix holding the eigenvalues of A. This decomposition is a sequence of a rigid body motion, a scale, followed by another rigid body motion.

Eigenfunction A function that gets mapped to a multiple of itself by some linear map defined on a function space.

Eigenvalue If a linear map takes some vector to itself multiplied by some constant, then that constant is an eigenvalue of the map.

Eigenvector A vector whose direction is unchanged by a linear map.

Elementary row operation This type of operation on a linear system includes exchanging rows, adding a multiple of one row to another, and multiplying a row by a scalar. Gauss elimination uses these operations to transform a linear system into a simpler one.

Ellipse A bounded conic section. When written in implicit form utilizing the matrix form, its 2×2 matrix has a positive determinant and two positive eigenvalues.

Equiangular polygon All interior angles at the vertices of a polygon are equal.

Equilateral polygon All sides of a polygon are of equal length.

Explicit equation Procedure to directly produce the value of a function or map.

Exterior angle A polygon is composed of edges. An exterior angle is the angle between two consecutive edges, measured on the outside of the polygon. Also called the turning angle.

Foot of a point The point on a line or a plane that is closest to a given point.

Frenet frame A 3D orthonormal coordinate system that exists at any point on a differentiable curve. It is defined by the tangent, normal, and binormal at a given point on the curve.

Gauss elimination The process of forward elimination, which transforms a linear system into an equivalent linear system with an upper triangular coefficient matrix, followed by back substitution.

Gauss-Jacobi iteration Solving a linear system by successively improving an initial guess for the solution vector by updating all solution components simultaneously.

Gauss-Seidel iteration Solving a linear system by successively improving an initial guess for the solution vector by updating each component as a new value is computed.

Generalized inverse Another name for the pseudoinverse. The concept of matrix inverse based on the SVD, which can be defined for nonsquare and singular matrices as well as square and nonsingular matrices.

Gram-Schmidt method A method for creating an orthonormal coordinate system from orthogonal projections of a given set of linearly independent vectors.

Homogeneous coordinates   Points in 2D affine space may be viewed as projections of points in 3D affine space, all being multiples of each other. The coordinates of any of these points are the homogeneous coordinates of the given point.

Homogeneous linear system   A linear system whose right-hand side consists of zeroes only.

Householder matrix   A special reflection matrix, which is designed to reflect a vector into a particular subspace while leaving some components unchanged. This matrix is symmetric, involutary, and orthogonal. It is key to the Householder method for solving a linear system.

Householder method   A method for solving an $m \times n$ linear system based on reflection matrices (rather than shears as in Gauss elimination). If the system is overdetermined, the Householder method results in the least squares solution.

Hyperbola   An unbounded conic section with two branches. When written in implicit form utilizing the matrix form, its 2×2 matrix has a negative determinant and a positive and a negative eigenvalue.

Idempotent   A map is idempotent if repeated applications of the map yield the same result as only one application. Example: projections.

Identity matrix   A square matrix with entries 1 on the diagonal and entries 0 elsewhere. This matrix maps every vector to itself.

Image   The result of a map. The preimage is mapped to the image.

Implicit equation   A definition of a multivariate function by specifying a relationship among its arguments.

Incenter   The center of a triangle's incircle, which is tangent to the three edges.

Inflection point   A point on a 2D curve where the curvature is zero.

Inner product   Given two elements of a linear space, their inner product is a scalar. The dot product is an example of an inner product. An inner product is formed from a vector product rule that satisfies symmetry, homogeneity, additivity, and positivity requirements. If two vectors are orthogonal, then their inner product is zero.

Interior angle   A polygon is composed of edges. An interior angle is formed by two consecutive edges inside the polygon.

Inverse matrix   A matrix maps a vector to another vector. The inverse matrix undoes this map.

Involutary matrix   A matrix for which the inverse is equal to the original matrix.

Kernel   The set of vectors being mapped to the zero vector by a linear map. Also called the null space.

Least squares approximation   The best approximation to an overdetermined linear system.

Line   Given two points in affine space, the set of all barycentric combinations is a line.

Line segment   A line, but with all coefficients of the barycentric combinations being nonnegative.

Linear combination   A weighted sum of vectors.

Linear functional   A linear map taking the elements of a linear space to the reals. Linear functionals are said to form the dual space of this linear space.

Linear interpolation   A weighted average of two points, where the weights sum to one and are linear functions of a parameter.

Linearity property   Preservation of linear combinations, thus incorporating the standard operations in a linear space: addition and scalar multiplication.

Linearly independent   A set of vectors is called linearly independent if none of its elements may be written as a linear combination of the remaining ones.

Linear map   A map of a linear space to another linear space such that linear relationships between vectors are not changed by the map.

Linear space   A set of vectors with the property that any linear combination of any two vectors is also in the set. Also called a vector space.

Linear system   The equations resulting from writing a given vector as an unknown linear combination of a given set of vectors.

Length   The magnitude of a vector.

Local coordinates   A specific coordinate system used to define a geometric object. This object may then be placed in a global coordinate system.

Major axis   An ellipse is symmetric about two axes that intersect at the center of the ellipse. The longer axis is the major axis.

Map   The process of changing objects. Example: rotating and scaling an object. The object being mapped is called the preimage, the result of the map is called the image.

Matrix   The coordinates of a linear map, written in a rectangular array of scalars.

Matrix norm   A characterization of a matrix, similar to a vector norm, which measures the "size" of the matrix.

Minor axis   An ellipse is symmetric about two axes that intersect at the center of the ellipse. The shorter axis is the minor axis.

Monomial curve   A curve represented in terms of the monomial polynomials.

Moore-Penrose generalized inverse   Another name for the pseudo-inverse.

***n*-gon**   An n-sided equiangular and equilateral polygon. Also called regular polygon.

Nonlinear map   A map that does not preserve linear relationships. Example: a perspective map.

Norm   A function that assigns a length to a vector.

Normal   A vector that is perpendicular to a line or a plane.

Normal equations   A linear system is transformed into a new, square linear system, called the normal equations, which will result in the least squares solution to the original system.

Null space   The set of vectors mapped to the zero vector by a linear map. The zero vector is always in this set. Also called the kernel.

Orthogonality Two vectors are orthogonal in an inner product space if their inner product vanishes.

Orthogonal matrix A matrix that leaves angles and lengths unchanged. The columns are unit vectors and the inverse is simply the transpose. Example: 2D rotation matrix.

Orthonormal A set of unit vectors is called orthonormal if any two of them are perpendicular.

Osculating plane At a point on a curve, the first and second derivative vectors span the osculating plane. The osculating circle, which is related to the curvature at this point, lies in this plane.

Overdetermined linear system A linear system with more equations than unknowns.

Parabola An unbounded conic section with one branch. When written in implicit form utilizing the matrix form, its 2×2 matrix has a zero determinant and has one zero eigenvalue.

Parallel Two planes are parallel if they have no point in common. Two vectors are parallel if they are multiples of each other.

Parallel projection A projection in which all vectors are projected in the same direction. If this direction is perpendicular to the projection plane it is an orthogonal projection, otherwise it is an oblique projection.

Parallelogram rule Two vectors define a parallelogram. The sum of these two vectors results in the diagonal of the parallelogram.

Parametric equation Description of an object by making each point/vector a function of a set of real numbers (the parameters).

Permutation matrix A matrix that exchanges rows or columns in a second matrix. It is a square and orthogonal matrix that has one entry of 1 in each row and entries of 0 elsewhere. It is used to perform pivoting in Gauss elimination.

Pivot The leading coefficient—first from the left—in a row of a matrix. In Gauss elimination, the pivots are typically the diagonal elements, as they play a central role in the algorithm.

Pivoting   The process of exchanging rows or columns so the pivot element is the largest element. Pivoting is used to improve numerical stability of Gauss elimination.

Plane   Given three points in affine space, the set of all barycentric combinations is a plane.

Point   A location, i.e., an element of an affine space.

Point cloud   A set of 3D points without any additional structure.

Point normal plane   A special representation of an implicit plane equation, formed by a point in the plane and the (unit) normal to the plane.

Polygon   The set of edges formed by connecting a set of points.

Polyline   Vertices connected by edges.

Positive definite matrix   If a quadratic form results in only nonnegative values for any nonzero argument, then its matrix is positive definite.

Principal components analysis (PCA)   A method for analyzing data by creating an orthogonal local coordinate frame—the principal axes—based on the dimensions with highest variance. This allows for an orthogonal transformation of the data to the coordinate axes for easier analysis.

Projection   An idempotent linear map that reduces the dimension of a vector.

Pseudoinverse   A generalized inverse that is based on the SVD. All matrices have a pseudoinverse, even nonsquare and singular ones.

Pythagorean theorem   The square of the length of the hypotenuse of a right triangle is equal to the sum of the squares of the lengths of the other two sides.

Quadratic form   A bivariate quadratic polynomial without linear or constant terms.

Range   A linear map maps from one space to another. The "to" space is the range.

Rank   The dimension of the image space of a map. Also the number of nonzero singular values.

Ratio   A measure of how three collinear points are distributed. If one is the midpoint of the other two, the ratio is 1.

Reflection matrix   A linear map that reflects a vector about a line (2D) or plane (3D). It is an orthogonal and involutary map.

Residual vector   The error incurred during the solution process of a linear system.

Rhombus   A parallelogram with all sides equal.

Rigid body motion   An affine map that leaves distances and angles unchanged, thus an object is not deformed. Examples: rotation and translation.

Rotation matrix   A linear map that rotates a vector through a particular angle around a fixed vector. It is an orthogonal matrix, thus the inverse is equal to the transpose and the determinant is one, making it a rigid body motion.

Row echelon form   Describes the inverted "staircase" form of a matrix, normally as a result of forward elimination and pivoting in Gauss elimination. The leading coefficient in a row (pivot) is to the right of the pivot in the previous row. Zero rows are the last rows.

Row space   The rows of a matrix form a set of vectors. Their span is the row space.

Scalar triple product   Computes the signed volume of three vectors using the cross product and dot product. Same as a 3×3 determinant.

Scaling matrix   A linear map with a diagonal matrix. If all diagonal elements are equal, it is a uniform scaling, which will uniformly shrink or enlarge a vector. If the diagonal elements are not equal, it is a nonuniform scaling. The inverse is a scaling matrix with inverted scaling elements.

Semi-major axis   One half the major axis of an ellipse.

Semi-minor axis   One half the minor axis of an ellipse.

Shear matrix   A linear map that translates a vector component proportionally to one or more other components. It has ones on the diagonal. It is area preserving. The inverse is a shear matrix with the off-diagonal elements negated.

Singular matrix   A matrix with zero determinant, thus it is not invertible.

Singular value decomposition (SVD)   The decomposition of a matrix into a sequence of a rigid body motion, a scale, followed by another rigid body motion, and the scale factors are the singular values.

Singular values   The singular values of a matrix A are the square roots of the eigenvalues of $A^{\mathrm{T}}A$.

Span   For a given set of vectors, its span is the set (space) of all vectors that can be obtained as linear combinations of these vectors.

Star   The triangles sharing one vertex in a triangulation.

Stationary vector   An eigenvector with 1 as its associated eigenvalue.

Stochastic matrix   A matrix for which the columns or rows sum to one.

Subspace   A set of linearly independent vectors defines a linear space. The span of any subset of these vectors defines a subspace of that linear space.

Symmetric matrix   A square matrix whose elements below the diagonal are mirror images of those above the diagonal.

Translation   An affine map that changes point locations by a constant vector.

Transpose matrix   The matrix whose rows are formed by the columns of a given matrix.

Triangle inequality   The sum of any two edge lengths in a triangle is larger than the third one.

Triangulation   Also called triangle mesh: a set of 2D or 3D points that is faceted into nonoverlapping triangles.

Trimmed surface   A parametric surface with parts of its domain removed.

Trivial reject   An inexpensive test in a process that eliminates unnecessary computations.

Turning angle   A polygon is composed of edges. An exterior angle formed by two consecutive edges is a turning angle.

Underdetermined linear system   A linear system with fewer equations than unknowns.

Unit vector   A vector whose length is 1.

Upper triangular matrix   A matrix with only zero entries below the diagonal.

Valence   The number of triangles in the star of a vertex.

Vector   An element of a linear space or, equivalently, the difference of two points in an affine space.

Vector norm   A measure for the magnitude of a vector. The Euclidean norm is the standard measure of length; others, such as the p-norms, are used as well.

Vector space   Linear space.

Zero vector   A vector of zero length. Every linear space contains a zero vector.

Selected Exercise Solutions

1a. The triangle vertex with coordinates $(0.1, 0.1)$ in the $[\mathbf{d}_1, \mathbf{d}_2]$-system is mapped to

$$x_1 = 0.9 \times 1 + 0.1 \times 3 = 1.2,$$
$$x_2 = 0.9 \times 2 + 0.1 \times 3 = 2.1.$$

The triangle vertex $(0.9, 0.2)$ in the $[\mathbf{d}_1, \mathbf{d}_2]$-system is mapped to

$$x_1 = 0.1 \times 1 + 0.9 \times 3 = 2.8,$$
$$x_2 = 0.8 \times 2 + 0.2 \times 3 = 2.2.$$

The triangle vertex $(0.4, 0.7)$ in the $[\mathbf{d}_1, \mathbf{d}_2]$-system is mapped to

$$x_1 = 0.6 \times 1 + 0.4 \times 3 = 1.8,$$
$$x_2 = 0.3 \times 2 + 0.7 \times 3 = 2.7.$$

1b. The coordinates $(2, 2)$ in the $[\mathbf{e}_1, \mathbf{e}_2]$-system are mapped to

$$u_1 = \frac{2-1}{3-1} = \frac{1}{2},$$
$$u_2 = \frac{2-2}{3-2} = 0.$$

3. The local coordinates $(0.5, 0, 0.7)$ are mapped to

$$x_1 = 1 + 0.5 \times 1 = 1.5,$$
$$x_2 = 1 + 0 \times 2 = 1,$$
$$x_3 = 1 + 0.7 \times 4 = 3.8.$$

4. We have simply moved Exercise 1a to 3D, so the first two coordinates are identical. Since the 3D triangle local coordinates are at the extents of the local frame, they will be at the extents of the global frame. Therefore, (u_1, u_2, u_3) is mapped to $(x_1, x_2, x_3) = (1.2, 2.1, 4)$, $(v_1, v_2, v_3$ is mapped to $(x_1, x_2, x_3) = (2.8, 2.2, 8)$, and (w_1, w_2, w_3) is mapped to $(x_1, x_2, x_3) = (1.8, 2.7, 4)$.

7. The local coordinates $(3/4, 1/2)$, which are at the midpoint of the two given local coordinate sets, has global coordinates

$$(6\frac{1}{2}, 5) = \frac{1}{2}(5, 2) + \frac{1}{2}(8, 8).$$

Its coordinates are the midpoint between the given global coordinate sets.

9. The aspect ratio tells us that

$$\Delta_1 = 2 \times \frac{4}{3} = \frac{8}{3}.$$

The global coordinates of $(1/2, 1/2)$ are

$$x_1 = 0 + \frac{1}{2} \cdot \frac{8}{3} = \frac{4}{3} \qquad x_2 = 0 + \frac{1}{2} \cdot 2 = 1.$$

10. Each coordinate will follow similarly, so let's work out the details for x_1. First, construct the ratio that defines the relationship between a coordinate u_1 in the NDC system and coordinate x_1 in the viewport

$$\frac{u_1 - (-1)}{1 - (-1)} = \frac{x_1 - \min_1}{\max_1 - \min_1}.$$

Solve for x_1, and the equations for x_2 follow similarly, namely

$$x_1 = \frac{(\max_1 - \min_1)}{2}(u_1 + 1) + \min_1,$$
$$x_2 = \frac{(\max_2 - \min_2)}{2}(u_2 + 1) + \min_2.$$

Chapter 2

2. The vectors $\mathbf{v}$ and $\mathbf{w}$ form adjacent sides of a parallelogram. The vectors $\mathbf{v} + \mathbf{w}$ and $\mathbf{v} - \mathbf{w}$ form the diagonals of this parallelogram, and an example is illustrated in Sketch 2.6.

4. The operations have the following results.

(a) vector

(b) point

(c) point

(d) vector

(e) vector

(f) point

5. The midpoint between $\mathbf{p}$ and $\mathbf{q}$ is

$$\mathbf{m} = \frac{1}{2}\mathbf{p} + \frac{1}{2}\mathbf{q}.$$

8. A triangle.

9. The length of the vector

$$\mathbf{v} = \begin{bmatrix} -4 \\ -3 \end{bmatrix}$$

is

$$\|\mathbf{v}\| = \sqrt{(-4)^2 + (-3)^2} = 5.$$

13. The distance between $\mathbf{p}$ and $\mathbf{q}$ is 1.

14. A unit vector has length one.

15. The normalized vector

$$\frac{\mathbf{v}}{\|\mathbf{v}\|} = \begin{bmatrix} -4/5 \\ -3/5 \end{bmatrix}.$$

17. The barycentric coordinates are $(1-t)$ and t such that $\mathbf{r} = (1-t)\mathbf{p} + t\mathbf{q}$. We determine the location of $\mathbf{r}$ relative to $\mathbf{p}$ and $\mathbf{q}$ by calculating $l_1 = \|\mathbf{r} - \mathbf{p}\| = 2\sqrt{2}$ and $l_2 = \|\mathbf{q} - \mathbf{r}\| = 4\sqrt{2}$. The barycentric coordinates must sum to one, so we need the total length $l_3 = l_1 + l_2 = 6\sqrt{2}$. Then the barycentric coordinates are $t = l_1/l_3 = 1/3$ and $(1 - t) = 2/3$. Check that this is correct:

$$\begin{bmatrix} 3 \\ 3 \end{bmatrix} = \frac{2}{3}\begin{bmatrix} 1 \\ 1 \end{bmatrix} + \frac{1}{3}\begin{bmatrix} 7 \\ 7 \end{bmatrix}.$$

19. No, they are linearly dependent.

21. Yes, $\mathbf{v}_1$ and $\mathbf{v}_2$ form a basis for $\mathbb{R}^2$ since they are linearly independent.

24. The dot product, $\mathbf{v} \cdot \mathbf{w} = 5 \times 0 + 4 \times 1 = 4$. Scalar product is another name for dot product and the dot product has the symmetry property, therefore $\mathbf{w} \cdot \mathbf{v} = 4$.

25. The angle between the vectors $\begin{bmatrix} 5 \\ 5 \end{bmatrix}$ and $\begin{bmatrix} 3 \\ -3 \end{bmatrix}$ is $90°$ by inspection. Sketch it and this will be clear. Additionally, notice $5 \times 3 + 5 \times -3 = 0$.

27. The angles fall into the following categories: θ_1 is obtuse, θ_2 is a right angle, and θ_3 is acute.

28. The orthogonal projection $\mathbf{u}$ of $\mathbf{w}$ onto $\mathbf{v}$ is determined by (2.21), or specifically,

$$\mathbf{u} = \frac{\begin{bmatrix} 1 \\ -1 \end{bmatrix} \cdot \begin{bmatrix} 3 \\ 2 \end{bmatrix}}{(\sqrt{2})^2} \cdot \begin{bmatrix} 1 \\ -1 \end{bmatrix} = \begin{bmatrix} 1/2 \\ -1/2 \end{bmatrix}.$$

Draw a sketch to verify. Therefore, the $\mathbf{u}^\perp$ that completes the orthogonal decomposition of $\mathbf{w}$ is

$$\mathbf{u}^\perp = \mathbf{w} - \mathbf{u} = \begin{bmatrix} 3 \\ 2 \end{bmatrix} - \begin{bmatrix} 1 \\ -1 \end{bmatrix} = \begin{bmatrix} 2 \\ 3 \end{bmatrix}.$$

Add $\mathbf{u}^\perp$ to your sketch to verify.

31. Equality of the Cauchy-Schwartz inequality holds when $\mathbf{v}$ and $\mathbf{w}$ are linearly dependent. A simple example:

$$\mathbf{v} = \begin{bmatrix} 1 \\ 0 \end{bmatrix} \quad \text{and} \quad \mathbf{w} = \begin{bmatrix} 3 \\ 0 \end{bmatrix}.$$

The two sides of Cauchy-Schwartz are then $(\mathbf{v}\cdot\mathbf{w})^2 = 9$ and $\|\mathbf{v}\|^2\|\mathbf{w}\|^2 = 1^2 \times 3^2 = 9$.

32. No, the triangle inequality states that $\|\mathbf{v} + \mathbf{w}\| \leq \|\mathbf{v}\| + \|\mathbf{w}\|$.

Chapter 3

1. The line is defined by the equation $\mathbf{l}(t) = \mathbf{p} + t(\mathbf{q} - \mathbf{p})$, thus

$$\mathbf{l}(t) = \begin{bmatrix} 0 \\ 1 \end{bmatrix} + t \begin{bmatrix} 4 \\ 1 \end{bmatrix}.$$

We can check that

$$\mathbf{l}(0) = \begin{bmatrix} 0 \\ 1 \end{bmatrix} + 0 \times \begin{bmatrix} 4 \\ 1 \end{bmatrix} = \mathbf{p},$$

$$\mathbf{l}(1) = \begin{bmatrix} 0 \\ 1 \end{bmatrix} + 1 \times \begin{bmatrix} 4 \\ 1 \end{bmatrix} = \mathbf{q}.$$

3. The parameter value $t = 2$ is outside of $[0, 1]$, therefore $\mathbf{l}(2)$ is not formed from a convex combination.

5. First form the vector

$$\mathbf{q} - \mathbf{p} = \begin{bmatrix} 2 \\ -1 \end{bmatrix},$$

then $\mathbf{a}$ is perpendicular to this vector, so let

$$\mathbf{a} = \begin{bmatrix} 1 \\ 2 \end{bmatrix}.$$

Next, calculate $c = -a_1 p_1 - a_2 p_2 = 2$, which makes the equation of the line $x_1 + 2x_2 + 2 = 0$.

Alternatively, we could have let

$$\mathbf{a} = \begin{bmatrix} -1 \\ -2 \end{bmatrix}.$$

Then the implicit equation of the line is $-x_1 - 2x_2 - 2 = 0$, which is simply equivalent to multiplying the previous equation by -1.

6. The point r_0 is not on the line $x_1 + 2x_2 + 2 = 0$, since $0 + 2 \times 0 + 2 = 2 \neq 0$. The point r_1 is on the line since $-4 + 2(1) + 2 = 0$. The point r_2 is not on the lines since $5 + 2(1) + 2 \neq 0$. The point r_3 is not on the line since $-3 + 2(-1) + 2 \neq 0$.

10. The explicit equation of the line $6x_1 + 3x_2 + 3 = 0$ is $x_2 = -2x_1 - 1$.

11. The slope is 4 and e_2-intercept is $x_2 = -1$.

14. The implicit equation of the line is $-3x_1 - 2x_2 + 5 = 0$. (Check that the given point, p, from the parametric form satisfies this equation.)

16. To form the implicit equation, we first find a vector parallel to the line,

$$\mathbf{v} = \begin{bmatrix} 2 \\ 2 \end{bmatrix} - \begin{bmatrix} -1 \\ 0 \end{bmatrix} = \begin{bmatrix} 3 \\ 2 \end{bmatrix}.$$

From $\mathbf{v}$ we select the components of $\mathbf{a}$ as $a_1 = -2$ and $a_2 = 3$, and then $c = -2$. Therefore, the implicit equation is $-2x_1 + 3x_2 - 2 = 0$. (Check that the two points forming the parametric form satisfy this equation.)

17. The parametric form of the line is $l(t) = \mathbf{p} + t\mathbf{v}$. Construct a vector that is perpendicular to $\mathbf{a}$:

$$\mathbf{v} = \begin{bmatrix} 2 \\ -3 \end{bmatrix},$$

find one point on the line:

$$\mathbf{p} = \begin{bmatrix} -1/3 \\ 0 \end{bmatrix},$$

and the definition is complete. (This selection of $\mathbf{p}$ and $\mathbf{v}$ follow the steps outlined in Section 3.5).

19. Let $\mathbf{w} = \mathbf{r} - \mathbf{p}$, then

$$\mathbf{w} = \begin{bmatrix} 5 \\ 0 \end{bmatrix}$$

and $\|\mathbf{w}\| = 5$. The cosine of the angle between $\mathbf{w}$ and $\mathbf{v}$ is $\cos \alpha = 1/\sqrt{2}$. The distance of $\mathbf{r}$ to the line is $d = \|\mathbf{w}\| \sin \alpha$, where $\sin \alpha = \sqrt{1 - \cos^2 \alpha}$, thus $d = 5/\sqrt{2}$.

21. Let $\mathbf{q}$ be the foot of the point $\mathbf{r}$ and $\mathbf{q} = \mathbf{p} + t\mathbf{v}$. Define $\mathbf{w} = \mathbf{r} - \mathbf{p}$, then

$$t = \frac{\mathbf{v} \cdot \mathbf{w}}{\|\mathbf{v}\|^2} = \frac{5}{2}.$$

The foot of the point is

$$\mathbf{q} = \begin{bmatrix} 0 \\ 0 \end{bmatrix} + \frac{5}{2} \begin{bmatrix} 1 \\ 1 \end{bmatrix} = \begin{bmatrix} 5/2 \\ 5/2 \end{bmatrix}.$$

(The distance between $\mathbf{r}$ and the foot of $\mathbf{r}$ is $5/\sqrt{2}$, confirming our solution to Exercise 19.)

24. The lines are identical!

27. The midpoint $\mathbf{r}$ is $1 : 1$ with respect to the points $\mathbf{p}$ and $\mathbf{q}$. This means that it has parameter value $t = 1/(1 + 1) = 1/2$ with respect to the line $\mathbf{l}(t) = \mathbf{p} + t(\mathbf{q} - \mathbf{p})$, and

$$\mathbf{r} = \begin{bmatrix} 0 \\ 1 \end{bmatrix} + \frac{1}{2} \begin{bmatrix} 4 \\ 1 \end{bmatrix} = \begin{bmatrix} 2 \\ 3/2 \end{bmatrix}.$$

The vector $\mathbf{v}^{\perp} = \begin{bmatrix} -1 \\ 4 \end{bmatrix}$ is perpendicular to $\mathbf{q} - \mathbf{p}$. The line $\mathbf{m}(t)$ is then defined as

$$\mathbf{m}(t) = \begin{bmatrix} 3/2 \\ 2 \end{bmatrix} + t \begin{bmatrix} -1 \\ 4 \end{bmatrix}.$$

Chapter 4

1. The linear combination is $\mathbf{w} = 2\mathbf{c}_1 + (1/2)\mathbf{c}_2$. In matrix form:

$$\begin{bmatrix} 1 & 0 \\ 0 & 2 \end{bmatrix} \begin{bmatrix} 2 \\ 1/2 \end{bmatrix} = \begin{bmatrix} 2 \\ 1 \end{bmatrix}.$$

3. Yes, $\mathbf{v} = A\mathbf{w}$ where

$$\mathbf{w} = \begin{bmatrix} 3 \\ -2 \end{bmatrix}.$$

6. The transpose of the given matrices are as follows:

$$A^{\mathrm{T}} = \begin{bmatrix} 0 & 1 \\ -1 & 0 \end{bmatrix}, \quad B^{\mathrm{T}} = \begin{bmatrix} 1 & -1 \\ -1 & 1/2 \end{bmatrix}, \quad \mathbf{v}^{\mathrm{T}} = \begin{bmatrix} 2 & 3 \end{bmatrix}.$$

7. The transpose of the 2×2 identity matrix I is (again)

$$I = \begin{bmatrix} 1 & 0 \\ 0 & 1 \end{bmatrix}.$$

8. From (4.9), we expect these two matrix sum-transpose results to be the same:

$$A + B = \begin{bmatrix} 1 & -2 \\ 0 & 1/2 \end{bmatrix}, \qquad [A + B]^{\mathrm{T}} = \begin{bmatrix} 1 & 0 \\ -2 & 1/2 \end{bmatrix};$$

$$A^{\mathrm{T}} = \begin{bmatrix} 0 & 1 \\ -1 & 0 \end{bmatrix}, \quad B^{\mathrm{T}} = \begin{bmatrix} 1 & -1 \\ -1 & 1/2 \end{bmatrix}, \quad A^{\mathrm{T}} + B^{\mathrm{T}} = \begin{bmatrix} 1 & 0 \\ -2 & 1/2 \end{bmatrix}.$$

And indeed they are.

9. The rank, or the number of linearly independent column vectors of B is two. We cannot find a scalar α such that

$$\begin{bmatrix} 1 \\ -1 \end{bmatrix} = \alpha \begin{bmatrix} -1 \\ 1/2 \end{bmatrix},$$

therefore, these vectors are linearly independent.

11. The zero matrix has rank zero.

14. The product $A\mathbf{v}$:

$$
\begin{array}{cc|c}
 & & 2 \\
 & & 3 \\
\hline
0 & -1 & -3 \\
1 & 0 & 2
\end{array}.
$$

The product $B\mathbf{v}$:

$$
\begin{array}{cc|c}
 & & 2 \\
 & & 3 \\
\hline
1 & -1 & -1 \\
-1 & 1/2 & -1/2
\end{array}.
$$

15. The matrix represents a uniform scaling,

$$
S = \begin{bmatrix} 3 & 0 \\ 0 & 3 \end{bmatrix}.
$$

19. This one is a little tricky! It is a rotation; notice that the determinant is one. See the discussion surrounding (4.15).

21. No. Simply check that the matrix multiplied by itself does not result in the matrix again:

$$
A^2 = \begin{bmatrix} -1 & 0 \\ 0 & -1 \end{bmatrix} \neq A.
$$

25. The determinant of A:

$$
|A| = \begin{vmatrix} 0 & -1 \\ 1 & 0 \end{vmatrix} = 0 \times 0 - (-1) \times 1 = 1.
$$

26. The determinant of B:

$$
|B| = \begin{vmatrix} 1 & -1 \\ -1 & 1/2 \end{vmatrix} = 1 \times 1/2 - (-1) \times (-1) = -1/2.
$$

28. The sum

$$
A + B = \begin{bmatrix} 1 & -2 \\ 0 & 1/2 \end{bmatrix}.
$$

The product $(A + B)\mathbf{v}$:

$$
\begin{array}{cc|c}
 & & 2 \\
 & & 3 \\
\hline
1 & -2 & -4 \\
0 & 1/2 & 3/2
\end{array}
$$

and

$$
A\mathbf{v} + B\mathbf{v} = \begin{bmatrix} -3 \\ 2 \end{bmatrix} + \begin{bmatrix} -1 \\ -1/2 \end{bmatrix} = \begin{bmatrix} -4 \\ 3/2 \end{bmatrix}.
$$

31. The matrix $A^2 = A \cdot A$ equals

$$
\begin{bmatrix} -1 & 0 \\ 0 & -1 \end{bmatrix}.
$$

Does this look familiar? Reflect on the discussion surrounding (4.15).

33. Matrix multiplication cannot increase the rank of the result, hence the rank of MN is one or zero.

Chapter 5

1. The linear system takes the form

$$
\begin{bmatrix} 2 & 6 \\ -3 & 0 \end{bmatrix} \begin{bmatrix} x_1 \\ x_2 \end{bmatrix} = \begin{bmatrix} 6 \\ 3 \end{bmatrix}.
$$

2. This system is inconsistent because there is no solution vector $\mathbf{u}$ that will satisfy this linear system.

4. The linear system

$$
\begin{bmatrix} 2 & 6 \\ -3 & 0 \end{bmatrix} \begin{bmatrix} x_1 \\ x_2 \end{bmatrix} = \begin{bmatrix} 6 \\ 3 \end{bmatrix}
$$

has the solution

$$
x_1 = \frac{\begin{vmatrix} 6 & 6 \\ 3 & 0 \end{vmatrix}}{\begin{vmatrix} 2 & 6 \\ -3 & 0 \end{vmatrix}} = -1, \qquad x_2 = \frac{\begin{vmatrix} 2 & 6 \\ -3 & 3 \end{vmatrix}}{\begin{vmatrix} 2 & 6 \\ -3 & 0 \end{vmatrix}} = \frac{4}{3}.
$$

7. The linear system

$$
\begin{bmatrix} 2 & 6 \\ -3 & 0 \end{bmatrix} \begin{bmatrix} x_1 \\ x_2 \end{bmatrix} = \begin{bmatrix} 6 \\ 3 \end{bmatrix}
$$

is transformed to

$$
\begin{bmatrix} 2 & 6 \\ 0 & 9 \end{bmatrix} \begin{bmatrix} x_1 \\ x_2 \end{bmatrix} = \begin{bmatrix} 6 \\ 12 \end{bmatrix}
$$

after one step of forward elimination. Now that the matrix is upper triangular, back substitution is used to find

$$
x_2 = 12/9 = 4/3,
$$

$$
x_1 = \left(6 - 6\frac{4}{3} \right) / 2 = -1.
$$

Clearly, the same solution as obtained by Cramer's rule in Exercise 4.

9. The Gauss elimination steps from Section 5.4 need to be modified because $a_{1,1} = 0$. Therefore, we add a pivoting step, which means that we exchange rows, resulting in

$$\begin{bmatrix} 2 & 2 \\ 0 & 4 \end{bmatrix} \begin{bmatrix} x_1 \\ x_2 \end{bmatrix} = \begin{bmatrix} 6 \\ 8 \end{bmatrix}.$$

Now the matrix is in upper triangular form and we use back substitution to find $x_2 = 2$ and $x_1 = 1$.

13. No, this system has only the trivial solution, $x_1 = x_2 = 0$.

14. We find the kernel of the matrix

$$C = \begin{bmatrix} 2 & 6 \\ 4 & 12 \end{bmatrix}$$

by first establishing the fact that $\mathbf{c}_2 = 3\mathbf{c}_1$. Therefore, all vectors of the form $v_1 = -3v_2$ are a part of the kernel. For example, $\begin{bmatrix} -3 \\ 1 \end{bmatrix}$ is in the kernel.

Alternatively, we could have set up a homogeneous system and used Gauss elimination to find the kernel: starting with

$$\begin{bmatrix} 2 & 6 \\ 4 & 12 \end{bmatrix} \mathbf{v} = \begin{bmatrix} 0 \\ 0 \end{bmatrix},$$

perform one step of forward elimination,

$$\begin{bmatrix} 2 & 6 \\ 0 & 0 \end{bmatrix} \mathbf{v} = \begin{bmatrix} 0 \\ 0 \end{bmatrix},$$

then the back substitution steps begin with arbitrarily setting $v_2 = 1$, and then $v_1 = -3$, which is the same result as above.

16. The inverse of the matrix A:

$$A^{-1} = \begin{bmatrix} 0 & -3/9 \\ 1/6 & 1/9 \end{bmatrix}.$$

Check that $AA^{-1} = I$.

19. An orthogonal matrix for which the column vectors are orthogonal and unit length. A rotation matrix such as

$$\begin{bmatrix} \cos 30° & -\sin 30° \\ \sin 30° & \cos 30° \end{bmatrix}$$

is an example.

21. Simply by inspection, we see that the map is a reflection about the $\mathbf{e}_1$-axis, thus the matrix is

$$A = \begin{bmatrix} 1 & 0 \\ 0 & -1 \end{bmatrix}.$$

Let's find this result using (5.19), then

$$V = \begin{bmatrix} 1 & 1 \\ 0 & 1 \end{bmatrix} \qquad V' = \begin{bmatrix} 1 & 1 \\ 0 & -1 \end{bmatrix}.$$

We find

$$V^{-1} = \begin{bmatrix} 1 & -1 \\ 0 & 1 \end{bmatrix},$$

and $A = V'V^{-1}$ results in the reflection already given.

Chapter 6

1. The point $\mathbf{q} = \begin{bmatrix} 2/3 \\ 4/3 \end{bmatrix}$. The transformed points:

$$\mathbf{r}' = \begin{bmatrix} 3 \\ 4 \end{bmatrix}, \quad \mathbf{s}' = \begin{bmatrix} 11/2 \\ 6 \end{bmatrix}, \quad \mathbf{q}' = \begin{bmatrix} 14/3 \\ 16/3 \end{bmatrix}.$$

The point $\mathbf{q}'$ is in fact equal to $(1/3)\mathbf{r}' + (2/3)\mathbf{s}'$.

3. The affine map takes the $\mathbf{x}_i$ to

$$\mathbf{x}_1' = \begin{bmatrix} -4 \\ 0 \end{bmatrix}, \quad \mathbf{x}_2' = \begin{bmatrix} -1 \\ 1 \end{bmatrix}, \quad \mathbf{x}_3' = \begin{bmatrix} 2 \\ 2 \end{bmatrix}.$$

The ratio of three collinear points is preserved by an affine map, thus the ratio of the $\mathbf{x}_i$ and the $\mathbf{x}_i'$ is $1 : 1$.

4. In order to rotate a point $\mathbf{x}$ around another point $\mathbf{r}$, construct the affine map

$$\mathbf{x}' = A(\mathbf{x} - \mathbf{r}) + \mathbf{r}.$$

In this exercise, we rotate $90°$, $\mathbf{x} = \begin{bmatrix} -2 \\ -2 \end{bmatrix}$, and $\mathbf{r} = \begin{bmatrix} -2 \\ 2 \end{bmatrix}$. The matrix

$$A = \begin{bmatrix} 0 & -1 \\ 1 & 0 \end{bmatrix},$$

thus $\mathbf{x}' = \begin{bmatrix} 2 \\ 2 \end{bmatrix}$. Be sure to draw a sketch!

7. The point $\mathbf{p}_0$ is on the line, therefore $\mathbf{p}_0' = \mathbf{p}_0$. Since the five points are uniformly spaced and collinear, we can apply the reflection affine map to $\mathbf{p}_4$ and then use linear interpolation (another, simpler affine map) to the other points. First, we want to find the foot of $\mathbf{p}_4$ on $l(t)$,

$$\mathbf{q}_4 = \mathbf{p}_0 + t\mathbf{v}.$$

We find t by projecting $\mathbf{p}_4$ onto the line. Let

$$\mathbf{w} = \mathbf{p}_4 - \mathbf{p}_0 = \begin{bmatrix} -4 \\ 0 \end{bmatrix},$$

then

$$t = \frac{\mathbf{v} \cdot \mathbf{w}}{\|\mathbf{v}\|^2} = 1,$$

and

$$\mathbf{q}_4 = \begin{bmatrix} 2 \\ 1 \end{bmatrix} + 1 \begin{bmatrix} -2 \\ 2 \end{bmatrix} = \begin{bmatrix} 0 \\ 3 \end{bmatrix}.$$

This point is the midpoint between $\mathbf{p}_4$ and $\mathbf{p}_4'$, thus

$$\mathbf{p}_4' = 2\mathbf{q}_4 - \mathbf{p}_4 = \begin{bmatrix} 2 \\ 5 \end{bmatrix}.$$

Now we use linear interpolation to find the other points:

$$\mathbf{p}_1' = \frac{3}{4}\mathbf{p}_0' + \frac{1}{4}\mathbf{p}_4' = \begin{bmatrix} 2 \\ 2 \end{bmatrix}, \quad \mathbf{p}_2' = \frac{1}{2}\mathbf{p}_0' + \frac{1}{2}\mathbf{p}_4' = \begin{bmatrix} 2 \\ 3 \end{bmatrix}, \quad \mathbf{p}_3' = \frac{1}{4}\mathbf{p}_0' + \frac{3}{4}\mathbf{p}_4' = \begin{bmatrix} 2 \\ 4 \end{bmatrix}.$$

8. We want to define A in the affine map

$$\mathbf{x}' = A(\mathbf{x} - \mathbf{a}_1) + \mathbf{a}_1'.$$

Let $V = [\mathbf{v}_2 \ \mathbf{v}_3]$ and $V' = [\mathbf{v}_2' \ \mathbf{v}_3']$, where

$$\mathbf{v}_2 = \mathbf{a}_2 - \mathbf{a}_1 = \begin{bmatrix} 1 \\ 0 \end{bmatrix}, \qquad \mathbf{v}_3 = \mathbf{a}_3 - \mathbf{a}_1 = \begin{bmatrix} 0 \\ 1 \end{bmatrix},$$

$$\mathbf{v}_2' = \mathbf{a}_2' - \mathbf{a}_1' = \begin{bmatrix} -1 \\ 1 \end{bmatrix}, \qquad \mathbf{v}_3' = \mathbf{a}_3' - \mathbf{a}_1' = \begin{bmatrix} -1 \\ 0 \end{bmatrix}.$$

The matrix A maps the $\mathbf{v}_i$ to the $\mathbf{v}_i'$, thus $AV = V'$, and

$$A = V'V^{-1} = \begin{bmatrix} -1 & -1 \\ 1 & 0 \end{bmatrix} \begin{bmatrix} 1 & 0 \\ 0 & 1 \end{bmatrix} = \begin{bmatrix} -1 & -1 \\ 1 & 0 \end{bmatrix}.$$

Check that the $\mathbf{a}_i$ are mapped to $\mathbf{a}_i'$. The point $\mathbf{x} = [1/2 \ 1/2]^{\mathrm{T}}$, the midpoint between $\mathbf{a}_2$ and $\mathbf{a}_3$, is mapped to $\mathbf{x} = [0 \ 1/2]^{\mathrm{T}}$, the midpoint between $\mathbf{a}_2'$ and $\mathbf{a}_3'$

9. Affine maps take collinear points to collinear points and preserve ratios, thus

$$\mathbf{x}' = \begin{bmatrix} 0 \\ 0 \end{bmatrix}.$$

11. Mapping a point $\mathbf{x}$ in NDC coordinates to $\mathbf{x}'$ in the viewport involves the following steps.

   1. Translate $\mathbf{x}$ by an amount that translates $\mathbf{l}_n$ to the origin.

2. Scale the resulting **x** so that the sides of the NDC box are of unit length.

3. Scale the resulting **x** so that the unit box is scaled to match the viewport's dimensions.

4. Translate the resulting **x** by l_v to align the scaled box with the viewport.

The sides of the viewport have the lengths $\Delta_1 = 20$ and $\Delta_2 = 10$. We can then express the affine map as

$$\mathbf{x}' = \begin{bmatrix} \Delta_1/2 & 0 \\ 0 & \Delta_2/2 \end{bmatrix} \left(\mathbf{x} - \begin{bmatrix} -1 \\ -1 \end{bmatrix} \right) + \begin{bmatrix} 10 \\ 10 \end{bmatrix}.$$

Applying this affine map to the NDC points in the question yields

$$\mathbf{x}'_1 = \begin{bmatrix} 10 \\ 10 \end{bmatrix}, \quad \mathbf{x}'_2 = \begin{bmatrix} 30 \\ 20 \end{bmatrix}, \quad \mathbf{x}'_3 = \begin{bmatrix} 15 \\ 17\frac{1}{2} \end{bmatrix}.$$

Of course we could easily "eyeball" $\mathbf{x}'_1$ and $\mathbf{x}'_2$, so they provide a good check for our affine map. Be sure to make a sketch.

12. No, affine maps do not transform perpendicular lines to perpendicular lines. A simple shear is a counterexample.

Chapter 7

2. Form the matrix $A - \lambda I$. The characteristic equation is

$$\lambda^2 - 2\lambda - 3 = 0.$$

The eigenvalues are $\lambda_1 = 3$ and $\lambda_2 = -1$. The eigenvectors are

$$\mathbf{r}_1 = \begin{bmatrix} -1/\sqrt{2} \\ 1/\sqrt{2} \end{bmatrix} \quad \text{and} \quad \mathbf{r}_2 = \begin{bmatrix} 1/\sqrt{2} \\ 1/\sqrt{2} \end{bmatrix}.$$

The dominant eigenvalue is $\lambda_1 = 3$.

4. The area of the parallelogram is 8 because $|A| = \lambda_1 \lambda_2 = 4 \times 2 = 8$.

6. The eigenvalues for the matrix in Example 7.1 are $\lambda_1 = 3$ and $\lambda_2 = 1$. Let $v = 1/\sqrt{2}$. For λ_1, we solve a homogenous equation, and find that we can choose either

$$\mathbf{r}_1 = \begin{bmatrix} v \\ v \end{bmatrix} \quad \text{or} \quad \mathbf{r}_1 = \begin{bmatrix} -v \\ -v \end{bmatrix}$$

as the eigenvector. For λ_2, we can choose either

$$\mathbf{r}_2 = \begin{bmatrix} v \\ -v \end{bmatrix} \quad \text{or} \quad \mathbf{r}_2 = \begin{bmatrix} -v \\ v \end{bmatrix}$$

as the eigenvector. The matrix of eigenvectors is $R = [\mathbf{r}_1 \quad \mathbf{r}_2]$, so this leads to the four possibilities for R:

$$R_1 = \begin{bmatrix} v & v \\ v & -v \end{bmatrix}, \quad R_2 = \begin{bmatrix} v & -v \\ v & v \end{bmatrix}, \quad R_3 = \begin{bmatrix} -v & v \\ -v & -v \end{bmatrix}, \quad R_4 = \begin{bmatrix} -v & -v \\ -v & v \end{bmatrix},$$

where the subscripts have no intrinsic meaning; they simply allow us to refer to each matrix.

It is easy to identify the rotation matrices by examining the diagonal elements. Matrix R_2 is a rotation by $45°$ and R_3 is a rotation by $270°$. Matrix R_1 is a reflection about $\mathbf{e}_1$ and a rotation by $45°$,

$$R_1 = \begin{bmatrix} v & -v \\ v & v \end{bmatrix} \begin{bmatrix} 1 & 0 \\ 0 & -1 \end{bmatrix}.$$

Matrix R_4 is a reflection about $\mathbf{e}_2$ and a rotation by $45°$,

$$R_4 = \begin{bmatrix} v & -v \\ v & v \end{bmatrix} \begin{bmatrix} -1 & 0 \\ 0 & 1 \end{bmatrix}.$$

Another way to identify the matrices: the reflection-rotation is a symmetric matrix and thus has real eigenvalues, whereas the rotation has complex eigenvalues. Sketching the $\mathbf{r}_1$ and $\mathbf{r}_2$ combinations is also helpful for identifying the action of the map. This shows us how $\mathbf{e}_1$ and $\mathbf{e}_2$ are mapped.

7. The matrix A is symmetric, therefore the eigenvalues will be real.

9. The eigenvalues of A are the roots of the characteristic polynomial

$$p(\lambda) = (a_{1,1} - \lambda)(a_{2,2} - \lambda) - a_{1,2}a_{2,1}$$

and since multiplication is commutative, the characteristic polynomial for A^{T} is identical to A's. Therefore the eigenvalues are identical.

11. The eigenvalues and corresponding eigenvectors for A are $\lambda_1 = 4$, $\mathbf{r}_1 = [0 \quad 1]^{\mathrm{T}}$, $\lambda_2 = 2$, $\mathbf{r}_2 = [1 \quad 0]^{\mathrm{T}}$.

The projection matrices are

$$P_1 = \mathbf{r}_1 \mathbf{r}_1^{\mathrm{T}} = \begin{bmatrix} 0 & 0 \\ 0 & 1 \end{bmatrix} \quad \text{and} \quad P_2 = \mathbf{r}_2 \mathbf{r}_2^{\mathrm{T}} = \begin{bmatrix} 0 & 1 \\ 0 & 0 \end{bmatrix}.$$

The action of the map on $\mathbf{x}$ is

$$A\mathbf{x} = 4P_1\mathbf{x} + 2P_2\mathbf{x} = \begin{bmatrix} 0 \\ 4 \end{bmatrix} + \begin{bmatrix} 2 \\ 0 \end{bmatrix} = \begin{bmatrix} 2 \\ 4 \end{bmatrix}.$$

Draw a sketch!

12. The quadratic form for C_1 is an ellipsoid and $\lambda_i = 4, 2$. The quadratic form for C_2 is an hyperboloid and $\lambda_i = 6.12, -2.12$. The quadratic form for C_3 is an paraboloid and $\lambda_i = 6, 0$.

13. The quadratic forms for the given C_i are

$$f_1(\mathbf{v}) = 3v_1^2 + 2v_1v_2 + 3v_2^2,$$
$$f_2(\mathbf{v}) = 3v_1^2 + 8v_1v_2 + 1v_2^2,$$
$$f_3(\mathbf{v}) = 3v_1^2 + 6v_1v_2 + 3v_2^2.$$

15. $C = \begin{bmatrix} 2 & 2 \\ 2 & 3 \end{bmatrix}$.

16. No, it is not a quadratic form because there is a linear term $2v_2$.

18. The 2×2 matrix A is symmetric, therefore the eigenvalues will be real. The eigenvalues will be positive since the matrix is positive definite, which we conclude from the fact that the determinant is positive.

20. Repeated linear maps do not change the eigenvectors, as demonstrated by (7.18), and the eigenvalues are easily computed. The eigenvalues of A are $\lambda_1 = 3$ and $\lambda_2 = -1$; therefore, for A^2 we have $\lambda_1 = 9$ and $\lambda_2 = 1$, and for A^3 we have $\lambda_1 = 27$ and $\lambda_2 = -1$.

Chapter 8

1. For the given vector

$$\mathbf{r} = \begin{bmatrix} 4 \\ 2 \\ 4 \end{bmatrix},$$

we calculate $\|\mathbf{r}\| = 6$, then

$$\frac{\mathbf{r}}{\|\mathbf{r}\|} = \begin{bmatrix} 2/3 \\ 1/3 \\ 2/3 \end{bmatrix}.$$

$\|2\mathbf{r}\| = 2\|\mathbf{r}\| = 12$.

3. The cross product of the vectors $\mathbf{v}$ and $\mathbf{w}$:

$$\mathbf{v} \wedge \mathbf{w} = \begin{bmatrix} 0 \\ -1 \\ 1 \end{bmatrix}.$$

5. The sine of the angle between $\mathbf{v}$ and $\mathbf{w}$:

$$\sin\theta = \frac{\|\mathbf{v} \wedge \mathbf{w}\|}{\|\mathbf{v}\|\|\mathbf{w}\|} = \frac{\sqrt{2}}{1 \times \sqrt{3}} = 0.82.$$

This means that $\theta = 54.7°$. Draw a sketch to double-check this for yourself.

7. Two 3D lines are skew if they do not intersect. They intersect if there exists s and t such that $l_1(t) = l_2(s)$, or

$$t \begin{bmatrix} 0 \\ 0 \\ 1 \end{bmatrix} - s \begin{bmatrix} -1 \\ -1 \\ 1 \end{bmatrix} = \begin{bmatrix} 1 \\ 1 \\ 1 \end{bmatrix} - \begin{bmatrix} 0 \\ 0 \\ 1 \end{bmatrix}.$$

Examining each coordinate, we find that $s = 1$ and $t = 1$ satisfy these three equations, thus the lines intersect and are not skew.

8. The point normal form of the plane through $\mathbf{p}$ with normal direction $\mathbf{r}$ is found by first defining the normal $\mathbf{n}$ by normalizing $\mathbf{r}$:

$$\mathbf{n} = \frac{\mathbf{r}}{\|\cdot\|} = \begin{bmatrix} 2/3 \\ 1/3 \\ 2/3 \end{bmatrix}.$$

The point normal form of the plane is

$$\mathbf{n}(\mathbf{x} - \mathbf{p}) = 0,$$

or

$$\frac{2}{3}x_1 + \frac{1}{3}x_2 + \frac{2}{3}x_3 - \frac{2}{3} = 0.$$

10. A parametric form of the plane P through the points $\mathbf{p}$, $\mathbf{q}$, and $\mathbf{r}$ is

$$\begin{aligned} P(s,t) &= \mathbf{p} + s(\mathbf{q} - \mathbf{p}) + t(\mathbf{r} - \mathbf{p}) \\ &= (1 - s - t)\mathbf{p} + s\mathbf{q} + t\mathbf{r} \\ &= (1 - s - t)\begin{bmatrix} 0 \\ 0 \\ 1 \end{bmatrix} + s\begin{bmatrix} 1 \\ 1 \\ 1 \end{bmatrix} + t\begin{bmatrix} 4 \\ 2 \\ 4 \end{bmatrix}. \cdot \cdot \end{aligned}$$

12. The length d of the projection of $\mathbf{w}$ bound to $\mathbf{q}$ is calculated by using two definitions of $\cos(\theta)$:

$$\cos(\theta) = \frac{d}{\|\mathbf{w}\|} \quad \text{and} \quad \cos(\theta) = \frac{\mathbf{v} \cdot \mathbf{w}}{\|\mathbf{v}\|\|\mathbf{w}\|}.$$

Solve for d,

$$d = \frac{\mathbf{v} \cdot \mathbf{w}}{\|\mathbf{v}\|} = \frac{\begin{bmatrix} 1 \\ 0 \\ 0 \end{bmatrix} \cdot \begin{bmatrix} 1 \\ 1 \\ 1 \end{bmatrix}}{\left\| \begin{bmatrix} 1 \\ 0 \\ 0 \end{bmatrix} \right\|} = 1.$$

The projection length has nothing to do with $\mathbf{q}$!

13. The distance h may be found using two definitions of $\sin(\theta)$:

$$\sin(\theta) = \frac{h}{\|\mathbf{w}\|} \quad \text{and} \quad \sin(\theta) = \frac{\|\mathbf{v} \wedge \mathbf{w}\|}{\|\mathbf{v}\|\|\mathbf{w}\|}.$$

Solve for h,

$$h = \frac{\|\mathbf{v} \wedge \mathbf{w}\|}{\|\mathbf{v}\|} = \frac{\left\| \begin{bmatrix} 0 \\ -1 \\ 1 \end{bmatrix} \right\|}{\left\| \begin{bmatrix} 1 \\ 0 \\ 0 \end{bmatrix} \right\|} = \sqrt{2}.$$

14. The cross product of parallel vectors results in the zero vector.

17. The volume V formed by the vectors $\mathbf{v}, \mathbf{w}, \mathbf{u}$ can be computed as the scalar triple product

$$V = \mathbf{v} \cdot (\mathbf{w} \wedge \mathbf{u}).$$

This is invariant under cyclic permutations, thus we can also compute V as

$$V = \mathbf{u} \cdot (\mathbf{v} \wedge \mathbf{w}),$$

which allows us to reuse the cross product from Exercise 1. Thus,

$$V = \begin{bmatrix} 0 \\ 0 \\ 1 \end{bmatrix} \cdot \begin{bmatrix} 0 \\ -1 \\ 1 \end{bmatrix} = 1.$$

18. This solution is easy enough to determine without formulas, but let's practice using the equations. First, project $\mathbf{w}$ onto $\mathbf{v}$, forming

$$\mathbf{u}_1 = \frac{\mathbf{v} \cdot \mathbf{w}}{\|\mathbf{v}\|^2} \mathbf{v} = \begin{bmatrix} 1 \\ 0 \\ 0 \end{bmatrix}.$$

Next form $\mathbf{u}_2$ so that $\mathbf{w} = \mathbf{u}_1 + \mathbf{u}_2$:

$$\mathbf{u}_2 = \mathbf{w} - \mathbf{u}_1 = \begin{bmatrix} 0 \\ 1 \\ 1 \end{bmatrix}.$$

To complete the orthogonal frame, we compute

$$\mathbf{u}_3 = \mathbf{u}_1 \wedge \mathbf{u}_2 = \begin{bmatrix} 0 \\ -1 \\ 1 \end{bmatrix}.$$

Draw a sketch to see what you have created. This frame is not orthonormal, but that would be easy to do.

19. We use barycentric coordinates to determine the color at a point inside the triangle given the colors at the vertices:

$$
\begin{aligned}
\mathbf{i_c} &= \frac{1}{3}\mathbf{i_p} + \frac{1}{3}\mathbf{i_q} + \frac{1}{3}\mathbf{i_r} \\
&= \frac{1}{3}\begin{bmatrix} 1 \\ 0 \\ 0 \end{bmatrix} + \frac{1}{3}\begin{bmatrix} 0 \\ 1 \\ 0 \end{bmatrix} + \frac{1}{3}\begin{bmatrix} 1 \\ 0 \\ 0 \end{bmatrix} \\
&= \begin{bmatrix} 2/3 \\ 1/3 \\ 0 \end{bmatrix},
\end{aligned}
$$

which is reddish-yellow.

1. The equation in matrix form is

Chapter 9

$$
\mathbf{v}' = \begin{bmatrix} 7 \\ 6 \\ 3 \end{bmatrix} = \begin{bmatrix} 1 & 2 & 0 \\ 1 & 0 & 3 \\ 1 & 0 & 0 \end{bmatrix} \begin{bmatrix} 3 \\ 2 \\ 1 \end{bmatrix}.
$$

2. The transpose of the given matrix is

$$
\begin{bmatrix} 1 & -1 & 2 \\ 5 & -2 & 3 \\ -4 & 0 & -4 \end{bmatrix}.
$$

3. The vector $\mathbf{u}$ is *not* an element of the subspace defined by $\mathbf{w}$ and $\mathbf{v}$ because we cannot find scalars s and t such that $\mathbf{u} = s\mathbf{w} + t\mathbf{v}$. This linear combination requires that $t = 0$ and $t = 1$. (We could check if the parallelepiped formed by the three vectors, $V = \mathbf{u} \cdot (\mathbf{v} \wedge \mathbf{w})$, is zero. Here we find that $V = 1$, confirming the result.)

5. No, $\mathbf{u}$ is by definition orthogonal to $\mathbf{v}$ and $\mathbf{w}$, therefore it is not possible to write $\mathbf{u}$ as a linear combination of $\mathbf{v}$ and $\mathbf{w}$.

7. No. If $\mathbf{v}$ and $\mathbf{w}$ were linearly dependent, then they would form a 1D subspace of $\mathbb{R}^3$.

10. The scale matrix is

$$
\begin{bmatrix} 2 & 0 & 0 \\ 0 & 1/4 & 0 \\ 0 & 0 & -4 \end{bmatrix}.
$$

This matrix changes the volume of a unit cube to be $2 \times 1/4 \times -4 = -2$.

12. The shear matrix is

$$
\begin{bmatrix} 1 & 0 & -a/c \\ 0 & 1 & -b/c \\ 0 & 0 & 1 \end{bmatrix}.
$$

Shears do not change volume, therefore the volume of the mapped unit cube is still 1.

14. To rotate about the vector $\begin{bmatrix} -1 \\ 0 \\ -1 \end{bmatrix}$, first form the unit vector $\mathbf{a} = \begin{bmatrix} -1/\sqrt{2} \\ 0 \\ -1/\sqrt{2} \end{bmatrix}$. Then, following (9.10), the rotation matrix is

$$\begin{bmatrix} \frac{1}{2}(1 + \frac{\sqrt{2}}{2}) & 1/2 & \frac{1}{2}(1 - \frac{\sqrt{2}}{2}) \\ -1/2 & \sqrt{2}/2 & 1/2 \\ \frac{1}{2}(1 - \frac{\sqrt{2}}{2}) & -1/2 & \frac{1}{2}(1 + \frac{\sqrt{2}}{2}) \end{bmatrix}.$$

The matrices for rotating about an arbitrary vector are difficult to verify by inspection. One test is to check the vector about which we rotated. You'll find that

$$\begin{bmatrix} -1 \\ 0 \\ -1 \end{bmatrix} \rightarrow \begin{bmatrix} -1 \\ 0 \\ -1 \end{bmatrix},$$

which is precisely correct.

16. The projection is defined as $P = AA^{\mathrm{T}}$, where $A = [\mathbf{u}_1 \quad \mathbf{u}_2]$, which results in

$$P = \begin{bmatrix} 1/2 & 1/2 & 0 \\ 1/2 & 1/2 & 0 \\ 0 & 0 & 1 \end{bmatrix}.$$

The action of P on a vector $\mathbf{v}$ is

$$\mathbf{v}' = \begin{bmatrix} 1/2(v_1 + v_2) \\ 1/2(v_1 + v_2) \\ v_3 \end{bmatrix}.$$

The vectors, $\mathbf{v}_1, \mathbf{v}_2, \mathbf{v}_3$ are mapped to

$$\mathbf{v}_1' = \begin{bmatrix} 1 \\ 1 \\ 1 \end{bmatrix}, \quad \mathbf{v}_2' = \begin{bmatrix} 1/2 \\ 1/2 \\ 0 \end{bmatrix}, \quad \mathbf{v}_3' = \begin{bmatrix} 0 \\ 0 \\ 0 \end{bmatrix}.$$

This is an orthogonal projection into the plane $x_1 - x_2 = 0$. Notice that the $\mathbf{e}_3$ component keeps its value since that component already lives in this plane.

The first two column vectors are identical, thus the matrix rank is 2 and the determinant is zero.

17. To find the projection direction, we solve the homogeneous system $P\mathbf{d} = \mathbf{0}$. This system gives us two equations to satisfy:

$$d_1 + d_2 = 0 \quad \text{and} \quad d_3 = 0,$$

which have infinitely many nontrivial solutions, $\mathbf{d} = [c \quad -c \quad 0]^{\mathrm{T}}$. All vectors $\mathbf{d}$ are mapped to the zero vector. The vector $\mathbf{v}_3$ in the previous exercise is part of the kernel, thus its image $\mathbf{v}_3' = \mathbf{0}$.

20. The matrices have the following determinants:

$$|A_1| = c_1|A|, \quad |A_2| = c_1 c_2 |A|, \quad |A_3| = c_1 c_2 c_3 |A|.$$

21. The determinant of the inverse is

$$|A^{-1}| = \frac{1}{|A|} = \frac{1}{2}.$$

22. The resulting matrix is

$$\begin{bmatrix} 2 & 3 & -4 \\ 3 & 9 & -4 \\ -1 & -9 & 4 \end{bmatrix}.$$

23. The matrix products are as follows:

$$AB = \begin{bmatrix} 5 & 3 & 2 \\ 0 & 2 & 0 \\ 2 & 2 & 1 \end{bmatrix} \quad \text{and} \quad BA = \begin{bmatrix} 2 & 0 & 0 \\ 4 & 3 & 4 \\ 1 & 2 & 3 \end{bmatrix}.$$

25. The inverse matrices are as follows:

$$\text{rotation:} \quad \begin{bmatrix} 1/\sqrt{2} & 0 & -1/\sqrt{2} \\ 0 & 1 & 0 \\ 1/\sqrt{2} & 0 & 1/\sqrt{2} \end{bmatrix},$$

$$\text{scale:} \quad \begin{bmatrix} 2 & 0 & 0 \\ 0 & 4 & 0 \\ 0 & 0 & 1/2 \end{bmatrix},$$

$$\text{projection:} \quad \text{no inverse exists.}$$

27. Only square matrices have an inverse, therefore this 3×2 matrix has no inverse.

29. The matrix $(A^\mathrm{T})^\mathrm{T}$ is simply A.

1. An affine map is comprised of a linear map and a translation.   Chapter 10

2. Construct the matrices in (10.6). For this problem, use the notation $A = YX^{-1}$, where

$$Y = \begin{bmatrix} \mathbf{y}_2 - \mathbf{y}_1 & \mathbf{y}_3 - \mathbf{y}_1 & \mathbf{y}_4 - \mathbf{y}_1 \end{bmatrix} = \begin{bmatrix} 1 & 1 & 1 \\ -1 & 0 & 0 \\ 0 & -1 & 1 \end{bmatrix}$$

and

$$X = \begin{bmatrix} \mathbf{x}_2 - \mathbf{x}_1 & \mathbf{x}_3 - \mathbf{x}_1 & \mathbf{x}_4 - \mathbf{x}_1 \end{bmatrix} = \begin{bmatrix} -1 & -1 & -1 \\ 1 & 0 & 0 \\ 0 & -1 & 1 \end{bmatrix}.$$

The first task is to find X^{-1}:

$$X^{-1} = \begin{bmatrix} 0 & 1 & 0 \\ -1/2 & -1/2 & -1/2 \\ -1/2 & -1/2 & 1/2 \end{bmatrix}.$$

Always check that $XX^{-1} = I$. Now A takes the form:

$$A = \begin{bmatrix} -1 & 0 & 0 \\ 0 & -1 & 0 \\ 0 & 0 & 1 \end{bmatrix},$$

which isn't surprising at all if you sketched the tetrahedra formed by the $\mathbf{x}_i$ and $\mathbf{y}_i$.

3. The point $\mathbf{p}$ is mapped to

$$\mathbf{p}' = \begin{bmatrix} -1 \\ -1 \\ 1 \end{bmatrix}.$$

We can find this point using the matrix from the previous exercise, $\mathbf{p}' = A\mathbf{p}$ or with barycentric coordinates. Clearly, the barycentric coordinates for $\mathbf{p}$ with respect to the $\mathbf{x}_i$ are $(1, 1, -1, 0)$, thus

$$\mathbf{p}' = 1\mathbf{y}_1 + 1\mathbf{y}_2 - 1\mathbf{y}_3 + 0\mathbf{y}_4.$$

6. As always, draw a sketch when possible! The plane that is used here is shown in Figure B.1. Since the plane is parallel to the $\mathbf{e}_2$ axis, only a side view is shown. The plane is shown as a thick line.

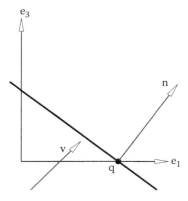

Figure B.1.
The plane for these exercises.

Construct the parallel projection map defined in (10.8):

$$\mathbf{x}' = \begin{bmatrix} 8/14 & 0 & -8/14 \\ 0 & 1 & 0 \\ -6/14 & 0 & 6/14 \end{bmatrix} \mathbf{x} + \begin{bmatrix} 6/14 \\ 0 \\ 6/14 \end{bmatrix}.$$

The $\mathbf{x}_i$ are mapped to

$$\mathbf{x}_1' = \begin{bmatrix} 1 \\ 0 \\ 0 \end{bmatrix}, \quad \mathbf{x}_2' = \begin{bmatrix} 3/7 \\ 1 \\ 3/7 \end{bmatrix}, \quad \mathbf{x}_3' = \begin{bmatrix} 1 \\ 0 \\ 0 \end{bmatrix}, \quad \mathbf{x}_4' = \begin{bmatrix} -2/14 \\ 0 \\ 12/14 \end{bmatrix}.$$

Since $\mathbf{x}_1$ is already in the plane, $\mathbf{x}_1' = \mathbf{x}_1$. The projection direction $\mathbf{v}$ is parallel to the $\mathbf{e}_2$-axis, thus this coordinate is unchanged in $\mathbf{x}_2$. Additionally, $\mathbf{v}$ projects equally into the $\mathbf{e}_1$- and $\mathbf{e}_3$-axes. Rationalize the results for $\mathbf{x}_3$ and $\mathbf{x}_4$ yourself.

8. Construct the perspective projection vector equation defined in (10.10). This will be different for each of the $\mathbf{x}_i$:

$$\mathbf{x}_1' = \frac{3/5}{3/5}\mathbf{x}_1 = \begin{bmatrix} 1 \\ 0 \\ 0 \end{bmatrix},$$

$$\mathbf{x}_2' = \frac{3/5}{0}\mathbf{x}_2,$$

$$\mathbf{x}_3' = \frac{3/5}{-4/5}\mathbf{x}_3 = \begin{bmatrix} 0 \\ 0 \\ 3/4 \end{bmatrix},$$

$$\mathbf{x}_4' = \frac{3/5}{4/5}\mathbf{x}_4 = \begin{bmatrix} 0 \\ 0 \\ 3/4 \end{bmatrix}.$$

There is no solution for $\mathbf{x}_2'$ because $\mathbf{x}_2$ is projected parallel to the plane.

12. First, translate so that a point $\mathbf{p}$ on the line is at the origin. Second, rotate $90°$ about the $\mathbf{e}_1$-axis with the matrix R. Third, remove the initial translation. Let's define

$$\mathbf{p} = \begin{bmatrix} 1 \\ 1 \\ 0 \end{bmatrix}, \qquad R = \begin{bmatrix} 1 & 0 & 0 \\ 0 & \cos(90°) & -\sin(90°) \\ 0 & \sin(90°) & \cos(90°) \end{bmatrix},$$

and then the affine map is defined as

$$\mathbf{q}' = R(\mathbf{q} - \mathbf{p}) + \mathbf{p} = \begin{bmatrix} 1 \\ 1 \\ -1 \end{bmatrix}.$$

13. Let S be the scale matrix and R be the rotation matrix, then the affine map is $\mathbf{x}' = RS\mathbf{x} + \mathbf{l}'$, or specifically

$$\mathbf{x}' = \begin{bmatrix} 2/\sqrt{2} & 0 & -2/\sqrt{2} \\ 0 & 2 & 0 \\ 2/\sqrt{2} & 0 & 2/\sqrt{2} \end{bmatrix} \mathbf{x} + \begin{bmatrix} -2 \\ -2 \\ -2 \end{bmatrix}.$$

The point $\mathbf{u}$ is mapped to

$$\mathbf{u}' = \begin{bmatrix} -2 \\ 0 \\ 0.83 \end{bmatrix}.$$

Chapter 11

1. First we find the point normal form of the implicit plane as

$$\frac{1}{3}x_1 + \frac{2}{3}x_2 - \frac{2}{3}x_3 - \frac{1}{3} = 0.$$

Substitute $\mathbf{p}$ into the plane equation to find the distance d to the plane, $d = -5/3 \approx -1.67$. Since the distance is negative, the point is on the opposite side of the plane as the normal direction.

4. The point on each line that corresponds to the point of closest proximity is found by setting up the linear system,

$$[\mathbf{x}_2(s_2) - \mathbf{x}_1(s_1)]\mathbf{v}_1 = 0,$$
$$[\mathbf{x}_2(s_2) - \mathbf{x}_1(s_1)]\mathbf{v}_2 = 0.$$

For the given lines, this linear system is

$$\begin{bmatrix} 3 & -1 \\ 1 & -3 \end{bmatrix} \begin{bmatrix} s_1 \\ s_2 \end{bmatrix} = \begin{bmatrix} 1 \\ -1 \end{bmatrix},$$

and the solution is $s_1 = 1/2$ and $s_2 = 1/2$. This corresponds to closest proximity points

$$\mathbf{x}_1(1/2) = \begin{bmatrix} 1/2 \\ 1/2 \\ 1/2 \end{bmatrix} \qquad \text{and} \quad \mathbf{x}_2(1/2) = \begin{bmatrix} 1/2 \\ 1/2 \\ 1/2 \end{bmatrix}.$$

The lines intersect!

5. Make a sketch! You will find that the line is parallel to the plane. The actual calculations would involve finding the parameter t on the line for the intersection:

$$t = \frac{(\mathbf{p} - \mathbf{q}) \cdot \mathbf{n}}{\mathbf{v} \cdot \mathbf{n}}.$$

In this exercise, $\mathbf{v} \cdot \mathbf{n} = 0$.

7. The vector $\mathbf{a}$ is projected in the direction $\mathbf{v}$, thus

$$\mathbf{a}' = \mathbf{a} + t\mathbf{v},$$

for some unknown t. The vector $\mathbf{a}'$ is in the plane P, thus $\mathbf{a}' \cdot \mathbf{n} = 0$. Substitute the expression for $\mathbf{a}'$ into the plane equation,

$$(\mathbf{a} + t\mathbf{v}) \cdot \mathbf{n} = 0,$$

from which we solve for t. Therefore, the vector $\mathbf{a}$ is projected to the vector

$$\mathbf{a}' = \mathbf{a} - \frac{\mathbf{a} \cdot \mathbf{n}}{\mathbf{v} \cdot \mathbf{n}} \mathbf{v}.$$

9. The parametric form of the plane containing the triangle is given by

$$\mathbf{x}(u, v) = \begin{bmatrix} 0 \\ 0 \\ 1 \end{bmatrix} + u \begin{bmatrix} 1 \\ 0 \\ -1 \end{bmatrix} + v \begin{bmatrix} 0 \\ 1 \\ -1 \end{bmatrix}.$$

Thus the intersection problem reduces to finding u, v, t in

$$\begin{bmatrix} 2 \\ 2 \\ 2 \end{bmatrix} + t \begin{bmatrix} -1 \\ -1 \\ -2 \end{bmatrix} = \begin{bmatrix} 0 \\ 0 \\ 1 \end{bmatrix} + u \begin{bmatrix} 1 \\ 0 \\ -1 \end{bmatrix} + v \begin{bmatrix} 0 \\ 1 \\ -1 \end{bmatrix}.$$

We rewrite this as a linear system:

$$\begin{bmatrix} 1 & 0 & 1 \\ 0 & 1 & 1 \\ -1 & -1 & 2 \end{bmatrix} \begin{bmatrix} u \\ v \\ t \end{bmatrix} = \begin{bmatrix} 2 \\ 2 \\ 1 \end{bmatrix}.$$

Since we have yet to discuss Gauss elimination for 3×3 matrices, let's use the hint to solve the system. The first two equations lead to $u = 2 - t$ and $v = 2 - t$. This allows us to use the first equation to find $t = 5/4$, then we can easily find that

$$\begin{bmatrix} u \\ v \\ t \end{bmatrix} = \begin{bmatrix} 3/4 \\ 3/4 \\ 5/4 \end{bmatrix}.$$

Since $1 - u - v = -1/2$, the intersection point is outside of the triangle.

10. The reflected direction is

$$\mathbf{v}' = \begin{bmatrix} 1/3 \\ 1/3 \\ -2/3 \end{bmatrix} + \frac{4}{3} \begin{bmatrix} 0 \\ 0 \\ 1 \end{bmatrix} = \begin{bmatrix} 1/3 \\ 1/3 \\ 2/3 \end{bmatrix}.$$

12. With a sketch, you can find the intersection point without calculation, but let's practice calculating the point. The linear system is:

$$\begin{bmatrix} 1 & 1 & 0 \\ 1 & 0 & 0 \\ 0 & 0 & 1 \end{bmatrix} \mathbf{x} = \begin{bmatrix} 1 \\ 1 \\ 4 \end{bmatrix},$$

and the solution is $\begin{bmatrix} 1 \\ 0 \\ 4 \end{bmatrix}$.

14. With a sketch you can find the intersection line without calculation, but let's practice calculating the line. The planes $x_1 = 1$ and $x_3 = 4$ have normal vectors

$$\mathbf{n}_1 = \begin{bmatrix} 1 \\ 0 \\ 0 \end{bmatrix} \quad \text{and} \quad \mathbf{n}_2 = \begin{bmatrix} 0 \\ 0 \\ 1 \end{bmatrix},$$

respectively. Form the vector

$$\mathbf{v} = \mathbf{n}_1 \wedge \mathbf{n}_2 = \begin{bmatrix} 0 \\ 1 \\ 0 \end{bmatrix},$$

and then a third plane is defined by $\mathbf{v} \cdot \mathbf{x} = 0$. Set up a linear system to intersect these three planes:

$$\begin{bmatrix} 1 & 0 & 0 \\ 0 & 0 & 1 \\ 0 & 1 & 0 \end{bmatrix} \mathbf{p} = \begin{bmatrix} 1 \\ 4 \\ 0 \end{bmatrix},$$

and the solution is

$$\begin{bmatrix} 1 \\ 0 \\ 4 \end{bmatrix}.$$

The intersection of the given two planes is the line

$$\mathbf{l}(t) = \begin{bmatrix} 1 \\ 0 \\ 4 \end{bmatrix} + t \begin{bmatrix} 0 \\ 1 \\ 0 \end{bmatrix}.$$

16. Set

$$\mathbf{b}_1 = \begin{bmatrix} 1 \\ 0 \\ 0 \end{bmatrix},$$

and then calculate

$$\mathbf{b}_2 = \begin{bmatrix} 1 \\ 1 \\ 0 \end{bmatrix} - (1) \begin{bmatrix} 1 \\ 0 \\ 0 \end{bmatrix} = \begin{bmatrix} 0 \\ 1 \\ 0 \end{bmatrix},$$

$$\mathbf{b}_3 = \begin{bmatrix} -1 \\ -1 \\ 1 \end{bmatrix} - (-1) \begin{bmatrix} 1 \\ 0 \\ 0 \end{bmatrix} - (-1) \begin{bmatrix} 0 \\ 1 \\ 0 \end{bmatrix} = \begin{bmatrix} 0 \\ 0 \\ 1 \end{bmatrix}.$$

Each vector is unit length, so we have an orthonormal frame.

Chapter 12

1. Refer to the system as $A\mathbf{u} = \mathbf{b}$. We do not expect a unique solution since $|A| = 0$. In fact, the $\mathbf{a}_i$ span the $[\mathbf{e}_1, \mathbf{e}_3]$-plane and $\mathbf{b}$ lives in $\mathbb{R}^3$, therefore the system is inconsistent because no solution exists.

3. The t_i must be unique. If $t_i = t_j$ then two rows of the matrix would be identical and result in a zero determinant.

4. The steps for transforming A to upper triangular are given along with the augmented matrix after each step.

Exchange row_1 and row_3.

$$\begin{bmatrix} 2 & 0 & 0 & 1 & 1 \\ 0 & 0 & 1 & -2 & 2 \\ 1 & 0 & -1 & 2 & -1 \\ 1 & 1 & 1 & 1 & -3 \end{bmatrix}$$

$\text{row}_3 \leftarrow \text{row}_3 - \frac{1}{2}\text{row}_1, \quad \text{row}_4 \leftarrow \text{row}_4 - \frac{1}{2}\text{row}_1$

$$\begin{bmatrix} 2 & 0 & 0 & 1 & 1 \\ 0 & 0 & 1 & -2 & 2 \\ 0 & 0 & -1 & 3/2 & -3/2 \\ 0 & 1 & 1 & 1/2 & -7/2 \end{bmatrix}$$

Exchange row_2 and row_4.

$$\begin{bmatrix} 2 & 0 & 0 & 1 & 1 \\ 0 & 1 & 1 & 1/2 & -7/2 \\ 0 & 0 & -1 & 3/2 & -3/2 \\ 0 & 0 & 1 & -2 & 2 \end{bmatrix}$$

$\text{row}_4 \leftarrow \text{row}_4 + \text{row}_3$

$$\begin{bmatrix} 2 & 0 & 0 & 1 & 1 \\ 0 & 1 & 1 & 1/2 & -7/2 \\ 0 & 0 & -1 & 3/2 & -3/2 \\ 0 & 0 & 0 & -1/2 & 1/2 \end{bmatrix}$$

This is the upper triangular form of the coefficient matrix. Back substitution results in the solution vector

$$\mathbf{v} = \begin{bmatrix} 1 \\ -3 \\ 0 \\ -1 \end{bmatrix}.$$

5. The steps for transforming A to upper triangular are given along with the augmented matrix after each step.

Exchange row_1 and row_2.

$$\begin{bmatrix} 1 & 0 & 0 & 0 \\ 0 & 0 & 1 & -1 \\ 1 & 1 & 1 & -1 \end{bmatrix}.$$

$\text{row}_3 \leftarrow \text{row}_3 - \text{row}_1$

$$\begin{bmatrix} 1 & 0 & 0 & 0 \\ 0 & 0 & 1 & -1 \\ 0 & 1 & 1 & -1 \end{bmatrix}.$$

Exchange row_2 and row_3.

$$\begin{bmatrix} 1 & 0 & 0 & 0 \\ 0 & 1 & 1 & -1 \\ 0 & 0 & 1 & -1 \end{bmatrix}.$$

This is the upper triangular form of the coefficient matrix. Back substitution results in the solution vector

$$\mathbf{v} = \begin{bmatrix} 0 \\ 0 \\ -1 \end{bmatrix}.$$

7. The linear system in row echelon form is

$$\begin{bmatrix} 1 & 2/3 & 0 \\ 0 & 1 & -2 \\ 0 & 0 & 1 \end{bmatrix} \mathbf{u} = \begin{bmatrix} 1/3 \\ 0 \\ 1/4 \end{bmatrix}.$$

9. The 5×5 permutation matrix that exchanges rows 3 and 4 is

$$\begin{bmatrix} 1 & 0 & 0 & 0 & 0 \\ 0 & 1 & 0 & 0 & 0 \\ 0 & 0 & 0 & 1 & 0 \\ 0 & 0 & 1 & 0 & 0 \\ 0 & 0 & 0 & 0 & 1 \end{bmatrix}.$$

Let's test the matrix: $[v_1, v_2, v_3, v_4, v_5]$ is mapped to $[v_1, v_2, v_4, v_3, v_5]$.

11. For a 3×3 matrix, $G = G_2 P_2 G_1 P_1$, and in this example, there is no pivoting for column 2, thus $P_2 = I$. The other matrices are

$$P_1 = \begin{bmatrix} 0 & 1 & 0 \\ 1 & 0 & 0 \\ 0 & 0 & 1 \end{bmatrix}, \quad G_1 = \begin{bmatrix} 1 & 0 & 0 \\ -1/2 & 1 & 0 \\ -1 & 0 & 1 \end{bmatrix}, \quad G_2 = \begin{bmatrix} 1 & 0 & 0 \\ 0 & 1 & 0 \\ 0 & 1 & 1 \end{bmatrix},$$

resulting in

$$G = \begin{bmatrix} 0 & 1 & 0 \\ 1 & -1/2 & 0 \\ 1 & -3/2 & 1 \end{bmatrix}.$$

Let's check this result by computing

$$GA = \begin{bmatrix} 0 & 1 & 0 \\ 1 & -1/2 & 0 \\ 1 & -3/2 & 1 \end{bmatrix} \begin{bmatrix} 2 & -2 & 0 \\ 4 & 0 & -2 \\ 4 & 2 & -4 \end{bmatrix} = \begin{bmatrix} 4 & 0 & -2 \\ 0 & -2 & 1 \\ 0 & 0 & -1 \end{bmatrix},$$

which is the final upper triangular matrix from the example.

13. For this homogeneous system, we apply forward elimination, resulting in

$$\begin{bmatrix} 4 & 1 & 2 \\ 0 & 1/2 & 0 \\ 0 & 0 & 0 \end{bmatrix} \mathbf{u} = \begin{bmatrix} 0 \\ 0 \\ 0 \end{bmatrix}.$$

Before executing back substitution, assign a nonzero value to u_3, so choose $u_3 = 1$. Back substitution results in $u_2 = 0$ and $u_1 = -1/2$.

16. Simply by looking at the matrix, it is clear that second and third column vectors are linearly dependent, thus it is not invertible.

17. This is a rotation matrix, and the inverse is simply the transpose.

$$\begin{bmatrix} \cos\theta & 0 & \sin\theta \\ 0 & 1 & 0 \\ -\sin\theta & 0 & \cos\theta \end{bmatrix}.$$

19. This matrix is not square, therefore it is not invertible.

21. The forward substitution algorithm for solving the lower triangular linear system $L\mathbf{y} = \mathbf{b}$ is as follows.

Forward substitution:
$y_1 = b_1$
For $j = 2, \ldots, n$,
$$y_j = b_j - y_1 l_{j,1} - \ldots - y_{j-1} l_{n,j-1}.$$

23. The determinant is equal to 5.

24. The rank is 3 since the determinant is nonzero.

26. The determinants that we need to calculate are

$$
\begin{vmatrix} 3 & 0 & 1 \\ 1 & 2 & 0 \\ 1 & 1 & 1 \end{vmatrix} = 5
$$

and

$$
\begin{vmatrix} 8 & 0 & 1 \\ 6 & 2 & 0 \\ 6 & 1 & 1 \end{vmatrix} = 10, \quad
\begin{vmatrix} 3 & 8 & 1 \\ 1 & 6 & 0 \\ 1 & 6 & 1 \end{vmatrix} = 10, \quad
\begin{vmatrix} 3 & 0 & 8 \\ 1 & 2 & 6 \\ 1 & 1 & 6 \end{vmatrix} = 10.
$$

The solution is $\begin{bmatrix} 2 \\ 2 \\ 2 \end{bmatrix}$.

28. This intersection problem was introduced as Example 11.4 and it is illustrated in Sketch 11.12. The linear system is

$$
\begin{bmatrix} 1 & 0 & 1 \\ 0 & 0 & 1 \\ 0 & 1 & 0 \end{bmatrix}
\begin{bmatrix} x_1 \\ x_2 \\ x_3 \end{bmatrix} =
\begin{bmatrix} 1 \\ 1 \\ 2 \end{bmatrix}.
$$

Solving it by Gauss elimination needs very few steps. We simply exchange rows two and three and then apply back substitution to find

$$
\begin{bmatrix} x_1 \\ x_2 \\ x_3 \end{bmatrix} =
\begin{bmatrix} 0 \\ 2 \\ 1 \end{bmatrix}.
$$

30. Use the explicit form of the line, $x_2 = ax_1 + b$. The overdetermined system is

$$
\begin{bmatrix} -2 & 1 \\ -1 & 1 \\ 0 & 1 \\ 1 & 1 \\ 2 & 1 \end{bmatrix}
\begin{bmatrix} a \\ b \end{bmatrix} =
\begin{bmatrix} 2 \\ 1 \\ 0 \\ 1 \\ 2 \end{bmatrix}.
$$

Following (12.21), we form the normal equations and the linear system becomes

$$
\begin{bmatrix} 10 & 0 \\ 0 & 5 \end{bmatrix}
\begin{bmatrix} a \\ b \end{bmatrix} =
\begin{bmatrix} 0 \\ 6 \end{bmatrix}.
$$

Thus the least squares line is $x_2 = 6/5$. Sketch the data and the line to convince yourself.

32. Use the explicit form of the line, $x_2 = ax_1 + b$. The linear system for two points is not overdetermined, however let's apply the least squares technique to see what happens. The system is

$$
\begin{bmatrix} -4 & 1 \\ 4 & 1 \end{bmatrix}
\begin{bmatrix} a \\ b \end{bmatrix} =
\begin{bmatrix} 1 \\ 3 \end{bmatrix}.
$$

The normal equations are

$$\begin{bmatrix} 32 & 0 \\ 0 & 2 \end{bmatrix} \begin{bmatrix} a \\ b \end{bmatrix} = \begin{bmatrix} 8 \\ 4 \end{bmatrix}.$$

Thus the least squares line is $x_2 = (1/4)x_1 + 2$. Let's evaluate the line at the given points:

$$x_2 = (1/4)(-4) + 2 = 1 \qquad x_2 = (1/4)(4) + 2 = 3$$

This is the interpolating line! This is what we expect since the solution to the original linear system must be unique. Sketch the data and the line to convince yourself.

1. The linear system after running through the algorithm for $j = 1$ is

$$\begin{bmatrix} -1.41 & -1.41 & -1.41 \\ 0 & -0.14 & -0.14 \\ 0 & -0.1 & 0.2 \end{bmatrix} \mathbf{u} = \begin{bmatrix} -1.41 \\ 0 \\ 0.3 \end{bmatrix}.$$

The linear system for $j = 2$ is

$$\begin{bmatrix} -1.41 & -1.41 & -1.41 \\ 0 & 0.17 & -0.02 \\ 0 & 0 & 0.24 \end{bmatrix} \mathbf{u} = \begin{bmatrix} -1.41 \\ -0.19 \\ 0.24 \end{bmatrix},$$

and from this one we can use back substitution to find the solution

$$\mathbf{u} = \begin{bmatrix} 1 \\ -1 \\ 1 \end{bmatrix}.$$

3. The Euclidean norm is also called the 2-norm, and for

$$\mathbf{v} = \begin{bmatrix} 1 \\ 1 \\ 1 \\ 1 \end{bmatrix},$$

$\|\mathbf{v}\|_2 = \sqrt{1^2 + 1^2 + 1^2 + 1^2} = 2$.

6. The new vector norm is $\max\{2|v_1|, |v_2|, \ldots, |v_n|\}$. First we show that the norm satisfies the properties in (13.6)–(13.9). Because of the absolute value function, $\max\{2|v_1|, |v_2|, \ldots, |v_n|\} \geq 0$ and $\max\{2|v_1|, |v_2|, \ldots, |v_n|\} = 0$ only for $\mathbf{v} = \mathbf{0}$. The third property is satisfied as follows.

$$\begin{aligned} \|c\mathbf{v}\| &= \max\{2|cv_1|, |cv_2|, \ldots, |cv_n|\} \\ &= |c| \max\{2|v_1|, |v_2|, \ldots, |v_n|\} \\ &= |c|\|\mathbf{v}\|. \end{aligned}$$

Chapter 13

Finally, the triangle inequality is satisfied as follows.

$$\begin{aligned}
\|\mathbf{v} + \mathbf{w}\| &= \max\{2|v_1 + w_1|, |v_2 + w_2|, \ldots, |v_n + w_n|\} \\
&\leq \max\{2|v_1| + 2|w_1|, |v_2| + |w_2|, \ldots, |v_n| + |w_n|\} \\
&\leq \max\{2|v_1|, |v_2|, \ldots, |v_n|\} + \max\{2|w_1|, |w_2|, \ldots, |w_n|\} \\
&= \|\mathbf{v}\| + \|\mathbf{w}\|.
\end{aligned}$$

The outline of all unit vectors takes the form of a rectangle with lower left vertex $\mathbf{l}$ and upper right vertex $\mathbf{u}$:

$$\mathbf{l} = \begin{bmatrix} -1/2 \\ -1 \end{bmatrix}, \qquad \mathbf{u} = \begin{bmatrix} 1/2 \\ 1 \end{bmatrix}.$$

8. The image vectors $A\mathbf{v}_i$ are given by

$$A\mathbf{v}_1 = \begin{bmatrix} 1 \\ 0 \end{bmatrix}, \quad A\mathbf{v}_2 = \begin{bmatrix} 1 \\ 1 \end{bmatrix}, \quad A\mathbf{v}_3 = \begin{bmatrix} -1 \\ 0 \end{bmatrix}, \quad A\mathbf{v}_4 = \begin{bmatrix} -1 \\ -1 \end{bmatrix}.$$

Their 2-norms are given by $1, \sqrt{2}, 1, \sqrt{2}$, thus $\max \|A\mathbf{v}_i\| = \sqrt{2} \approx 1.41$ which is a reasonable guess for the true value 1.62. As we increase the number of $\mathbf{v}_i$ (to 10, say), we will get much closer to the true norm.

9. No, because for any real number c, $\det cA = c^n \det A$. This violates (13.15).

10. One is the smallest possible condition number for a nonsingular matrix A since

$$\kappa(A) = \sigma_1/\sigma_n,$$

and $\sigma_n \leq \sigma_1$ by definition. (For a singular matrix, $\sigma_n = 0$.)

11. To find the condition number of A, we calculate its singular values by first forming

$$A^{\mathrm{T}}A = \begin{bmatrix} 1.49 & -0.79 \\ -0.79 & 1.09 \end{bmatrix}.$$

The characteristic equation $|A^{\mathrm{T}}A - \lambda' I| = 0$ is

$$\lambda'^2 - 2.58\lambda' + 1 = 0,$$

and its roots are

$$\lambda'_1 = 2.10 \quad \text{and} \quad \lambda'_2 = 0.475.$$

The singular values of A are $\sigma_1 = \sqrt{2.10} = 1.45$ and $\sigma_2 = \sqrt{0.475} = 0.69$. Thus the condition number of the matrix A is $\kappa(A) = \sigma_1/\sigma_n = 2.10$.

13. With the given projection matrix A, we form

$$A^T A = \begin{bmatrix} 1 & 0 \\ 0 & 0 \end{bmatrix}.$$

This symmetric matrix has eigenvalues $\lambda_1' = 1$ and $\lambda_2' = 0$, thus A has singular values $\sigma_1 = 1$ and $\sigma_2 = 0$. The condition number $\kappa(A) = 1/0 = \infty$. This confirms what we already know from Section 5.9: a projection matrix is not invertible.

15. For any i,

$$\mathbf{u}^i = \begin{bmatrix} 1/i \\ 1 \\ -1/i \end{bmatrix}.$$

Hence there is a limit, namely $[0, 1, 0]^T$.

17. We have

$$D = \begin{bmatrix} 4 & 0 & 0 \\ 0 & 8 & 0 \\ 0 & 0 & 2 \end{bmatrix} \quad \text{and} \quad R = \begin{bmatrix} 0 & 0 & -1 \\ 2 & 0 & 2 \\ 1 & 0 & 0 \end{bmatrix}.$$

Hence

$$\mathbf{u}^{(2)} = \begin{bmatrix} 0.25 & 0 & 0 \\ 0 & 0.125 & 0 \\ 0 & 0 & 0.5 \end{bmatrix} \left(\begin{bmatrix} 2 \\ -2 \\ 0 \end{bmatrix} - \begin{bmatrix} 0 & 0 & -1 \\ 2 & 0 & 2 \\ 1 & 0 & 0 \end{bmatrix} \begin{bmatrix} 0 \\ 0 \\ 0 \end{bmatrix} \right) = \begin{bmatrix} 0.5 \\ -0.25 \\ 0 \end{bmatrix}.$$

Next:

$$\mathbf{u}^{(3)} = \begin{bmatrix} 0.25 & 0 & 0 \\ 0 & 0.125 & 0 \\ 0 & 0 & 0.5 \end{bmatrix} \left(\begin{bmatrix} 2 \\ -2 \\ 0 \end{bmatrix} - \begin{bmatrix} 0 & 0 & -1 \\ 2 & 0 & 2 \\ 1 & 0 & 0 \end{bmatrix} \begin{bmatrix} 0.5 \\ -0.25 \\ 0 \end{bmatrix} \right) = \begin{bmatrix} 0.5 \\ -0.375 \\ -0.25 \end{bmatrix}.$$

Similarly, we find

$$\mathbf{u}^{(4)} = \begin{bmatrix} 0.438 & -0.313 & -0.25 \end{bmatrix}^T.$$

The true solution is

$$\mathbf{u} = \begin{bmatrix} 0.444 & -0.306 & -.222 \end{bmatrix}^T.$$

19. The first three iterations yield the vectors

$$\begin{bmatrix} -0.5 \\ -0.25 \\ -0.25 \\ -0.5 \end{bmatrix}, \quad \begin{bmatrix} 0.187 \\ 0.437 \\ 0.4375 \\ 0.187 \end{bmatrix}, \quad \begin{bmatrix} -0.156 \\ 0.093 \\ 0.0937 \\ -0.156 \end{bmatrix}.$$

The actual solution is

$$\begin{bmatrix} -0.041 \\ 0.208 \\ 0.208 \\ -0.041 \end{bmatrix}.$$

Chapter 14

1. Yes, $\mathbf{r}$, a linear combination of $\mathbf{u}, \mathbf{v}, \mathbf{w}$, is also in $\mathbb{R}^4$ because of the linearity property.

2. Yes, C is an element of $\mathcal{M}_{3\times3}$ because the rules of matrix arithmetic are consistent with the linearity property.

4. No. If we multiply an element of that set by -1, we produce a vector that is not in the set, a violation of the linearity condition.

5. No, $\mathbf{v} = \mathbf{u} + \mathbf{w}$.

7. Yes, $\mathbf{r}$ is a linear combination of $\mathbf{u}, \mathbf{v}, \mathbf{w}$, namely $\mathbf{r} = 2\mathbf{u} + 3\mathbf{w}$, therefore it is in the subspace defined by these vectors.

10. The matrix A is 2×4. Since the $\mathbf{v}_i$ are mapped to the $\mathbf{w}_i$ via a linear map, we know that the same linear relationship holds among the image vectors,
$$\mathbf{w}_4 = 3\mathbf{w}_1 + 6\mathbf{w}_2 + 9\mathbf{w}_3.$$

11. Forward elimination produces the matrix
$$\begin{bmatrix} 1 & 2 & 0 \\ 0 & 0 & 1 \\ 0 & 0 & 0 \\ 0 & 0 & 0 \end{bmatrix}$$
from which we conclude the rank is 2.

14. This inner product satisfies the symmetry and positivity requirements, however it fails homogeneity and additivity. Let's look at homogeneity:
$$\langle \alpha\mathbf{v}, \mathbf{w} \rangle = (\alpha v_1)^2 w_1^2 + (\alpha v_2)^2 w_2^2 + (\alpha v_3)^2 w_3^2$$
$$= \alpha^2 \langle \mathbf{v}, \mathbf{w} \rangle$$
$$\neq \alpha \langle \mathbf{v}, \mathbf{w} \rangle.$$

16. Yes, $\langle A, B \rangle$ satisfies the properties (14.3)–(14.6) of an inner product. Symmetry: This is easily satisfied since products, $a_{i,j}b_{i,j}$, are commutative.

Positivity:
$$\langle A, A \rangle = a_{1,1}^2 + a_{1,2}^2 + \ldots + a_{3,3}^2 \geq 0,$$
and equality is achieved if all $a_{i,j} = 0$. If A is the zero matrix, all $a_{i,j} = 0$, then $\langle A, B \rangle = 0$.

Homogeneity and Additivity: If C is also in this space, then
$$\langle \alpha A + \beta B, C \rangle = (\alpha a_{1,1} + \beta b_{1,1})c_{1,1} + \ldots + (\alpha a_{3,3} + \beta b_{3,3})c_{3,3}$$
$$= \alpha a_{1,1}c_{1,1} + \beta b_{1,1}c_{1,1} + \ldots + \alpha a_{3,3}c_{3,3} + \beta b_{3,3}c_{3,3}$$
$$= \alpha \langle A, C \rangle + \beta \langle B, C \rangle.$$

17. Yes. We have to check the four defining properties (14.3)–(14.6). Each is easily verified. Symmetry: For instance $p_0 q_0 = q_0 p_0$, so this property is easily verified.

Positivity: $\langle p, p \rangle = p_0^2 + p_1^2 + p_2^2 \geq 0$ and $\langle p, p \rangle = 0$ if $p_0 = p_1 = p_2 = 0$. If $p(t) = 0$ for all t, then $p_0 = p_1 = p_2 = 0$, and clearly then $\langle p, p \rangle = 0$.

Homogeneity and Additivity: We can tackle these together,

$$\langle \alpha r + \beta p, q \rangle = \alpha \langle r, q \rangle + \beta \langle p, q \rangle,$$

by first noting that if r is also a quadratic polynomial, then

$$\alpha r + \beta p = \alpha r_0 + \beta p_0 + (\alpha r_1 + \beta p_1)t + (\alpha r_2 + \beta p_2)t^2.$$

This means that

$$\begin{aligned} \langle \alpha r + \beta p, q \rangle &= (\alpha r_0 + \beta p_0)q_0 + (\alpha r_1 + \beta p_1)q_1 + (\alpha r_2 + \beta p_2)q_2 \\ &= \alpha r_0 q_0 + \beta p_0 q_0 + \alpha r_1 q_1 + \beta p_1 q_1 + \alpha r_2 q_2 + \beta p_2 q_2 \\ &= \alpha \langle r, q \rangle + \beta \langle p, q \rangle. \end{aligned}$$

19. The Gram-Schmidt method produces

$$\mathbf{b}_1 = \begin{bmatrix} 1/\sqrt{2} \\ 1/\sqrt{2} \\ 0 \end{bmatrix}, \quad \mathbf{b}_2 = \begin{bmatrix} -1/\sqrt{2} \\ 1/\sqrt{2} \\ 0 \end{bmatrix}, \quad \mathbf{b}_3 = \begin{bmatrix} 0 \\ 0 \\ -1 \end{bmatrix}.$$

Knowing that the $\mathbf{b}_i$ are normalized and checking that $\mathbf{b}_i \cdot \mathbf{b}_j = 0$, we can be confident that this is an orthonormal basis. Another tool we have is the determinant, which will be one,

$$\begin{vmatrix} \mathbf{b}_1 & \mathbf{b}_2 & \mathbf{b}_3 \end{vmatrix} = 1$$

21. One such basis is given by the four matrices

$$\begin{bmatrix} 1 & 0 \\ 0 & 0 \end{bmatrix}, \begin{bmatrix} 0 & 1 \\ 0 & 0 \end{bmatrix}, \begin{bmatrix} 0 & 0 \\ 1 & 0 \end{bmatrix}, \begin{bmatrix} 0 & 0 \\ 0 & 1 \end{bmatrix}.$$

23. No. The linearity conditions are violated. For example $\Phi(-\mathbf{u}) = \Phi(\mathbf{u})$, contradicting the linearity condition, which would demand $\Phi(\alpha \mathbf{u}) = \alpha \Phi(\mathbf{u})$ with $\alpha = -1$.

1. First we find the eigenvalues by looking for the roots of the characteristic equations

$$\det[A - \lambda I] = \begin{vmatrix} 2 - \lambda & 1 \\ 1 & 2 - \lambda \end{vmatrix} = 0, \tag{B.1}$$

Chapter 15

which is $\lambda^2 - 4\lambda + 3 = 0$ and when factored becomes $(\lambda - 3)(\lambda - 1) = 0$. This tells us that the eigenvalues are $\lambda_1 = 3$ and $\lambda_2 = 1$.

The eigenvectors $\mathbf{r}_i$ are found by inserting each λ_i into $[A - \lambda_i I]\mathbf{r} = \mathbf{0}$. We find that

$$\mathbf{r}_1 = \begin{bmatrix} 1 \\ 1 \end{bmatrix} \quad \text{and} \quad \mathbf{r}_2 = \begin{bmatrix} -1 \\ 1 \end{bmatrix}.$$

3. Since this is an upper triangular matrix, the eigenvalues are on the diagonal, thus they are $4, 3, 2, 1$.

4. The eigenvalue is 2 because $A\mathbf{r} = 2\mathbf{r}$.

6. The rank of A is two since one eigenvalue is zero. A matrix with a zero eigenvalue is singular, thus the determinant must be zero.

8. The vector $\mathbf{e}_2$ is an eigenvector since $R\mathbf{e}_2 = \mathbf{e}_2$ and the corresponding eigenvalue is 1.

10. The dominant eigenvalue is $\lambda_1 = 3$.

12. We have

$$\mathbf{r}^{(2)} = \begin{bmatrix} 5 \\ 3 \\ 5 \end{bmatrix}, \quad \mathbf{r}^{(3)} = \begin{bmatrix} 25 \\ 13 \\ 21 \end{bmatrix}, \quad \mathbf{r}^{(4)} = \begin{bmatrix} 121 \\ 55 \\ 93 \end{bmatrix}.$$

The last two give the ratios

$$\frac{r_1^{(4)}}{r_1^{(3)}} = 4.84, \quad \frac{r_2^{(4)}}{r_2^{(3)}} = 4.23, \quad \frac{r_3^{(4)}}{r_3^{(3)}} = 4.43.$$

The true dominant eigenvalue is 4.65.

13. We have

$$\mathbf{r}^{(2)} = \begin{bmatrix} 0 \\ -1 \\ 6 \end{bmatrix}, \quad \mathbf{r}^{(3)} = \begin{bmatrix} 48 \\ -13 \\ 2 \end{bmatrix}, \quad \mathbf{r}^{(4)} = \begin{bmatrix} -368 \\ -17 \\ 410 \end{bmatrix}.$$

The last two give the ratios

$$\frac{r_1^{(4)}}{r_1^{(3)}} = -7.66, \quad \frac{r_2^{(4)}}{r_2^{(3)}} = 1.31, \quad \frac{r_3^{(4)}}{r_3^{(3)}} = 205.$$

This is not close yet to revealing the true dominant eigenvalue, -13.02.

14. Stochastic: B, D. Not stochastic: A has a negative element. One column of C does not sum to one.

16. Figure B.2 illustrates the directed graph.

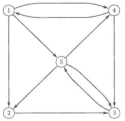

Figure B.2.
Graph showing the connectivity defined by C.

The corresponding stochastic matrix is

$$D = \begin{bmatrix} 0 & 0 & 0 & 1/2 & 0 \\ 1/3 & 0 & 0 & 0 & 1/3 \\ 0 & 1 & 0 & 1/2 & 1/3 \\ 1/3 & 0 & 0 & 0 & 1/3 \\ 1/3 & 0 & 1 & 0 & 0 \end{bmatrix}.$$

19. The map $Lf = f''$, has eigenfunctions $\sin(kx)$ for $k = 1, 2, \ldots$ since

$$\frac{d^2 \sin(kx)}{dx^2} = -k\frac{d \cos(kx)}{dx} = -k^2 \sin(kx),$$

and the corresponding eigenvalues are $-k^2$.

Chapter 16

1. First we form
$$A^{\mathrm{T}}A = \begin{bmatrix} 1 & 0 \\ 0 & 16 \end{bmatrix},$$
and identify the eigenvalues as $\lambda_i = 16, 1$. The corresponding normalized eigenvectors are the columns of
$$V = \begin{bmatrix} 0 & 1 \\ 1 & 0 \end{bmatrix}.$$
The singular values of A are $\sigma_1 = \sqrt{16} = 4$ and $\sigma_2 = \sqrt{1} = 1$, thus
$$\Sigma = \begin{bmatrix} 4 & 0 \\ 0 & 1 \end{bmatrix}.$$
Next we form
$$AA^{\mathrm{T}} = \begin{bmatrix} 1 & 0 \\ 0 & 16 \end{bmatrix},$$
and as this matrix is identical to $A^{\mathrm{T}}A$, we can construct $U = V$ without much work at all. The SVD of A is complete: $A = U\Sigma V^{\mathrm{T}}$.

2. The eigendecomposition is identical to the SVD because the matrix is symmetric and positive definite.

4. The semi-major axis length is 2 and the semi-minor axis length is 1/2. This matrix is symmetric and positive definite, so the singular values are identical to the eigenvalues, which are clearly $\lambda_i = 2, 1/2$. In general, these lengths are the singular values.

7. Since A is a diagonal matrix, the eigenvalues are easily identified to be $\lambda_i = -2, 1, 1$, hence $\det A = -2 \times 1 \times 1 = -2$. To find A's singular values we find the eigenvalues of
$$A^{\mathrm{T}}A = \begin{bmatrix} 4 & 0 & 0 \\ 0 & 1 & 0 \\ 0 & 0 & 1 \end{bmatrix},$$

which are $4, 1, 1$. Therefore, the singular values of A are $\sigma_i = 2, 1, 1$ and $|\det A| = 2 \times 1 \times 1 = 2$. The two equations do indeed return the same value for $|\det A|$.

9. Since the first and last column vectors are linearly dependent, this matrix has rank 2, thus we know one eigenvalue, $\lambda_3 = 0$, and $\det A = 0$. Similarly, a rank deficient matrix as this is guaranteed to have $\sigma_3 = 0$, confirming that $\det A = 0$.

10. The pseudoinverse is defined in (16.7) as $A^\dagger = V\Sigma^\dagger U^{\mathrm{T}}$. The SVD of $A = U\Sigma V^{\mathrm{T}}$ is comprised of

$$U = \begin{bmatrix} -1 & 0 & 0 \\ 0 & 1 & 0 \\ 0 & 0 & 1 \end{bmatrix}, \qquad \Sigma = \begin{bmatrix} 2 & 0 \\ 0 & 1 \\ 0 & 0 \end{bmatrix}, \qquad V = \begin{bmatrix} 0 & 1 \\ 1 & 0 \end{bmatrix}.$$

To form the pseudoinverse of A, we find

$$\Sigma^\dagger = \begin{bmatrix} 1/2 & 0 & 0 \\ 0 & 1 & 0 \end{bmatrix},$$

then

$$A^\dagger = \begin{bmatrix} 0 & 1 & 0 \\ -0.5 & 0 & 0 \end{bmatrix}.$$

12. First we form $A^{\mathrm{T}}A = [4]$; this is a 1×1 matrix. The eigenvalue is 4 and the corresponding eigenvector is 1, so $V = [1]$. Next we form

$$AA^{\mathrm{T}} = \begin{bmatrix} 4 & 0 & 0 \\ 0 & 0 & 0 \\ 0 & 0 & 0 \end{bmatrix},$$

which has eigenvalues $\lambda_i = 4, 0, 0$ and the corresponding eigenvectors form the columns of

$$U = \begin{bmatrix} 1 & 0 & 0 \\ 0 & 1 & 0 \\ 0 & 0 & 1 \end{bmatrix};$$

note that the last two columns of U form a basis for the null space of A^{T}. The next step is to form

$$\Sigma^\dagger = \begin{bmatrix} 1/2 & 0 & 0 \end{bmatrix},$$

and then the pseudoinverse is

$$A^\dagger = V\Sigma^\dagger U^{\mathrm{T}} = \begin{bmatrix} 1/2 & 0 & 0 \end{bmatrix}.$$

Check (16.8) and (16.9) to be sure this solution satisfies the properties of the pseudoinverse.

14. The least squares system is $A\mathbf{x} = \mathbf{b}$. We proceed with the enumerated steps.

1. Compute the SVD of A:

$$U = \begin{bmatrix} 0 & 1 & 0 \\ 0.89 & 0 & -0.45 \\ 0.45 & 0 & 0.89 \end{bmatrix}, \quad \Sigma = \begin{bmatrix} 2.24 & 0 \\ 0 & 1 \\ 0 & 0 \end{bmatrix}, \quad V = \begin{bmatrix} 0 & 1 \\ 1 & 0 \end{bmatrix}.$$

The pseudoinverse of Σ is

$$\Sigma^\dagger = \begin{bmatrix} 0.45 & 0 & 0 \\ 0 & 1 & 0 \end{bmatrix}.$$

2. Compute

$$\mathbf{z} = U^T \mathbf{b} = \begin{bmatrix} 0.89 \\ 0 \\ -0.45 \end{bmatrix}.$$

3. Compute

$$\mathbf{y} = \Sigma^\dagger \mathbf{z} = \begin{bmatrix} 0.4 \\ 0 \end{bmatrix}.$$

4. Compute the solution

$$\mathbf{x} = V\mathbf{y} = \begin{bmatrix} 0 \\ 0.4 \end{bmatrix}.$$

We can check this against the normal equation approach:

$$A^T A \mathbf{v} = A^T \mathbf{b},$$

$$\begin{bmatrix} 1 & 0 \\ 0 & 5 \end{bmatrix} \mathbf{v} = \begin{bmatrix} 0 \\ 2 \end{bmatrix},$$

and we can see that the solution is the same as that from the pseudo-inverse.

17. Form the covariance matrix by first creating

$$X^T X = \begin{bmatrix} 12 & 8 \\ 8 & 12 \end{bmatrix},$$

and then divide this matrix by the number of points, 7, thus the co-variance matrix is

$$C = \begin{bmatrix} 1.71 & 1.14 \\ 1.14 & 1.71 \end{bmatrix}.$$

Find the (normalized) eigenvectors of C and make them the columns of V. Looking at a sketch of the points and from our experience so far, it is clear that

$$V = \begin{bmatrix} 0.707 & -0.707 \\ 0.707 & 0.707 \end{bmatrix}.$$

The data points transformed into the principal components coordinate system are

$$\hat{X} = XV = \begin{bmatrix} -2.82 & 0 \\ -1.41 & 0 \\ 0 & 0 \\ 1.41 & 0 \\ 2.82 & 0 \\ 0 & 1.41 \\ 0 & -1.41 \end{bmatrix}.$$

Chapter 17

1a. First of all, draw a sketch. Before calculating the barycentric coordinates (u, v, w) of the point

$$\begin{bmatrix} 0 \\ 1.5 \end{bmatrix},$$

notice that this point is on the edge formed by $\mathbf{p}_1$ and $\mathbf{p}_3$. Thus, the barycentric coordinate $v = 0$.

The problem now is simply to find u and w such that

$$\begin{bmatrix} 0 \\ 1.5 \end{bmatrix} = u \begin{bmatrix} 1 \\ 1 \end{bmatrix} + w \begin{bmatrix} -1 \\ 2 \end{bmatrix}$$

and $u + w = 1$. This is simple enough to see, without computing! The barycentric coordinates are $(1/2, 0, 1/2)$.

1b. Add the point

$$\mathbf{p} = \begin{bmatrix} 0 \\ 0 \end{bmatrix}$$

to the sketch from the previous exercise. Notice that $\mathbf{p}$, $\mathbf{p}_1$, and $\mathbf{p}_2$ are collinear. Thus, we know that $w = 0$.

The problem now is to find u and v such that

$$\begin{bmatrix} 0 \\ 0 \end{bmatrix} = u \begin{bmatrix} 1 \\ 1 \end{bmatrix} + v \begin{bmatrix} 2 \\ 2 \end{bmatrix}$$

and $u + v = 1$. This is easy to see: $u = 2$ and $v = -1$. If this wasn't obvious, you would calculate

$$u = \frac{\|\mathbf{p}_2 - \mathbf{p}\|}{\|\mathbf{p}_2 - \mathbf{p}_1\|},$$

then $v = 1 - u$.

Thus, the barycentric coordinates of $\mathbf{p}$ are $(2, -1, 0)$.

If you were to write a subroutine to calculate the barycentric coordinates, you would not proceed as we did here. Instead, you would calculate the area of the triangle and two of the three sub-triangle areas. The third barycentric coordinate, say w, can be calculated as $1 - u - v$.

1c. To calculate the barycentric coordinates (i_1, i_2, i_3) of the incenter, we need the lengths of the sides of the triangles:

$$s_1 = 3, \quad s_2 = \sqrt{5}, \quad s_3 = \sqrt{2}.$$

The sum of the lengths, or circumference c, is approximately $c = 6.65$. Thus the barycentric coordinates are

$$\left(\frac{3}{6.65}, \frac{\sqrt{5}}{6.65}, \frac{\sqrt{2}}{6.65} \right) = (0.45, 0.34, 0.21).$$

Always double-check that the barycentric coordinates sum to one. Additionally, check that these barycentric coordinates result in a point in the correct location:

$$0.45 \begin{bmatrix} 1 \\ 1 \end{bmatrix} + 0.34 \begin{bmatrix} 2 \\ 2 \end{bmatrix} + 0.21 \begin{bmatrix} -1 \\ 2 \end{bmatrix} = \begin{bmatrix} 0.92 \\ 1.55 \end{bmatrix}.$$

Plot this point on your sketch, and this looks correct! Recall the incenter is the intersection of the three angle bisectors.

1d. Referring to the circumcenter equations from Section 17.3, first calculate the dot products

$$d_1 = \begin{bmatrix} 1 \\ 1 \end{bmatrix} \cdot \begin{bmatrix} -2 \\ 1 \end{bmatrix} = -1,$$

$$d_2 = \begin{bmatrix} -1 \\ -1 \end{bmatrix} \cdot \begin{bmatrix} -3 \\ 0 \end{bmatrix} = 3,$$

$$d_3 = \begin{bmatrix} 2 \\ -1 \end{bmatrix} \cdot \begin{bmatrix} 3 \\ 0 \end{bmatrix} = 6,$$

then $D = 18$. The barycentric coordinates (cc_1, cc_2, cc_3) of the circumcenter are

$$cc_1 = -1 \times 9/18 = -1/2,$$
$$cc_2 = 3 \times 5/18 = 5/6,$$
$$cc_3 = 6 \times 2/18 = 2/3.$$

The circumcenter is

$$\frac{-1}{2} \begin{bmatrix} 1 \\ 1 \end{bmatrix} + \frac{5}{6} \begin{bmatrix} 2 \\ 2 \end{bmatrix} + \frac{2}{3} \begin{bmatrix} -1 \\ 2 \end{bmatrix} = \begin{bmatrix} 0.5 \\ 2.5 \end{bmatrix}.$$

Plot this point on your sketch. Construct the perpendicular bisectors of each edge to verify.

1e. The centroid of the triangle is simple:

$$\mathbf{c} = \frac{1}{3} \left(\begin{bmatrix} 1 \\ 1 \end{bmatrix} + \begin{bmatrix} 2 \\ 2 \end{bmatrix} + \begin{bmatrix} -1 \\ 2 \end{bmatrix} \right) = \begin{bmatrix} 2/3 \\ 1\frac{2}{3} \end{bmatrix}.$$

5. The unit normal is $[0 \ \ 0 \ \ 1]^\mathrm{T}$. It is easy to determine since the triangle lives in the $[\mathbf{e}_1, \mathbf{e}_2]$-plane.

6. Form two vectors that span the triangle,

$$\mathbf{v} = \begin{bmatrix} 0 \\ 1 \\ 0 \end{bmatrix} \quad \text{and} \quad \mathbf{w} = \begin{bmatrix} -1 \\ 0 \\ 1 \end{bmatrix},$$

then the normal is formed as

$$\mathbf{n} = \frac{\mathbf{v} \wedge \mathbf{w}}{\| \cdot \|} = \frac{1}{\sqrt{2}} \begin{bmatrix} 1 \\ 0 \\ 1 \end{bmatrix}.$$

Chapter 18

3. A rhombus is equilateral but not equiangular.

5. The winding number is 0.

6. The area is 3.

8. The estimate normal to the polygon, using the methods from Section 18.7, is

$$\mathbf{n} = \frac{1}{\sqrt{6}} \begin{bmatrix} 1 \\ 1 \\ 2 \end{bmatrix}.$$

9. The outlier is $\mathbf{p}_4$; it should be

$$\begin{bmatrix} 0 \\ 3 \\ -3/2 \end{bmatrix}.$$

Three ideas for planarity tests were given in Section 18.8. There is not just one way to solve this problem, but one is not suitable for finding the outlier. If we use the "average normal test," which calculates the centroid of all points, then it will lie outside the plane since it is calculated using the outlier. Either the "volume test" or the "plane test" could be used.

For so few points, we can practically determine the true plane and the outlier. In "real world" applications, where we would have thousands of points, the question above is ill-posed. We would be inclined to calculate an average plane and then check if any points deviate significantly from this plane. Can you think of a good method to construct such an average plane?

Chapter 19

1. The matrix form for a circle in standard position is

$$\mathbf{x}^\mathrm{T} \begin{bmatrix} \lambda & 0 \\ 0 & \lambda \end{bmatrix} \mathbf{x} - r^2 = 0.$$

2. The ellipse is

$$\frac{1}{25}x_1^2 + \frac{1}{4}x_2^2 = 1.$$

4. The equation of the ellipse is

$$2x_1^2 + 10x_2^2 - 4 = 0.$$

It is fairly clear that rotating by $90°$ simply exchanges λ_1 and λ_2; however, as practice, we could construct the rotation:

$$A = \begin{bmatrix} 0 & -1 \\ 1 & 0 \end{bmatrix} \begin{bmatrix} 10 & 0 \\ 0 & 2 \end{bmatrix} \begin{bmatrix} 0 & 1 \\ -1 & 0 \end{bmatrix} = \begin{bmatrix} 2 & 0 \\ 0 & 10 \end{bmatrix}.$$

Thus the rotated ellipse is $10x_1^2 + 2x_2^2 - 4 = 0$.

5. The implicit equation of the hyperbola is

$$4x_1^2 + 4x_2^2 + 12x_1x_2 - 4 = 0.$$

In matrix form, the hyperbola is

$$\mathbf{x}^{\mathrm{T}} \begin{bmatrix} 4 & 6 \\ 6 & 4 \end{bmatrix} \mathbf{x} - 4 = 0.$$

7. The implicit equation of the conic in expanded form is

$$-5x_1^2 - 6x_2^2 - 2x_1x_2 + 2x_1 + 3 = 0.$$

9. The eigendecomposition is $A = RDR^{\mathrm{T}}$, where

$$R = \begin{bmatrix} 1/\sqrt{2} & -1/\sqrt{2} \\ 1/\sqrt{2} & 1/\sqrt{2} \end{bmatrix} \quad \text{and} \quad D = \begin{bmatrix} 6 & 0 \\ 0 & 4 \end{bmatrix}.$$

11. A conic in matrix form is given by (19.10) and in this case,

$$A = \begin{bmatrix} 1 & -1 \\ -1 & 0 \end{bmatrix}.$$

The $\det A < 0$, thus the conic is a hyperbola.

13. A conic in matrix form is given by (19.10) and in this case,

$$A = \begin{bmatrix} 2 & 0 \\ 0 & 0 \end{bmatrix}.$$

The $\det A = 0$, thus the conic is a parabola.

15. Write the circle in matrix form,

$$\mathbf{x}^\mathrm{T} \begin{bmatrix} 1 & 0 \\ 0 & 1 \end{bmatrix} \mathbf{x} - 2\mathbf{x}^\mathrm{T} \begin{bmatrix} 3 \\ -1 \end{bmatrix} + 6 = 0.$$

The translation is the solution to the linear system

$$\begin{bmatrix} 1 & 0 \\ 0 & 1 \end{bmatrix} \mathbf{v} = \begin{bmatrix} 3 \\ -1 \end{bmatrix},$$

which is

$$\mathbf{v} = \begin{bmatrix} 3 \\ -1 \end{bmatrix}.$$

To write the circle in standard position, we need $c = \mathbf{v}^\mathrm{T} A \mathbf{v} - 6 = 4$. Therefore, the circle in standard form is

$$\mathbf{x}^\mathrm{T} \begin{bmatrix} 1 & 0 \\ 0 & 1 \end{bmatrix} \mathbf{x} - 4 = 0.$$

Divide this equation by 4 so we can easily see the linear map that takes the circle to the ellipse,

$$\mathbf{x}^\mathrm{T} \begin{bmatrix} 1/4 & 0 \\ 0 & 1/4 \end{bmatrix} \mathbf{x} - 1 = 0.$$

Thus we need the scaling

$$\begin{bmatrix} 8 & 0 \\ 0 & 16 \end{bmatrix}.$$

17. An ellipse is the result because a conic's type is invariant under affine maps.

Chapter 20

1. The solution is shown in Figure B.3.

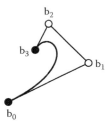

Figure B.3.
The curve for Exercise 1 in Chapter 20.

2. The evaluation point at $t = 1/2$ is calculated as follows:

$$\mathbf{d}_0 = \begin{bmatrix} 0 \\ 0 \end{bmatrix}$$

$$\mathbf{d}_1 = \begin{bmatrix} 6 \\ 3 \end{bmatrix} \quad \begin{bmatrix} 3 \\ 1.5 \end{bmatrix}$$

$$\mathbf{d}_2 = \begin{bmatrix} 3 \\ 6 \end{bmatrix} \quad \begin{bmatrix} 4.5 \\ 3 \end{bmatrix} \quad \begin{bmatrix} 3.75 \\ 2.25 \end{bmatrix}$$

$$\mathbf{d}_3 = \begin{bmatrix} 2 \\ 4 \end{bmatrix} \quad \begin{bmatrix} 2.5 \\ 5 \end{bmatrix} \quad \begin{bmatrix} 3.5 \\ 4 \end{bmatrix} \quad \begin{bmatrix} 3.625 \\ 3.125 \end{bmatrix}.$$

3. The first derivative is

$$\dot{\mathbf{d}}(0) = 3 \left(\begin{bmatrix} 6 \\ 3 \end{bmatrix} - \begin{bmatrix} 0 \\ 0 \end{bmatrix} \right) = 3 \begin{bmatrix} 6 \\ 3 \end{bmatrix}.$$

The second derivative is

$$\ddot{\mathbf{d}}(0) = 3 \times 2 \left(\begin{bmatrix} 3 \\ 6 \end{bmatrix} - 2 \begin{bmatrix} 6 \\ 3 \end{bmatrix} + \begin{bmatrix} 0 \\ 0 \end{bmatrix} \right) = 6 \begin{bmatrix} -9 \\ 0 \end{bmatrix}.$$

4. The de Casteljau scheme for the evaluation at $t = 1/2$ is given in the solution to Exercise 2. Thus the control polygon corresponding to $0 \leq t \leq 1/2$ is given by the diagonal of the scheme:

$$\begin{bmatrix} 0 \\ 0 \end{bmatrix}, \quad \begin{bmatrix} 3 \\ 1.5 \end{bmatrix}, \quad \begin{bmatrix} 3.75 \\ 2.25 \end{bmatrix}, \quad \begin{bmatrix} 3.625 \\ 3.125 \end{bmatrix}.$$

The polygon for $1/2 \leq t \leq 1$ is given by the bottom row, reading from right to left:

$$\begin{bmatrix} 3.625 \\ 3.125 \end{bmatrix}, \quad \begin{bmatrix} 3.5 \\ 4 \end{bmatrix}, \quad \begin{bmatrix} 2.5 \\ 5 \end{bmatrix}, \quad \begin{bmatrix} 2 \\ 4 \end{bmatrix}.$$

Try to sketch the algorithm into Figure. B.3.

5. The monomial coefficients $\mathbf{a}_i$ are

$$\mathbf{a}_0 = \begin{bmatrix} 0 \\ 0 \end{bmatrix}, \quad \mathbf{a}_1 = \begin{bmatrix} 18 \\ 9 \end{bmatrix}, \quad \mathbf{a}_2 = \begin{bmatrix} -27 \\ 0 \end{bmatrix}, \quad \mathbf{a}_3 = \begin{bmatrix} 11 \\ -5 \end{bmatrix}.$$

For additional insight, compare these vectors to the point and derivatives of the Bézier curve at $t = 0$.

6. The minmax box is given two points, which are the "lower left" and "upper right" extents determined by the control polygon,

$$\begin{bmatrix} 0 \\ 0 \end{bmatrix} \quad \text{and} \quad \begin{bmatrix} 6 \\ 6 \end{bmatrix}.$$

7. First, for the curvature at $t = 0$ we find

$$\dot{\mathbf{d}}(0) = \begin{bmatrix} 18 \\ 9 \end{bmatrix}, \quad \ddot{\mathbf{d}}(0) = \begin{bmatrix} -54 \\ 0 \end{bmatrix}, \quad |\dot{\mathbf{d}} \ \ddot{\mathbf{d}}| = 486.$$

Thus

$$\kappa(0) = \frac{|\dot{\mathbf{d}} \ \ddot{\mathbf{d}}|}{\|\dot{\mathbf{d}}\|^3} = \frac{486}{8150.46} = 0.06. \tag{B.2}$$

Next, for the curvature at $t = 1/2$ we find

$$\dot{\mathbf{d}}\left(\frac{1}{2}\right) = \begin{bmatrix} -0.75 \\ 5.25 \end{bmatrix}, \quad \ddot{\mathbf{d}}\left(\frac{1}{2}\right) = \begin{bmatrix} -21 \\ 3 \end{bmatrix}, \quad |\dot{\mathbf{d}} \ \ddot{\mathbf{d}}| = 108.$$

Thus

$$\kappa\left(\frac{1}{2}\right) = \frac{|\dot{\mathbf{d}} \ \ddot{\mathbf{d}}|}{\|\dot{\mathbf{d}}\|^3} = \frac{108}{148.88} = 0.72. \tag{B.3}$$

Examining your sketch of the curve, notice that the curve is flatter at $t = 0$, and this is reflected by these curvature values.

8. The Frenet frame at $t = 1/2$ is

$$\mathbf{f}_1 = \begin{bmatrix} 0.14 \\ 0.99 \\ 0 \end{bmatrix}, \quad \mathbf{f}_2 = \begin{bmatrix} 0 \\ 0 \\ 1 \end{bmatrix}, \quad \mathbf{f}_3 = \begin{bmatrix} 0.99 \\ -0.14 \\ 0 \end{bmatrix}.$$

9. We will evaluate the curve at $t = 2$ using the de Casteljau algorithm. Let's use the triangular schematic to guide the evaluation.

$$\mathbf{d}_0 = \begin{bmatrix} 0 \\ 0 \end{bmatrix}$$

$$\mathbf{d}_1 = \begin{bmatrix} 6 \\ 3 \end{bmatrix} \quad \begin{bmatrix} 12 \\ 6 \end{bmatrix}$$

$$\mathbf{d}_2 = \begin{bmatrix} 3 \\ 6 \end{bmatrix} \quad \begin{bmatrix} 0 \\ 9 \end{bmatrix} \quad \begin{bmatrix} -12 \\ 12 \end{bmatrix}$$

$$\mathbf{d}_3 = \begin{bmatrix} 2 \\ 4 \end{bmatrix} \quad \begin{bmatrix} 1 \\ 2 \end{bmatrix} \quad \begin{bmatrix} 2 \\ -5 \end{bmatrix} \quad \begin{bmatrix} 16 \\ -22 \end{bmatrix}.$$

We have moved fairly far from the polygon!

10. To achieve tangent continuity, the curve $\mathbf{c}(t)$ must have

$$\mathbf{c}_3 = \mathbf{d}_0,$$

and $\mathbf{c}_2$ must be on the line formed by $\mathbf{d}_0$ and $\mathbf{d}_1$:

$$\mathbf{c}_2 = \mathbf{c}_0 + c[\mathbf{d}_0 - \mathbf{d}_1]$$

for positive c. Let's choose $c = 1$, then

$$\mathbf{c}_2 = \begin{bmatrix} -6 \\ -3 \end{bmatrix}.$$

We are free to choose $\mathbf{c}_1$ and $\mathbf{c}_0$ anywhere!

Bibliography

[1] Adobe Systems Inc. *PostScript Language Tutorial and Cookbook.* Addison-Wesley Publishing Company, Inc., 1985.

[2] W. Boehm and H. Prautzsch. *Geometric Concepts for Geometric Design.* A K Peters Ltd., 1992.

[3] H. M. S. Coxeter. *Introduction to Geometry.* Wiley Text Books, second edition, 1989.

[4] M. de Berg, M. van Kreveld, M. Overmars, and O. Schwarzkopf. *Computational Geometry Algorithms and Applications.* Berlin: Springer-Verlag, 1997.

[5] M. Escher and J. Locher. *The Infinite World of M.C. Escher.* New York: Abradale Press/Harry N. Abrams, Inc., 1971.

[6] G. Farin. *NURBS: From Projective Geometry to Practical Use.* A K Peters, Ltd., second edition, 1999.

[7] G. Farin and D. Hansford. *The Essentials of CAGD.* A K Peters, Ltd., 2000.

[8] R. Goldman. Triangles. In A. Glassner, editor, *Graphics Gems, Volume 1*, pages 20–23. Academic Press, 1990.

[9] M. Goossens, F. Mittelbach, and A. Samarin. *The LaTeX Companion.* Reading, MA: Addison-Wesley Publishing Company, Inc., 1994.

[10] D. Hearn and M. Baker. *Computer Graphics with OpenGL, 3/E*. Prentice-Hall, 2003.

[11] L. Johnson and R. Riess. *Numerical Analysis*. Addison-Wesley Publishing Company, Inc., second edition, 1982.

[12] E. Kästner. *Erich Kästner erzaehlt Die Schildbürger*. Cecilie Dressler Verlag, 1995.

[13] L. Lamport. *LaTeX User's Guide and Reference Manual*. Addison-Wesley Publishing Company, Inc., 1994.

[14] P. Shirley. *Fundamentals of Computer Graphics*. A K Peters, Ltd., 2002.

[15] D. Shreiner, M. Woo, J. Neider, and T. Davis. *OpenGL Programming Guide*. Addison-Wesley Publishing Company, Inc., fourth edition, 2004.

[16] G. Strang. *Linear Algebra and Its Applications*. Thomson Brooks/Cole, fourth edition, 2006.

Index